Approaches to Global Sustainability, Markets, and Governance

Scopus indexed *Approaches to Global Sustainability, Markets, and Governance* book series takes a fresh and global approach to issues of corporate social responsibility, regulation, governance, and sustainability. It encompasses such issues as: environmental sustainability and managing the resources of the world; geopolitics and sustainability; global markets and their regulation; governance and the role of supranational bodies; sustainable production and resource acquisition; society and sustainability.

Although primarily a business and management series, it is interdisciplinary and includes contributions from the social sciences, technology, engineering, politics, philosophy, and other disciplines. It focuses on the issues at a meta-level, and investigates the ideas, organisation, and infrastructure required to address them.

The series is grounded in the belief that any global consideration of sustainability must include such issues as governance, regulation, geopolitics, the environment, and economic activity in combination to recognise the issues and develop solutions for the planet. At present such global meta-analysis is rare as current research assumes that the identification of local best practice will lead to solutions, and individual disciplines act in isolation rather than being combined to identify truly global issues and solutions.

This book series is part of the Social Responsibility Research Network (srrnet.org), a worldwide organisation with several thousand members. The network organises annual conferences, held at various locations around the world and many books come from these conferences. Submissions to the current conference are invited and details can be found at srrnet.org. Additionally submissions of chapters, or book proposals are always welcome at info@srrnet.org. If you are interested in contributing a monograph or curated volume to the series, please contact the Publisher: Mamun Mishra (mamun.mishra@springernature.com).

Rubee Singh · David Crowther
Editors

Transition Towards a Sustainable Future

Net Zero Policies and Environmental Sustainability

Springer

Editors
Rubee Singh
Institute of Business Management
GLA University
Mathura, Uttar Pradesh, India

David Crowther
Social Responsibility Research Network
UK

ISSN 2520-8772 ISSN 2520-8780 (electronic)
Approaches to Global Sustainability, Markets, and Governance
ISBN 978-981-97-5755-8 ISBN 978-981-97-5756-5 (eBook)
https://doi.org/10.1007/978-981-97-5756-5

This Springer imprint is published by the registered company Springer Nature Singapore Pte Ltd.
The registered company address is: 152 Beach Road, #21-01/04 Gateway East, Singapore 189721, Singapore

Preface

The book delves into the capacity of developing countries and developed countries to shift their strategical plan towards net zero policies. It focuses on the emergence of net zero policies in various industries and its strategic agenda for mitigating the effects of carbon emissions. In order to achieve environmental sustainability for future generations, it utilises emerging environmentally friendly technologies that contribute to a sustainable environment. Furthermore, it focuses on the obstacles that nations all over the world face when it comes to achieving sustainable development in a timely manner and provides perspectives on technological strategies that can assist in accomplishing this goal. Several significant aspects of the book are the importance of renewable energy sources, the promotion of eco-friendly innovation, and the creation of sustainable infrastructure for a sustainable future. It is essential to develop strong policy frameworks and international partnerships in order to assist worldwide nations during their transformation process.

The book is focused on emphasising the importance of education and knowledge in promoting environmental culture, as well as enabling people to engage in activities that promote a net zero emissions environment within communities in developing countries. It is the story of a future where world nations take the lead in sustainability, serving as a model for the rest of the world, in which nations across the globe take the lead in sustainable development. To achieve the net zero goals effectively, each nation should assess its current status and make an effort to improve its capabilities in order to assess its current status and make an effort to improve its capabilities. In order for the future of our planet to be truly sustainable, a comprehensive plan must take into consideration the particular challenges and opportunities that exist globally, while also utilising the technological advancements of the past to create a sustainable future.

As a part of addressing the global mandate, the book sheds light on various advancements in net zero policies, environmental sustainability, and emerging technologies in order to foster sustainability and also seeks to highlight the urgency of responding to the sustainable development goals (UN-SDGs), as part of the global mandate of ensuring sustainability. This book aims to offer a unique perspective in terms of exploring and integrating the innovative multidisciplinary ideas, techniques, and tools necessary to implement net zero policies, innovative ideas, technologies, and tools for environmental sustainability in order to demonstrate a unique perspective and to provide a unique perspective on how to implement them. Developing the sustainable development goals requires addressing all aspects of the development process, including food security, environmental protection, renewable energy, and economic stability, as well as making advances in the fields of information technology and telecommunications.

The purpose of this book is to give readers an in-depth understanding of how net zero policies are helping in the transition of sustainable technologies and promoting environmental sustainability for future generations for a period of time to come. There are a number of important issues that will be explored in this book such as how to navigate net zero policies and emerging technologies in the era of sustainability, as well as recommendations for adopting net zero policies in the future. There will be information provided to the readers concerning net zero emissions, as well as emerging technologies which drive environmental sustainability and strengthen the local economy. There is a potential that the book will discuss the role that policy frameworks, financing mechanisms, and international collaboration will play in facilitating the adoption of net zero policies, sustainable technologies, and policies that facilitate sustainability on a global scale, as well as encouraging global collaboration towards achieving sustainability on a global scale. This book explains net zero policies and outlines what businesses can do to reduce their carbon footprint directly and through their wider influence in order to make a sustainable commitment to the planet, and it also points out some of the obstacle's businesses may have to overcome in order to achieve a sustainable goal. We intend to reach out to a broad audience that includes undergraduates, postgraduates, practitioners, leading industry experts, and researchers in the fields of sustainable supply chains, environmental sciences, management, sustainable business management, and social sciences. Furthermore, this book is useful for industrial professionals who are engaged in projects related to net zero emissions, sustainable development, environment sustainability, business management, and entrepreneurship as well as their roles within the global economy.

Throughout this book, 14 chapters are explored that touch on net zero policies, environmental sustainability, and transition of sustainable technologies from the perspective of the organisation, the consumer, and the policy maker. It is a compilation of chapters written by both emerging and eminent authors from different parts of the world, including Sri Lanka, Nepal, Saudi Arabia, the UK, Germany, Dubai, Malaysia, and India. It provides a valuable and insightful compendium aimed at sustainability researchers, businesses, educators, policymakers, and anyone else interested in socio-environmental issues and sustainable development principles.

We hope this edited book will be capable to quench the thrust of the readers who are really interested in transition of emerging sustainable technologies in the field of net zero emissions, climate change, environmental sustainability.

Mathura, India — Rubee Singh
UK — David Crowther

Acknowledgements

We would like to express our gratitude to all those who have contributed to the creation of this book. Firstly, we thank the researchers and experts who have shared their knowledge and insights with us, providing valuable information on the current state of *Transition Towards a Sustainable Future: Net Zero Policies and Environmental Sustainability*. We also extend our appreciation to the organisations and institutions that have supported our work, including government agencies, non-governmental organisations, and academic institutions.

Finally, we would like to acknowledge the invaluable contributions of our colleagues and editors, who have worked tirelessly to ensure that this book is informative, engaging, and accessible to a wide range of readers. We hope that this book will inspire further research and action towards a net zero policies and emerging technologies for sustainable future to promote environmental sustainability as per United Nations agenda. We would like to express our sincere gratitude to the experts and stakeholders who contributed their valuable insights and perspectives to this book. Their contributions have been instrumental in shaping our understanding of the challenges and opportunities involved in achieving net zero policies with use of sustainable technologies at globally. We hope that this book will serve as a useful resource for policymakers, practitioners, and researchers working towards a net zero future for the continent.

We are immensely thankful to ***Book Series-Editors, Prof. David Crowther, Dr. Shahla Seifi***, and the entire team at Springer Nature for bringing out this publication in a very professional and timely manner.

Rubee Singh
David Crowther

About This Book

This book provides the latest research on net zero policies, environmental sustainability and its practical implications to adopt sustainable technologies for businesses in current scenario. A number of experts in academia have contributed to the content of this edited volume in order to explore the practical implications of the sustainability transition for net zero policies and utilising sustainable technologies to promote environmental sustainability. It is very crucial that businesses take environmental sustainability seriously so that they can make credible efforts to ensure sustainability for the next generation without damaging the existing resources, while reducing their greenhouse gas emissions. Developing and developed countries should develop actions, strategies, practices, and policies to sustain environmental sustainability. This book aims to offer a unique perspective in terms of exploring and integrating the innovative multidisciplinary ideas, techniques, and tools necessary to implement net zero policies. The purpose of this book is to give readers an in-depth understanding of how net zero policies are helping transition of sustainable technologies and promoting environmental sustainability for future generations for a period of time to come.

There is no doubt that navigating the complexities of the net zero policies in navigating the sustainability transition can be quite challenging. However, this book aims to provide a comprehensive guide to help users navigate the complexities of sustainable transition, drawing on the expertise of scholars in sustainable supply chains, environmental sciences, business management, and social sciences. There are a number of experts whose contributions are included in the book, and it examines the challenges that businesses face in reducing their carbon emissions as well as the potential green growth opportunities that they can leverage in order to achieve net zero policies. A must-read for students, research scholars, and industrial professionals involved in sustainable development research, eco-friendly business management, and emerging technologies. This book offers a thorough concept, roadmap, and analysis of various tactics and sustainable technological advancements that could propel the strategical journey of developed and developing countries towards net zero policies to reduce carbon emissions as per urgency need of United Nations

Agenda to make the world more sustainable. The book brings together a wealth of new research insights, ideas, and new directions for the implementation of sustainable transitions for net zero policies.

Contents

Editors and Contributors

About the Editors

Rubee Singh B.Sc., MBA, Ph.D., Post-Doctoral Researcher (D.Lit.)
Assistant Professor, GLA University, Mathura, Uttar Pradesh, India.
Pursuing- Post Doc (D.Lit.)- Kumaun University, Nainital, Uttarakhand, India.

Rubee Singh is Assistant Professor of Human Resource at Institute of Business Management, GLA University India. She is Post-Doctoral Researcher (D.Lit.) at Kumaun University, Nainital, India. She holds a Doctorate in Management (HR) from Noida International University India and an MBA in HR from Dr. A. P. J. Abdul Kalam Technical University India. She has more than ten years of experience in research and teaching. Apart from teaching and research, she has handled many administrative positions in premier institutes affiliated with A.K.T.U. She has conducted a number of FDPs, conferences, workshops, and HR conclaves in the areas of sustainability, CSR, AI, HR, human factors, women empowerment, and human rights. She is a United Nations volunteer and a contributing author for Voices of Youth (UNICEF). She is an honorary life member of Women Economic Forum and National Secretary of IATM (Youth Wing) International Anti-Terrorism Movement India. Dr. Singh is Managing Editor of IJARSH India. She is serving on the Editorial Board of more than 30 international journals. She serves as Advisory Board Member of the *Heliyon Journal*, Elsevier, *Research in Globalisation*, and reviewer of many reputed journals like *European Journal of Innovation Management* (EJIM), *British Food Journal*, *Sustainability* etc. She is life member of International Association of Engineers, Thailand, IMRTC USA, ISRD UK, etc. She is a bestselling author with significant experience as an excellent Editor, Eminent researcher, great academician, Global Ambassador (Children's) and Social Activist. She has presented many papers at national and international conferences, published 20 academic books including Springer, Elsevier, De Gruyter Nova Science Publishers, and more than 75+ research papers in SCI, ABDC-A, Scopus, and refereed/UGC index journals. Her areas of research are sustainability, digital HRM, sustainable HRM, circular economy, digital CE, net zero emissions, climate change, Industry 4.0, artificial intelligence, blockchain technology, and corporate social responsibility. Being a thought leader, her books are honoured by being added to the library, Ohio Christian University, University of California, Corporacion Universitaria Unisabanta, UMass Chan Medical School, Tunix Community College, USA, Aston University UK, Dow University of Health Sciences, Pakistan, University of Port Harcourt, Nigeria. Her book titled *Training & Development* was added to the MBA syllabus of "The Institute of Aeronautical Engineering" (Jawaharlal Nehru Technological University, Hyderabad). Dr. Singh has been honoured with many prestigious awards, which include Most Influential Woman Leaders in 2022, Best Researchers Award of the Year-2022–23 by GLA University, India, Child Ambassador Award of The Year 2022, Iconic Faculty Award 2019, International Young Researchers Award-2021, Young Scientist Award 2020, and Nari Ratan Award 2023, India Book of Records-2020 and Asia Book of Records Holder for 2021. The list of awards and appreciation is long, but Dr. Singh has always prioritised her work over recognition, making her unique and commendable in her field.

David Crowther B.A., MBA, M.Ed., Ph.D., PGCE, FCMA, CGMA, CPFA, MCMI
Professor, University of Bedfordshire, UK.
London School of Commerce, UK.
Ansted University, Malaysia.
Shanghai University, China.
www.davideacrowther.org; www.srrnet.org

David Crowther is an expert in various fields: he is a specialist in business administration, a games theoretician, and a psychologist while also a qualified accountant who had worked as an accountant, systems specialist, and general manager in local government, industry and commerce for over 20 years. After a number of years in the financial services sector, including a spell as divisional managing director during which he set up and ran a credit card scheme, he decided to leave the business world and become an academic. In 1994, he joined Aston University and there obtained a Ph.D. in 1999 for research into corporate social performance. In 2001, he became the first Professor of Corporate Social Responsibility in the world. He now holds various positions as a professor as well as being visiting professor in various institutions. He gives advice as Chief Research Fellow at the Lithuanian Institute of Agrarian Economics and has consulted extensively around the world. At Ansted University he is also a member of the Board and President of the Sustainability Research Institute. He has consulted widely including to US Presidential candidates, UN, EU, and governments. Prof. David has published more than 60 books and has also contributed several hundred articles to academic, business, and professional journals and to edited book collections. He has also spoken widely at conferences and seminars worldwide and acted as a consultant to a wide range of government, professional, and commercial organisations. In 2003, he organised the first-ever CSR conference and as a result of this he established the Social Responsibility Research Network and also launched

Social Responsibility Journal, the premier journal in its field. In 2010 he established the Organisation Governance Network and conference series. He is listed in Who's Who in the World and various other directories. His research is into sustainability, governance, and corporate social responsibility with a particular emphasis on the relationship between social, environmental, and financial performance. His latest project has been to edit and produce the Palgrave major reference work on CSR.

Contributors

Azamat Ali School of Liberal and Creative Arts (Journalism and Mass Communication), Lovely Professional University, Phagwara, Punjab, India

A. Arun Kumar ICFAI Law School, ICFAI Foundation for Higher Education (Deemed to be University Under Sec 3 of UGC Act, 1956), Hyderabad, Telangana, India

D. M. Arvind Mallik Associate Professor, Department of MBA, PES Institute of Technology and Management, Shivamogga, Karnataka, India

Vanita Bhardwaj Department of Engineering, Vivekananda Institute of Professional Studies, New Delhi, India

David Crowther Social Responsibility Research Network, London, UK

Narayanage Jayantha Dewasiri Department of Accountancy and Finance, Faculty of Management Studies, Sabaragamuwa University of Sri Lanka, Belihuloya, Sri Lanka;
Adjunct Faculty, University Center for Research and Development (UCRD), Chandigarh University, Mohali, Punjab, India

Jacinta Dsilva SEE Research Center, SEE Institute, Dubai, United Arab Emirates

W. K. A. C. Gnanapala Department of Tourism Management, Faculty of Management Studies, Sabaragamuwa University of Sri Lanka, Belihuloya, Sri Lanka

Dewanarayana Hiththaragedara Prasad Manuranga Gunathilaka Head of Quality Assurance, Environmental, Health and Safety, International Distillers Limited, Kaduwela, Sri Lanka

Abid Haleem Department of Mechanical Engineering, Jamia Millia Islamia, New Delhi, India

Vaibhav Jain Department of Mechanical Engineering, Maharaja Agrasen Institute of Technology, New Delhi, India

Amit Joshi Department of Management Studies, Kumaun University, Nainital, Uttarakhand, India

Shahbaz Khan Department of Industrial Engineering, University of Tabuk, Tabuk, Saudi Arabia

Sonal Khurana Department of Engineering, Vivekananda Institute of Professional Studies, New Delhi, India

Kishore Kumar School of Management, Bennett University, Greater Noida, India

Vikas Kumar Faculty of Business Law and Social Sciences, Birmingham City University, Birmingham, UK

Bikram Paul Singh Lehri University School of Business, Chandigarh University, Mohali, India

Bisma Mannan Department of Mechanical and Marine Engineering, University of Plymouth, Plymouth, UK

Udgam Mishra Tribhuvan University, Kathmandu, Nepal

Daravath Vikas Nayak Loyola Academy (Degree and PG College), Secunderabad, Telangana, India

Seema Rani Department of Commerce, Kurukshetra University, Kurukshetra, India

Mananage Shanika Hansini Rathnasiri Department of Marketing Management, Faculty of Management Studies, Sabaragamuwa University of Sri Lanka, Belihuloya, Sri Lanka

Kamalesh Ravesangar Department of Accounting and Business, Tunku Abdul Rahman University of Management and Technology, Kuala Lumpur, Malaysia

V. Rema Department of Data Sciences and Analytics, School of Social Sciences, Ramaiah University of Applied Sciences, Bengaluru, Karnataka, India

Anna Rostomyan Department of Psychology, Media University of Applied Sciences, Berlin, Germany

Smitha Sambrani Department of Business Management, University College of Commerce and Business Management, Osmania University, Hyderabad, Telangana, India

Mohamed Irfan Shaikh SEE Research Center, SEE Institute, Dubai, United Arab Emirates

Ankita Sharma Christ University, Bengaluru, India

Sameer Shekhar Department of Commerce (KSEC), Kalinga Institute of Industrial Technology (KIIT), Deemed to be University, Bhubaneshwar, Odisha, India

Rubee Singh Department of Management Studies, Kumaun University, Nainital, Uttarakhand, India;
Institute of Business Management, GLA University, Mathura, Uttar Pradesh, India

T. Sowmyya Forensic Science, Department of Chemistry, University College of Science, Osmania University, Hyderabad, Telangana, India

Rakesh Suryadevara ICFAI Law School, ICFAI Foundation for Higher Education (Deemed to be University Under Sec 3 of UGC Act, 1956), Hyderabad, Telangana, India

Lalit Mohan Tewari Department of Botany, DSB Campus, Kumaun University, Nainital, Uttarakhand, India

Fahad Ziaul Ain Usmani SEE Research Center, SEE Institute, Dubai, United Arab Emirates

Mohit Yadav Jindal Global Business School, Jindal Global University, Haryana, India

Part I
Net Zero Policies and Environmental Sustainability

Chapter 1
Sustainable Strategies for Net Zero: An Asia Pacific Perspective

Narayanage Jayantha Dewasiri

Abstract The Asia Pacific region stands at a critical juncture in its quest for sustainability, with the imperative of achieving net zero emissions looming large on the global agenda. This chapter explores the region's multifaceted landscape of net zero initiatives, examining success stories, challenges faced, policy recommendations, and future directions. From China's rapid expansion of renewable energy capacity to Japan's ambitious targets for 100% renewable energy, the region showcases remarkable progress alongside persistent challenges such as grid integration, regulatory barriers, and social inequalities. Strategic recommendations for policymakers include strengthening policy frameworks, enhancing international collaboration, and promoting social equity to accelerate progress toward net zero emissions. Innovations in renewable energy technologies, smart grids, circular economy solutions, and climate resilience are shaping the future of sustainable development in the region. A long-term vision for a sustainable future emphasizes achieving carbon neutrality, conserving biodiversity, promoting social equity, and fostering green economic growth. Ultimately, the journey toward net zero emissions in Asia requires collective action, innovation, and commitment from governments, businesses, and communities to pave the way toward a brighter and greener future.

Keywords Sustainable strategies · Net zero economy · Asia Pacific · Sustainable future · Sustainability

N. J. Dewasiri (✉)
Department of Accountancy and Finance, Faculty of Management Studies, Sabaragamuwa University of Sri Lanka, Belihuloya, Sri Lanka
e-mail: jayantha@mgt.sab.ac.lk

Adjunct Faculty, University Center for Research and Development (UCRD), Chandigarh University, Mohali, Punjab, India

R. Singh and D. Crowther (eds.), *Transition Towards a Sustainable Future*, Approaches to Global Sustainability, Markets, and Governance,
https://doi.org/10.1007/978-981-97-5756-5_1

1.1 Introduction

1.1.1 Overview of Net Zero and Sustainability

Achieving net zero emissions is essential for mitigating climate change and stabilizing global temperatures, representing a fundamental shift in how economies operate and grow. Haszeldine et al. (2018) stated that the net zero balances the amount of greenhouse gases extracted from and released into the atmosphere. This balance is crucial for stabilizing global temperatures and mitigating the severe impacts of climate change. Significant emissions reductions from renewable energy sources, improved energy efficiency, and carbon capture and storage technologies are needed to reach net zero (Becattini et al., 2021). It also demands changes in consumption patterns, industrial processes, and transportation systems. Net zero is more than an environmental goal; it represents a fundamental shift in economic operations, aiming to decouple economic growth from carbon emissions.

Sustainability, while including the goal of net zero, encompasses a broader perspective (Al Kez et al., 2024). It emphasizes how crucial it is to satisfy current wants without sacrificing the capacity of the next generations to satisfy their own. This comprehensive approach integrates environmental health, economic prosperity, and social equity (Sopiana & Harahap, 2023). Sustainability involves conserving natural resources, reducing pollution, and promoting biodiversity while ensuring that economic and social systems are robust, resilient, and inclusive (Mensah, 2019). Together, net zero and sustainability represent a comprehensive strategy (Vulturius et al., 2024) for addressing the pressing environmental challenges of our time. They underscore the imperative of aligning economic development with ecological limits to ensure a livable planet for future generations.

1.1.2 Importance of the Asia Pacific Region in Global Sustainability Efforts

The Asia Pacific region is critical in global sustainability due to its substantial economic influence, large population, and rich biodiversity (Yang et al., 2020). This region encompasses many countries, from highly industrialized nations like Japan and South Korea to rapidly developing economies like China and India and, more minor, developing island nations in the Pacific. These countries contribute significantly to global greenhouse gas emissions due to rapid industrialization, urbanization, and increasing energy demand. China, for instance, is the global leading emitter of carbon dioxide, while India is projected to be a significant driver of future emissions growth (Ortega-Ruiz et al., 2022).

Nonetheless, there is great potential for the Asia Pacific region to support global sustainability. Renewable energy sources, including solar, wind, and hydropower, are plentiful in Asia–Pacific and crucial for reducing reliance on fossil fuels (Mamat et al., 2019). Australia is a global leader in solar power, and nations like China and India are making significant advancements in deploying renewable energy (Prăvălie et al., 2019). Furthermore, the region's various ecosystems—coral reefs and tropical rainforests—are essential to preserving biodiversity and global carbon sequestration.

The Asia Pacific's commitment to achieving net zero is vital for achieving global climate goals, particularly those delineated in the Paris Agreement (Zhang et al., 2021). The region's actions can significantly influence global emissions trajectories and set a precedent for sustainable development practices worldwide. Furthermore, the Asia Pacific's unique challenges—such as varying levels of economic development, political systems, and cultural contexts—provide valuable lessons on the multifaceted approach needed to achieve sustainability. By addressing these challenges through innovative policies, technological advancements, and regional cooperation, the Asia Pacific has the potential to provide an example of how environmental sustainability and economic prosperity may coexist.

1.1.3 Purpose and Scope of the Chapter

This chapter investigates the sustainable strategies implemented across the Asia Pacific region to achieve net zero emissions. This chapter outlines the region's efforts toward a sustainable future by examining the current state of net zero policies, showcasing successful case studies, and analyzing the roles of governments, private sectors, and communities. It also identifies the challenges and barriers specific to the Asia Pacific, offering strategic recommendations for overcoming them. The scope of this study includes a comprehensive analysis of policy frameworks, technological innovations, and community engagements, highlighting the unique contributions and approaches of various countries within the region. Ultimately, this chapter seeks to underscore the critical importance of the Asia Pacific in global sustainability efforts and provide a roadmap for achieving net zero emissions through collaborative and innovative strategies.

1.2 Methodology

This chapter utilizes a systematic literature review to investigate sustainable strategies for achieving net zero in the Asia Pacific. This method involves a rigorous process of identifying, evaluating, and synthesizing research on a particular topic. The goal is to provide a thorough overview of existing knowledge, highlight research gaps, and offer insights for future research. To ensure a robust and transparent review process, we used the Preferred Reporting Items for Systematic Reviews and Meta-Analyses (PRISMA) method, following the methodologies outlined by Dewasiri et al. (2022), Rathnasiri and De Silva (2023), Jayawardena and Dewasiri (2023), Karunarathna et al. (2024), Dewasiri et al. (2023), and Dewasiri et al. (2024).

PRISMA is a widely recognized methodology that enhances the clarity and completeness of reporting in systematic reviews. The PRISMA approach involves a four-phase process: Identification, Screening, Eligibility, and Inclusion.

1.2.1 Identification

The researcher designed a comprehensive strategy to identify relevant samples. Several academic databases, including Scopus and Google Scholar, were utilized for this purpose. The search involved a combination of keywords such as "net zero," "sustainability," "Asia Pacific," "renewable energy," "carbon emissions," and "policy." Beyond database searches, the researcher manually reviewed the reference lists of pertinent articles and reviews to ensure no relevant studies were missed.

1.2.2 Screening

The initial search yielded 1041 articles. Duplicates were removed, resulting in 938 unique articles. Titles and abstracts of these articles were initially checked against the criteria developed for the inclusion and exclusion. Accordingly, the articles were included only if they focused on net zero strategies, sustainability practices, or policy analysis within the Asia Pacific region. Exclusion criteria included articles not available in English, those not peer-reviewed, and those unrelated to the primary research focus.

1.2.3 Eligibility

The full texts of 143 articles that passed the screening stage were assessed for eligibility. This assessment thoroughly evaluated the articles' relevance, methodology, and contributions to the topic. Articles that did not provide sufficient empirical data or were not directly relevant to the Asia Pacific context were excluded.

1.2.4 Inclusion

After the eligibility assessment, seventy-seven (77) articles were deemed suitable for inclusion in the review. These articles were selected based on their methodological rigor, relevance, and contributions to understanding sustainable strategies for net zero in the Asia Pacific region.

1.2.5 Data Extraction and Analysis

Data from the chosen articles were systematically extracted using a standardized form. Critical information collected included the study objectives, methodology, findings, and conclusions. Subsequently, the retrieved data were combined to find recurring themes, patterns, and gaps in the literature. A thematic analysis was conducted to categorize the findings into relevant themes. The analysis aimed to produce a set of strategies implemented across the Asia Pacific to achieve net zero and the challenges and opportunities identified in the literature.

1.2.6 Quality Assessment

A quality assessment was conducted on the selected studies using predefined criteria to ensure their reliability and validity. The criteria evaluated included the clarity of research objectives, the appropriateness of methodologies, the robustness of data analysis, and the validity of conclusions. Only studies meeting these quality standards were included in the final synthesis, ensuring the review's findings were based on high-quality evidence. Hence, the systematic literature review, guided by the PRISMA approach, provides a comprehensive and reliable synthesis of current knowledge on sustainable strategies for achieving net zero in the Asia Pacific region. The insights gained from this review are critical for informing policy, guiding future research, and supporting the region's efforts toward a sustainable future.

1.3 Findings and Discussion

This section reveals a detailed examination of the current state of net zero policies in Asia, uncovering the diverse commitments and strategies that characterize the region's approach to achieving sustainable, low-carbon futures. The analysis begins with an outline of the net zero targets made by various countries, identifying key leaders like China, India, Japan, and South Korea and comparing their approaches. It then discusses the sustainable strategies for achieving net zero, including adopting and integrating renewable energy, energy efficiency measures, technological innovations, and sustainable urban planning. The section highlights success stories, challenges faced, and lessons learned from net zero initiatives in the region by presenting case studies. The role of governments and policymakers is scrutinized, emphasizing policy frameworks, regulatory measures, and international collaborations that support net zero goals. Furthermore, the section explores the essential contributions of the private sector and community engagement in driving sustainability initiatives. Concluding with future directions and recommendations, it offers strategic insights for policymakers, identifies emerging trends, and outlines a long-term vision for a sustainable future in the Asia Pacific. Through this comprehensive analysis, the chapter underscores the critical importance of the region's efforts in the global pursuit of a net zero economy.

1.3.1 Current State of Net Zero Policies in the Asia Pacific

The Asia Pacific region, characterized by its vast economic diversity and significant environmental footprint, plays a crucial role in global exertions to minimize carbon emissions. As countries in this region confront the realities of climate change, many have set ambitious targets and are developing comprehensive strategies to transition toward sustainable, low-carbon economies. These varied commitments reflect each nation's unique economic, social, and political contexts. From the technological innovations in Japan and South Korea to the renewable energy advancements in China and India, the Asia Pacific is a vibrant landscape of diverse approaches to sustainability (Baloch et al., 2021). This section focuses on the current net zero commitments across the region, highlighting key countries leading the movement and comparing the different strategies employed to reach their environmental goals. Through this examination, we understand the region's critical role in the global pursuit of a sustainable future.

1.3.1.1 Overview of Net Zero Commitments in the Region

With its diverse economies and significant environmental impact, the Asia Pacific region has seen a range of net zero commitments from various countries (Zhang et al., 2021). These commitments are part of broader national strategies to overcome challenges in climate change and promote sustainability. Many countries in the region have set ambitious targets to achieve net zero targets by 2050, aligning with global climate goals established under the Paris Agreement.

The largest producer of greenhouse gases in the world, China, has pledged to become carbon neutral by 2060, as announced in 2020 (Davidson et al., 2021). This pledge represents a significant shift in China's climate policy and significantly impacts global efforts to reduce emissions. India, another leading emitter, targets to reach net zero by 2070, prioritizing the expansion of renewable energy and improvements in energy efficiency (Zhong et al., 2023). Meanwhile, Thailand has set its target of net zero by 2050 (Rajbhandari et al., 2024).

Japan and South Korea promised to attain net zero emissions by 2050 (Lau & Tsai, 2023). Their plans include reducing reliance on fossil fuels, enhancing energy efficiency, and promoting clean energy technologies. Although Australia has been slower in formalizing its net zero target, it has committed to achieving it by 2050, focusing on technological innovations and carbon capture and storage (Martin-Roberts et al., 2021).

Smaller nations in the region, such as Fiji and the Maldives, have also committed to net zero (Soomauroo et al., 2020). These nations are often motivated by the existential threat climate change poses to their environments and economies. These nations advocate for global climate action and lead by example in implementing sustainable practices.

1.3.1.2 Key Countries Leading the Net Zero Movement

Several nations in the Asia–Pacific area stand out for having taken proactive steps to reach net zero emissions:

China: China, the largest emitter of greenhouse gases, has set up a crucial obligation to achieve carbon neutrality by 2060. The country is making substantial investments in renewable energy, focusing on solar and wind power, and is at the forefront of global electric vehicle adoption (Liu et al., 2022). China's 14th Five-Year Plan prioritizes green development, with significant initiatives aimed at transitioning from coal to cleaner energy sources (Shang & Lv, 2023).

Japan: Japan's net zero by 2050 pledge is supported by its "Green Growth Strategy," which focuses on hydrogen energy, offshore wind power, and carbon recycling (Kawai et al., 2022; Oshiro et al., 2018). The country is also advancing nuclear energy and energy efficiency technologies as part of its comprehensive approach to decarburization.

South Korea: South Korea's "Green New Deal" targets a net zero economy by 2050 through significant investments in renewable energy, smart grids, and green infrastructure (Boafo et al., 2022). The country also promotes green finance and fosters industries supporting a low-carbon economy.

India: India has set a target to reach a net zero economy by 2070, supported by ambitious renewable energy goals (Bhattacharyya et al., 2022). This includes a goal of 500 GW of non-fossil fuel capacity by 2030 (Singh et al., 2023). Initiatives such as the International Solar Alliance and promoting sustainable urban development highlight India's commitment to a green transition.

Australia: Australia's path to net zero by 2050 focuses on technological advancements, comprising hydrogen production, carbon capture and storage, and renewable energy projects (Geroe, 2022). The country is also enhancing its climate resilience and adaptation strategies.

1.3.1.3 Comparison of Different Approaches Within the Region

The approaches to achieving net zero vary widely across the Asia Pacific, reflecting the region's diverse economic structures, energy dependencies, and policy environments:

Renewable Energy Adoption: Countries like China, India, and Australia heavily invest in renewable energy sources, each leveraging their unique geographical advantages. China focuses on large-scale solar and wind projects (Sun et al., 2017), India emphasizes solar energy and decentralized renewable solutions (Rathore et al., 2019), and Australia capitalizes on its vast solar and wind potential (Salahuddin et al., 2018).

Energy Efficiency: Japan and South Korea prioritize energy efficiency in industrial processes, buildings, and transportation (Tolliver et al., 2021). Their advanced technologies and stringent regulations support significant energy savings and emissions reductions.

Technological Innovations: The region emphasizes technological solutions. South Korea and Japan are leaders in hydrogen energy and electric vehicle technologies (Li & Kimura, 2021). Australia is pioneering carbon capture and storage (Thorpe, 2012), and China is at the forefront of renewable energy technologies and innovative grid systems (Li et al., 2020a, 2020b).

Policy Frameworks: Different policy frameworks reflect varying degrees of government intervention and market mechanisms (Rathnasiri et al., 2024). China employs a centralized, top-down approach with significant state investment in green technologies (Li & Du, 2021). In contrast, Australia relies more on market-based mechanisms and private sector involvement, supported by government incentives.

Community and International Engagement: Smaller nations, like Fiji and the Maldives, focus on community-based approaches and international cooperation (Soomauroo et al., 2020). These countries actively participate in global climate discussions and advocate for more excellent international support for climate adaptation and mitigation.

Overall, the Asia Pacific region's diverse strategies toward net zero reflect its complex economic landscape and varying capacities. The collective efforts of these countries are crucial for the universal combat of climate change, offering valuable lessons and innovative solutions that can be adapted and implemented worldwide.

1.3.2 Sustainable Strategies for Achieving Net Zero

Pursuing net zero emissions in the Asia Pacific region necessitates implementing sustainable strategies that holistically address energy production, consumption, and urban development. Renewable energy adoption and integration are foundational pillars, leveraging the region's abundant natural resources to conversion from fossil fuels. Concurrently, energy efficiency and conservation measures are vital in reducing energy demand and optimizing resource utilization. Technological innovations and green technologies drive transformative changes across sectors, facilitating more efficient and environmentally friendly processes (Rathnasiri et al., 2024). Moreover, sustainable urban planning and infrastructure development are imperative for creating resilient, low-carbon cities that support sustainable lifestyles. Asia Pacific can achieve net zero emissions through a comprehensive approach encompassing

Table 1.1 Themes identified as sustainable strategies through the SLR

#	Themes identified through SLR
1	Renewable energy adoption and integration
2	Energy efficiency and conservation measures
3	Technological innovations and green technologies
4	Sustainable urban planning and infrastructure development
5	Policy recommendations
6	Regional coordination
7	Public engagement and education

Source Authors' own

these strategies while fostering economic growth, environmental stewardship, and social equity. We conducted a thematic analysis to identify sustainable strategies for achieving net zero. These themes included renewable energy adoption, energy efficiency measures, technological innovations, policy frameworks, and community engagement, as indicated in Table 1.1.

1.3.2.1 Renewable Energy Adoption and Integration

Renewable energy adoption and integration are fundamental pillars in achieving net zero emissions in Asia (Zhang et al., 2021). To diversify their energy mix and lessen reliance on fossil fuels, nations should enhance their investments in solar, wind, hydro, and other renewable energy sources (Shahsavari & Akbari, 2018). By harnessing the abundant renewable resources available, such as sunlight and wind, nations can significantly decrease their carbon footprint while ensuring energy security and affordability (Bhuiyan, 2022). Integrating renewable energy into existing grids and infrastructure requires robust planning and investment in innovative grid technologies and solutions for energy storage to mitigate intermittency and variability, thereby maximizing the contribution of renewables to the overall energy supply.

1.3.2.2 Energy Efficiency and Conservation Measures

Murshed (2020) stressed that energy efficiency and conservation measures complement renewable energy adoption by reducing overall energy demand and improving resource utilization. Governments and industries throughout Asia are introducing a range of initiatives to improve energy efficiency in buildings, transportation, industry, and appliances (Lo, 2014). These efforts involve adopting energy-efficient technologies, building codes, standards, and motivations to inspire energy-saving initiatives. Additionally, promoting energy conservation practices through public awareness campaigns and educational programs fosters a culture of sustainability and responsible energy consumption, further contributing to net zero objectives.

1.3.2.3 Technological Innovations and Green Technologies

The shift to net zero is happening faster than ever, thanks to technological advancements and green technology, which are causing revolutionary changes in many industries (Rathnasiri et al., 2024). Progressions in electrification, carbon capture and storage, renewable energy generation, and sustainable manufacturing are increasing operations' environmental friendliness and efficiency. Governments and businesses are spending more money on R&D to facilitate the implementation of creative solutions that lower emissions and improve environmental sustainability (Jayarathne et al., 2021). Asia Pacific nations may open up new avenues for reaching net zero emissions while fostering economic growth and competitiveness by fostering an atmosphere that supports technology innovation through funding, cooperation, and legislation.

1.3.2.4 Infrastructure Development and Sustainable Urban Planning (SUP)

SUP and infrastructure development are critical for creating livable, resilient, and low-carbon regional cities (De-Jong et al., 2015). Rapid urbanization in the region poses challenges and opportunities for sustainable development (Chen et al., 2022). Emissions in urban areas can be significantly decreased by implementing integrated planning techniques that prioritize compact, mixed-use development, adequate public transportation, green spaces, and energy-efficient structures. Additionally, investing in climate-resilient infrastructure and nature-based solutions, like green roofs and urban forests, enhances cities' ability to adapt to climate change impacts while mitigating greenhouse gas emissions (Boateng et al., 2023). Countries may build more inclusive, equitable, and environmentally sustainable communities for future generations by integrating sustainability principles into urban design and infrastructure construction.

1.3.2.5 Policy Recommendations

Comprehensive policy frameworks are essential for overcoming the identified challenges and accelerating progress toward net zero emissions (Rathnasiri et al., 2024). Governments in the Asia Pacific can expedite the development and execution of robust policy instruments that incentivize renewable energy deployment, promote energy efficiency and conservation, stimulate technological innovation, and support sustainable urban planning and infrastructure development. Green building regulations, energy efficiency standards, carbon pricing schemes, renewable energy targets, and sustainable transportation policies are a few examples of policy measures (Lu et al., 2020). Furthermore, promoting regional cooperation and resource-sharing can amplify the impact of individual endeavors and facilitate the exchange of best practices, technology transfer, and capacity building among neighboring countries (Rafique & Rossi, 2023).

1.3.2.6 Regional Cooperation

Emphasizing the importance of regional cooperation and shared resources is paramount in achieving net zero goals in any region (Wimbadi & Djalante, 2020). Collaboration among countries enables the pooling of expertise, resources, and investments to address common challenges and capitalize on collective opportunities (Rathnasiri et al., 2024). Regional initiatives, such as cross-border energy networks, collaborative research initiatives, and capacity development programs, can facilitate the conversion to an interconnected, sustainable energy system. By promoting dialogue, cooperation, and solidarity, countries in the Asia Pacific can amplify their impact and accelerate progress toward achieving net zero emissions regionally and globally.

1.3.2.7 Public Engagement and Education

Public awareness and education are crucial in garnering support for net zero policies (Al Kez et al., 2024) and fostering behavioral change at the Individual and community levels. Governments, civil society organizations, and educational institutions should prioritize public engagement initiatives that enhance awareness of the urgency of climate actions, the paybacks of renewable energy, energy efficiency, and sustainable lifestyles. Enlightening programs targeting schools, workplaces, and communities can empower individuals with the knowledge, skills, and tools to adopt sustainable practices, make informed choices, and advocate for policy change. Public–private engagement (Ugwu et al., 2024) and education efforts can mobilize collective action and drive the evolution toward a net zero future in Asia by fostering a culture of sustainability and environmental stewardship.

1.3.3 Case Studies from the Asia Pacific

The Asia Pacific region presents a rich tapestry of net zero initiatives, including noteworthy examples from China, Japan, South Korea, Australia, and India. These case studies offer valuable insights into the diverse approaches taken by nations in their quest for sustainable, low-carbon futures. Australia, for instance, showcases innovative solutions, such as its progression toward renewable energy and carbon capture and storage initiatives. Meanwhile, India's ambitious renewable energy targets and decentralized energy systems exemplify the country's commitment to green growth and energy access. By examining these case studies alongside those from other key regional players, we gain a holistic understanding of the challenges, successes, and lessons learned in pursuing net zero emissions. Through a critical analysis of policy implementation and its impact, this section aims to illuminate pathways toward achieving sustainability goals in the Asia Pacific and beyond.

1.3.3.1 Success Stories of Net Zero Initiatives

China: China's rapid expansion of renewable energy capability, predominantly in solar and wind power, has remarkably reduced carbon emissions. The country's developments in clean energy infrastructure and stringent environmental regulations have paved the way toward becoming a global leader in renewable energy adoption. Notable projects include the development of the world's largest solar farm in Qinghai Province (Qaisrani et al., 2021) and the construction of offshore wind farms along the country's coastlines (DeCastro et al., 2019).

Japan: Following the 2011 nuclear disaster, Japan's Fukushima Prefecture has emerged as a pioneer in renewable energy (Schreurs, 2021). The region has specific goals to achieve 100% renewable energy by 2040, leveraging solar, wind, and biomass resources. Through community-based initiatives and public–private partnerships, Fukushima has revitalized its economy while tumbling its usage of fossil fuels and nuclear power (Feldhoff, 2016).

South Korea: South Korea's Green New Deal initiative (Kim, 2023) has catalyzed investments in renewable energy and sustainable infrastructure, positioning the country for a green transition. Success stories include the development of smart cities like Songdo (lim et al., 2023) and expanding electric vehicle charging infrastructure nationwide (Moon et al., 2018). Additionally, South Korea's commitment to carbon neutrality by 2050 has spurred innovation and investment in clean technologies.

Australia: Australia is generating much of its electricity from renewable sources, demonstrating how far it has come to adopt renewable energy (Li et al., 2020a, 2020b). The country's vast solar and wind resources have fueled the growth of utility-scale renewable projects, such as the Hornsdale Power Reserve, one of the world's largest lithium-ion battery storage facilities (Rangarajan et al., 2023). Despite policy uncertainty and grid integration challenges, Australia's renewable energy sector continues to thrive, driving emissions reductions and economic growth.

India: India is committed to promoting green growth and energy access, as seen by its ambitious goals for renewable energy and decentralized energy systems. The nation has emerged as a global leader in deploying renewable energy by expanding solar and wind power capacity and using cutting-edge financing techniques like reverse auctions and green bonds. Success stories include the development of the Gujarat Solar Park and the Saubhagya scheme, which aims to provide electricity access to all households (Stock, 2021). However, challenges related to grid reliability, land acquisition, and regulatory barriers persist, underscoring the need for continued policy support and investment in sustainable energy infrastructure.

1.3.3.2 Challenges Faced and Lessons Learned

China: Despite significant progress, China faces challenges integrating renewable energy into its grid and addressing energy inefficiencies in traditional industries. Additionally, the transition from coal remains complex and politically sensitive, requiring careful planning and stakeholder engagement (Van-Voss & Rafaty, 2022).

Japan: Fukushima's transition to renewable energy has encountered obstacles such as grid constraints and opposition from vested interests (Abin, 2021). The region's experience highlights the significance of community engagement, capacity building, and supportive policy guidelines in overcoming barriers to sustainable development.

South Korea: South Korea's rapid urbanization and industrialization have led to environmental degradation and pollution, posing challenges to its net zero aspirations. A comprehensive strategy that balances social justice, environmental preservation, and economic growth is needed to address these issues.

Australia: Despite its numerous resources in renewable energy, policy uncertainty and regulatory challenges have hindered Australia's evolution to a net zero economy (Gunningham & Bowman, 2016). More excellent policy stability and investment certainty are essential to unlocking renewable energy's full potential and achieving emissions reduction targets.

India: India faces challenges related to grid reliability, land acquisition, and regulatory barriers in scaling renewable energy deployment (Elavarasan et al., 2020). The decentralization of energy systems and innovative financing mechanisms have overcome some challenges. Still, sustained policy support and investment are needed to accelerate progress toward net zero emissions.

1.3.3.3 Analysis of Policy Implementation and Impact

China: China's Renewable Energy Law and National Emissions Trading System have driven renewable energy deployment and emissions reduction (Wang et al., 2024). However, efficient implementation and enforcement methods are necessary to ensure compliance and achieve the intended results.

Japan: Japan's feed-in tariff scheme and renewable energy auctions have incentivized private investment in clean energy projects (Chu & Takeuchi, 2022). Regulatory reforms and market liberalization are needed to foster competition and innovation in the renewable energy sector.

South Korea: South Korea's carbon pricing mechanism and renewable energy targets have stimulated development in low-carbon tools and technologies (Xu et al., 2023). Nevertheless, regulatory barriers and vested interests challenge the shift to sustainability through renewable energy.

Australia: Australia's renewable energy policies, such as the Renewable Energy Target and state-level renewable energy objectives, have sparked significant investment in renewable energy infrastructure (Nelson et al., 2022). However, investors are now cautious, and the shift to a low-carbon economy has been hampered by uneven policy support and a lack of long-term planning.

India: India's ambitious renewable energy targets and policy initiatives like the National Solar Mission have spurred investment and deployment of renewable energy projects (Chandel et al., 2016). However, regulatory challenges, grid integration issues, and financial constraints have slowed progress, underscoring the need for comprehensive policy reforms and investment incentives.

These case studies offer valuable insights into the successes, challenges, and lessons learned from net zero initiatives in the Asia Pacific. Through an analysis of policy implementation and its impact, countries in the region can identify the best opportunities for practices for a sustainable future.

1.3.4 Role of Governments and Policymakers

Governments and policymakers are pivotal in driving the transition toward net zero emissions, shaping policy frameworks, implementing regulatory measures, and fostering international collaboration (Fagerberg & Normann, 2023).

1.3.4.1 Policy Frameworks Supporting Net Zero Goals

Effective policy frameworks are essential for providing incentives and guidance to businesses, industries, and communities to reach a net zero economy (Rathnasiri et al., 2024). Countries across the Asia Pacific region increasingly adopt ambitious targets and strategies to align with global climate goals. These frameworks often include renewable energy targets, carbon pricing mechanisms, emission reduction goals, and energy efficiency standards. By providing clear direction and regulatory certainty, governments could create a favorable environment for initiatives in clean energy technologies and sustainable practices.

1.3.4.2 Regulatory Measures and Incentives

Regulatory measures and incentives are crucial in incentivizing emissions reduction and fostering innovation in clean technologies (Del Río & Kiefer, 2022). Governments can implement policies such as carbon pricing, emissions trading schemes, feed-in tariffs, and tax incentives to encourage businesses and individuals to transition toward low-carbon alternatives. Additionally, regulatory standards for energy efficiency, building codes, and vehicle emissions can drive market transformation and promote sustainable practices across sectors (Jayarathne et al., 2021). Governments can spur investment and innovation by aligning regulatory frameworks with net zero objectives while securing fair and equitable treatment for all stakeholders.

1.3.4.3 International Collaboration and Agreements

International collaboration and agreements are vital for addressing global climate challenges and achieving net zero emissions on a global scale. Asia Pacific region increasingly recognizes the importance of working together to share knowledge, resources, and best practices in climate action. Countries can set aggressive goals for lowering emissions and commit to collective action within multilateral agreements like the Paris Agreement (Rathnasiri et al., 2024). Bilateral and regional partnerships further enhance technology transfer, capacity building, and climate finance cooperation.

1.3.5 Private Sector and Community Engagement

Private sector companies and community organizations are integral in driving the transition toward net zero emissions, contributing through corporate responsibility initiatives, sustainability efforts, and grassroots community engagement (Mishra et al., 2022).

1.3.5.1 Corporate Responsibility and Sustainability Initiatives

A growing number of businesses in the Asia Pacific area are realizing how important it is to include sustainability and corporate responsibility in their operations. Businesses can reduce their carbon footprint, lessen their adverse effects on the environment, and strengthen their social and economic resiliency by implementing sustainability into their corporate plans. This means adopting energy-efficient practices, reducing waste and emissions, utilizing renewable energy sources, and investing in environmentally friendly supply chains (Mishra et al., 2024). Companies can also contribute to community development, environmental conservation, and poverty alleviation through corporate social responsibility programs, demonstrating their commitment to sustainable development.

1.3.5.2 Role of NGOs and Community Organizations

Non-governmental organizations (NGOs) and community officialdoms play a vital role in mobilizing grassroots support and driving change at the local level (Rigolon &

Gibson, 2021). These organizations often work closely with communities, advocating for environmental protection, social justice, and sustainable development. NGOs engage in various activities, including raising awareness, conducting research, providing education and training, and advocating for policy reform. NGOs and community organizations elevate the voices of underprivileged groups and guarantee that local viewpoints are incorporated into decision-making processes by enabling communities to tackle climate change and environmental issues.

1.3.6 Public Awareness and Participation

Public awareness and participation are essential for fostering a culture of sustainability and driving collective action toward net zero emissions (Constantino et al., 2022). Governments, businesses, and civil society organizations must engage with the public through education campaigns, outreach activities, and participatory decision-making processes. By increasing understanding of the effects of climate change, the benefits of renewable energy, and the need for conservation efforts, people may make educated decisions and adopt sustainable habits in their daily lives. Through initiatives such as citizen science projects, community consultations, and public hearings, public participation in climate action ensures that diverse perspectives are considered and solutions are inclusive and equitable. Public awareness and participation are critical drivers of social change and essential for achieving net zero emissions in Asia.

1.4 Future Directions and Recommendations

As the Asia Pacific region continues its journey toward a sustainable future, it is essential to consider future directions and provide recommendations to guide policymakers, businesses, and communities in achieving net zero emissions.

1.4.1 Strategic Recommendations for Policymakers

Policymakers have a crucial role in setting the strategic direction for climate action and creating an enabling environment for sustainable development. Policymakers are essential When setting the direction for climate action and creating an atmosphere that supports sustainable development. Some key proposals are regulatory standards for energy efficiency and emissions reduction, carbon pricing mechanisms, and more robust policy frameworks supporting net zero aims. Improving global cooperation is also essential since it facilitates the sharing information, resources, and best practices for mitigating and adapting to climate change. It is recommended that policymakers give top priority to participating in regional partnerships and global agreements in order to accelerate the transition to net zero emissions. In addition, governments ought to provide incentives for investment in clean energy technologies and research to propel technical progress and ease the shift to a low-carbon economy. Promoting social equity is crucial, to be sure. Action on climate change should be inclusive and consider marginalized people's needs and viewpoints. Prioritizing social equality programs and making sure that sustainable development projects serve all facets of society should be top priorities for policymakers.

1.4.2 Innovations and Emerging Trends

Innovations and new trends are shaping the Asia Pacific region's future of sustainable development (Rathnasiri et al., 2024). Developing renewable energy technology, such as solar photovoltaics, wind turbines, and energy storage systems, is a significant area of concentration since it lowers costs and raises efficiency, making renewable energy more widely available and reasonably priced than ever. The production, distribution, and consumption of energy are all undergoing radical change thanks to innovative grid technologies and energy management systems, which also make it possible to integrate renewable energy sources more seamlessly and increase total energy efficiency. Furthermore, the shift to a circular economy, in which resources are recycled, regenerated, and reused, is gaining popularity as a viable substitute for conventional linear economic models that minimize waste, conserve resources, and have a minimal negative environmental impact. Furthermore, there is a growing understanding of the significance of climate resilience and adaptation strategies due to the frequency and severity of climate-related disasters. Innovations in disaster preparedness, climate-resilient infrastructure, and community resilience are crucial for creating a sustainable and resilient future.

1.4.3 Long-Term Vision for a Sustainable Future

Making decisions and motivating group action require a long-term vision for a sustainable future that can be achieved through four distinct objectives. Achieving carbon neutrality- a state in which greenhouse gas emissions and removals are balanced- is a crucial component of this strategy since it helps stabilize the climate and avert the worst effects of climate change. Furthermore, preserving and reviving biodiversity is essential for upholding the resilience and health of ecosystems, promoting food security, and lessening the effects of climate change. In addition, a sustainable future must prioritize social justice and inclusion, guaranteeing that everyone can participate in decision-making and access necessities like food, clean water, and air. The shift toward a green economy, in which economic expansion is independent

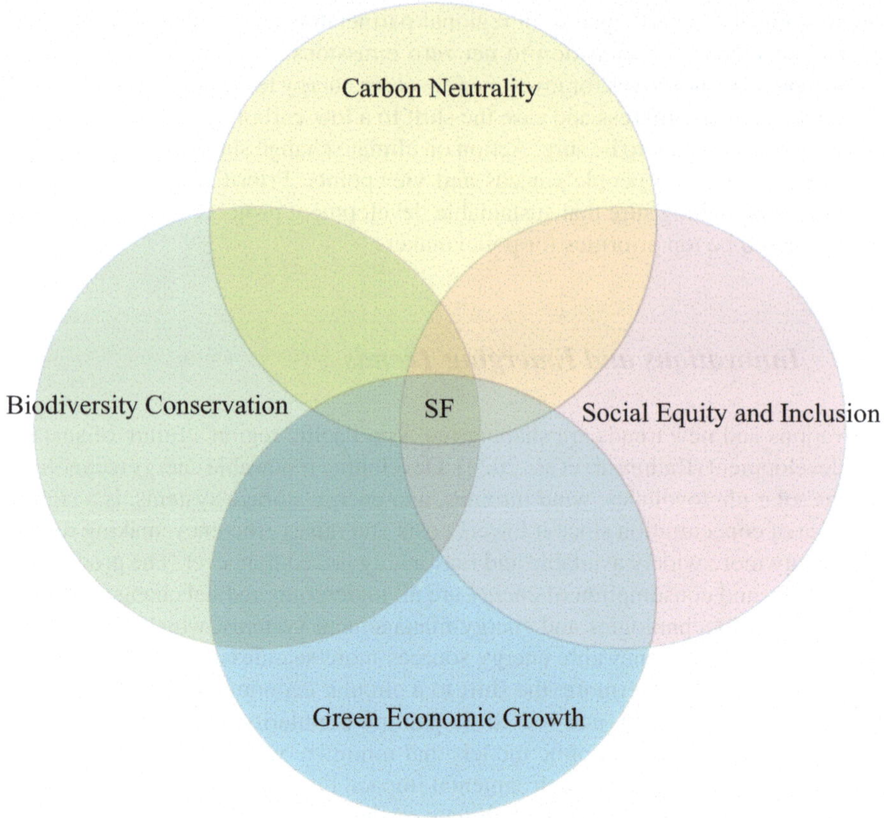

Fig. 1.1 Long-term vision for a sustainable future (SF). *Source* Authors' own

of environmental deterioration, is vital for accomplishing sustainable development objectives and enhancing the standard of living for everyone (Fig. 1.1).

By embracing these recommendations and aligning with a long-term vision for a sustainable future, the Asia Pacific region can chart a course toward prosperity, resilience, and environmental stewardship for future generations.

1.5 Conclusion

In conclusion, achieving net zero emissions in Asia is multifaceted and requires collaboration, innovation, and bold policy action. This chapter has explored various aspects of net zero initiatives, including success stories, challenges faced, policy recommendations, and future directions. Here, we summarize the key points and outline the path forward for the region.

1.6 Summary of Key Points

Success Stories: From China's rapid expansion of renewable energy capacity to Japan's ambitious targets for 100% renewable energy, the Asia Pacific region has demonstrated remarkable advancement for a sustainable future.

Challenges Faced: Despite successes, issues such as assimilating renewable energy into the grid, overcoming regulatory barriers, and addressing social inequalities persist and require concerted efforts from governments, businesses, and communities.

Policy Recommendations: Strategic recommendations for policymakers include strengthening policy frameworks, enhancing international collaboration, investing in innovation, and promoting social equity to accelerate progress toward net zero emissions.

Innovations and Emerging Trends: Innovations in renewable energy technologies, smart grids, circular economy solutions, and climate resilience are shaping the future of sustainable development in the region.

Long-Term Vision: A long-term vision for a sustainable future includes achieving carbon neutrality, conserving biodiversity, promoting social equity, and fostering green economic growth.

Hence, the way forward involves collective action and commitment from governments, businesses, and communities. The region can overcome challenges and achieve net zero emissions by embracing renewable energy, advancing innovation, promoting social inclusion, and prioritizing environmental stewardship.

1.6.1 Final Thoughts on Achieving Net Zero

Achieving net zero emissions is not merely a goal but a moral imperative for the Asia Pacific region and the world. It requires transformative changes in energy systems, economic structures, and societal values. By working together toward a shared sustainability vision, the Asia Pacific region can lead toward a brighter, greener future for future generations. In conclusion, while the road ahead may be challenging, the rewards of achieving net zero emissions are profound. The Asia Pacific region can pave the way toward a more resilient, prosperous, and equitable world by seizing opportunities, overcoming obstacles, and staying committed to sustainability.

References

Abin, B. A. (2021). Japan's energy security. Opportunities and challenges of renewables in post-Fukushima energy scenario. *Journal of Polity & Society, 13*, 23–47.

Al Kez, D., Lowans, C., & Foley, A. (2024). Sustainable development in third level programs: Distilling a pathway to a true net-zero education. *Sustainability, 16*(5), 1998.

Baloch, Z. A., Tan, Q., Khan, M. Z., Alfakhri, Y., & Raza, H. (2021). Assessing energy efficiency in the Asia-Pacific region and the mediating role of environmental pollution: Evidence from a super-efficiency model with a weighting preference scheme. *Environmental Science and Pollution Research, 28*(35), 48581–48594.

Becattini, V., Gabrielli, P., & Mazzotti, M. (2021). Role of carbon capture, storage, and utilization to enable a net-zero-CO_2-emissions aviation sector. *Industrial & Engineering Chemistry Research, 60*(18), 6848–6862.

Bhattacharyya, R., Singh, K. K., Grover, R. B., & Bhanja, K. (2022). Estimating minimum energy requirement for transitioning to a net-zero, developed India in 2070. *Current Science,* 517–527.

Bhuiyan, M. R. A. (2022). Overcome future environmental challenges through sustainable and renewable energy resources. *Micro & Nano Letters, 17*(14), 402–416.

Boafo, F. E., Kim, J. H., Ahn, J. G., Kim, S. M., Kim, J. T., & Zhang, L. (2022). Study on thermal characteristics and electrical performance of a hybrid building integrated photovoltaic (BIPV) system combined with vacuum insulation panel (VIP). *Energy and Buildings, 277*, 112574.

Boateng, E. A., Asibey, M. O., Cobbinah, P. B., Adutwum, I. O., & Blija, D. K. (2023). Enabling nature-based solutions: Innovating urban climate resilience. *Journal of Environmental Management, 332*, 117433.

Chandel, S. S., Shrivastva, R., Sharma, V., & Ramasamy, P. (2016). Overview of the renewable energy sector initiatives under India's national action plan on climate change. *Renewable and Sustainable Energy Reviews, 54*, 866–873.

Chen, M., Chen, L., Cheng, J., & Yu, J. (2022). Identifying interlinkages between urbanization and Sustainable Development Goals. *Geography and Sustainability, 3*(4), 339–346.

Chu, L., & Takeuchi, K. (2022). The non-operating solar projects: Examining the impact of the feed-in tariff amendment in Japan. *Energy Policy, 160*, 112712.

Constantino, S. M., Sparkman, G., Kraft-Todd, G. T., Bicchieri, C., Centola, D., Shell-Duncan, B., & Weber, E. U. (2022). Scaling up change: A critical review and practical guide to harnessing social norms for climate action. *Psychological Science in the Public Interest, 23*(2), 50–97.

Davidson, M., Karplus, V. J., Zhang, D., & Zhang, X. (2021). Policies and institutions to support carbon neutrality in China by 2060. *Economics of Energy & Environmental Policy, 10*(2), 7–17.

DeCastro, M., Salvador, S., Gómez-Gesteira, M., Costoya, X., Carvalho, D., Sanz-Larruga, F. J., & Gimeno, L. (2019). Europe, China, and the United States: Three different approaches to the development of offshore wind energy. *Renewable and Sustainable Energy Reviews, 109*, 55–70.

De-Jong, M., Joss, S., Schraven, D., Zhan, C., & Weijnen, M. (2015). Sustainable–smart–resilient–low carbon–eco–knowledge cities; making sense of a multitude of concepts promoting sustainable urbanization. *Journal of Cleaner Production, 109*, 25–38.

Del Río, P., & Kiefer, C. P. (2022). Which policy instruments promote innovation in renewable electricity technologies? A critical review of the literature with a focus on auctions. *Energy Research & Social Science, 89*, 102501.

Dewasiri, N. J., Baker, H. K., Banda, Y. W., & Rathnasiri, M. S. H. (2022). The dividend decision model: A possible solution for the dividend puzzle. In *Exploring the latest trends in management literature* (pp. 249–267). Emerald Publishing Limited. https://doi.org/10.1108/978-1-80262-277-520231021

Dewasiri, N. J., Dharmarathna, D. G., & Choudhary, M. (2024). Leveraging artificial intelligence for enhanced risk management in banking: A systematic literature review. In R. Singh, S. Khan, A. Kumar, & V. Kumar (Eds.), *Artificial intelligence enabled management: An emerging economy perspective* (pp. 197–213). De Gruyter. https://doi.org/10.1515/9783111172408

Dewasiri, N. J., Karunarathne, K. S. S. N., Menon, S., Jayarathne, P. G. S. A., & Rathnasiri, M. S. H. (2023). Fusion of artificial intelligence and blockchain in the banking industry: Current application, adoption, and future challenges. In A. Saini, & V. Garg (Ed.), *Transformation for sustainable business and management practices: Exploring the spectrum of Industry 5.0* (pp. 293–307). Emerald Publishing Limited. https://doi.org/10.1108/978-1-80262-277-520231021

Elavarasan, R. M., Shafiullah, G. M., Padmanaban, S., Kumar, N. M., Annam, A., Vetrichelvan, A. M., & Holm-Nielsen, J. B. (2020). A comprehensive review on renewable energy development, challenges, and policies of leading Indian states with an international perspective. *IEEE Access, 8*, 74432–74457.

Fagerberg, J., & Normann, H. E. (2023). Innovation policy, regulation and the transition to net zero. In *Handbook of innovation and regulation* (pp. 163–191). Edward Elgar Publishing.

Feldhoff, T. (2016). Asset-based community development in the energy sector: Energy and regional policy lessons from community power in Japan. *International Planning Studies, 21*(3), 261–277.

Geroe, S. (2022). 'Technology not taxes': A viable Australian path to net zero emissions? *Energy Policy, 165*, 112945.

Gunningham, N., & Bowman, M. (2016). Energy regulation for a low carbon economy: Obstacles and opportunities. *Environmental and Planning Law Journal, 33*(2), 118–142.

Haszeldine, R. S., Flude, S., Johnson, G., & Scott, V. (2018). Negative emissions technologies and carbon capture and storage to achieve the Paris Agreement commitments. *Philosophical Transactions of the Royal Society A: Mathematical, Physical and Engineering Sciences, 376*(2119), 20160447.

Jayarathne, P. G. S. A., Sachithra, V., & Dewasiri, N. J. (2021). Sustainable practices through competitive capabilities in the apparel industry: Sri Lankan experience. *FIIB Business Review*.https://doi.org/10.1177/23197145211060398

Jayawardena, N. S., & Dewasiri, N. J. (2023). Food acquisition and consumption issues of South Asian countries: A systematic literature review and future research agenda. *FIIB Business Review*. https://doi.org/10.1177/23197145231194113

Karunarathna, K. S. S. N., Dewasiri, N. J., Singh, R., & Rathnasiri, M. S. H. (2024). What does artificial intelligence–Powered ChatGPT bring to academia? A review. In R. Singh, S. Khan, A. Kumar, & V. Kumar (Eds.), *Artificial intelligence enabled management: An emerging economy perspective* (pp. 3–13). De Gruyter. https://doi.org/10.1515/9783111172408

Kawai, E., Ozawa, A., & Leibowicz, B. D. (2022). Role of carbon capture and utilization (CCU) for decarbonization of industrial sector: A case study of Japan. *Applied Energy, 328*, 120183.

Lau, H. C., & Tsai, S. C. (2023). Global decarbonization: Current status and what it will take to achieve net zero by 2050. *Energies, 16*(23), 7800.

Li, H. X., Edwards, D. J., Hosseini, M. R., & Costin, G. P. (2020a). A review on renewable energy transition in Australia: An updated depiction. *Journal of Cleaner Production, 242*, 118475.

Li, J., & Du, Y. (2021). Spatial effect of environmental regulation on green innovation efficiency: Evidence from prefectural-level cities in China. *Journal of Cleaner Production, 286*, 125032.

Li, Y., & Kimura, S. (2021). Economic competitiveness and environmental implications of hydrogen energy and fuel cell electric vehicles in ASEAN countries: The current and future scenarios. *Energy Policy, 148*, 111980.

Li, Z., Zhang, W., Zhang, R., & Sun, H. (2020b). Development of renewable energy multi-energy complementary hydrogen energy system (A case study in China): A review. *Energy Exploration & Exploitation, 38*(6), 2099–2127.

Lim, Y., Edelenbos, J., & Gianoli, A. (2023). Dynamics in the governance of intelligent cities: Insights from South Korean smart cities. *International Journal of Urban Sciences, 27*(1), 183–205.

Liu, Z., Deng, Z., He, G., Wang, H., Zhang, X., Lin, J., & Liang, X. (2022). Challenges and opportunities for carbon neutrality in China. *Nature Reviews Earth & Environment, 3*(2), 141–155.

Lo, K. (2014). A critical review of China's rapidly developing renewable energy and energy efficiency policies. *Renewable and Sustainable Energy Reviews, 29*, 508–516.

Lu, Y., Khan, Z. A., Alvarez-Alvarado, M. S., Zhang, Y., Huang, Z., & Imran, M. (2020). A critical review of sustainable energy policies for the promotion of renewable energy sources. *Sustainability, 12*(12), 5078.

Mamat, R., Sani, M. S. M., Khoerunnisa, F., & Kadarohman, A. (2019). Target and demand for renewable energy across 10 ASEAN countries by 2040. *The Electricity Journal, 32*(10), 106670.

Martin-Roberts, E., Scott, V., Flude, S., Johnson, G., Haszeldine, R. S., & Gilfillan, S. (2021). Carbon capture and storage at the end of a lost decade. *One Earth, 4*(11), 1569–1584.

Mensah, J. (2019). Sustainable development: Meaning, history, principles, pillars, and implications for human action: Literature review. *Cogent Social Sciences, 5*(1), 1653531.

Mishra, R., Naik, B. K. R., Raut, R. D., & Paul, S. K. (2022). Circular economy principles in community energy initiatives through stakeholder perspectives. *Sustainable Production and Consumption, 33*, 256–270.

Mishra, U., Rathnasiri, M. S. H., & Dewasiri, N. J. (2024). Net-zero economy and energy sustainability: A Nepalese perspective. In R. Singh, S. Khan, A. Kumar, A. Luthra, & H. Chokshi (Eds.), *Net zero economy, corporate social responsibility and sustainable value creation. CSR, sustainability, ethics & governance* (pp. 145–161). Springer. https://doi.org/10.1007/978-3-031-55779-8_9

Moon, H., Park, S. Y., Jeong, C., & Lee, J. (2018). Forecasting electricity demand of electric vehicles by analyzing consumers' charging patterns. *Transportation Research Part D: Transport and Environment, 62*, 64–79.

Murshed, M. (2020). An empirical analysis of the non-linear impacts of ICT-trade openness on renewable energy transition, energy efficiency, clean cooking fuel access and environmental sustainability in South Asia. *Environmental Science and Pollution Research, 27*(29), 36254–36281.

Nelson, T., Nolan, T., & Gilmore, J. (2022). What's next for the Renewable Energy Target–resolving Australia's integration of energy and climate change policy? *Australian Journal of Agricultural and Resource Economics, 66*(1), 136–163.

Ortega-Ruiz, G., Mena-Nieto, A., Golpe, A. A., & García-Ramos, J. E. (2022). CO_2 emissions and causal relationships in the six largest world emitters. *Renewable and Sustainable Energy Reviews, 162*, 112435.

Oshiro, K., Masui, T., & Kainuma, M. (2018). Transformation of Japan's energy system to attain net-zero emission by 2050. *Carbon Management, 9*(5), 493–501.

Prăvălie, R., Patriche, C., & Bandoc, G. (2019). Spatial assessment of solar energy potential at a global scale. A geographical approach. *Journal of Cleaner Production, 209*, 692–721.

Qaisrani, M. A., Wei, J., & Khan, L. A. (2021). Potential and transition of concentrated solar power: A case study of China. *Sustainable Energy Technologies and Assessments, 44*, 101052.

Rafique, S., & Rossi, A. (2023). Harnessing cross-border energy resources for regional development. *Journal of Regional Connectivity and Development, 2*(2), 305–316.

Rajbhandari, S., Winyuchakrit, P., Pradhan, B. B., Chaichaloempreecha, A., Pita, P., & Limmeechokchai, B. (2024). Thailand's net-zero emissions by 2050: Analysis of economy-wide impacts. *Sustainability Science, 19*(1), 189–202.

Rangarajan, A., Foley, S., & Trück, S. (2023). Assessing the impact of battery storage on Australian electricity markets. *Energy Economics, 120*, 106601.

Rathnasiri, M. S. H., & De Silva, V. (2023). The role of shopping values, cognitive and affective evaluations on the relationship between store environment and store patronage intention: A future research agenda, Sri Lanka. *Journal of Management Studies, 5*(1), 100–120.

Rathnasiri, M. S. H., Dewasiri, N. J., & Kumar, A. A. (2024). Policy framework and implementation strategies for Sri Lanka's transition to a net-zero economy. In R. Singh, S. Khan, A. Kumar, S. Luthra, & H. Chokshi (Eds.), *Net zero economy, corporate social responsibility and sustainable value creation. CSR, sustainability, ethics & governance* (pp. 43–60). Springer. https://doi.org/10.1007/978-3-031-55779-8_3

Rathore, P. K. S., Chauhan, D. S., & Singh, R. P. (2019). Decentralized solar rooftop photovoltaic in India: On the path of sustainable energy security. *Renewable Energy, 131*, 297–307.

Rigolon, A., & Gibson, S. (2021). The role of non-governmental organizations in achieving environmental justice for green and blue spaces. *Landscape and Urban Planning, 205*, 103970.

Salahuddin, M., Alam, K., Ozturk, I., & Sohag, K. (2018). The effects of electricity consumption, economic growth, financial development and foreign direct investment on CO_2 emissions in Kuwait. *Renewable and Sustainable Energy Reviews, 81*, 2002–2010.

Schreurs, M. A. (2021). Reconstruction and revitalization in Fukushima a decade after the "triple disaster" struck: Striving for sustainability and a new future vision. *International Journal of Disaster Risk Reduction, 53*, 102006.

Shahsavari, A., & Akbari, M. (2018). Potential of solar energy in developing countries for reducing energy-related emissions. *Renewable and Sustainable Energy Reviews, 90*, 275–291.

Shang, W. L., & Lv, Z. (2023). Low carbon technology for carbon neutrality in sustainable cities: A survey. *Sustainable Cities and Society, 92*, 104489.

Singh, K., Meena, R. S., Kumar, S., Dhyani, S., Sheoran, S., Singh, H. M., Pathak, V. V., Khalid, Z., Singh, A., Chopra, K., & Bajar, S. (2023). India's renewable energy research and policies to phase down coal: Success after Paris agreement and possibilities post-Glasgow Climate Pact. *Biomass and Bioenergy, 177*, 106944.

Soomauroo, Z., Blechinger, P., & Creutzig, F. (2020). Unique opportunities of island states to transition to a low-carbon mobility system. *Sustainability, 12*(4), 1435.

Sopiana, Y., & Harahap, M. A. K. (2023). Sustainable urban planning: A holistic approach to balancing environmental conservation, economic development, and social well-being. *West Science Interdisciplinary Studies, 1*(02), 43–53.

Stock, R. (2021). Illuminant intersections: Injustice and inequality through electricity and water infrastructures at the Gujarat Solar Park in India. *Energy Research & Social Science, 82*, 102309.

Sun, B., Yu, Y., & Qin, C. (2017). Should China focus on the distributed development of wind and solar photovoltaic power generation? A comparative study. *Applied Energy, 185*, 421–439.

Thorpe, A. (2012). Too little, too soon? An assessment of Australian carbon capture and storage legislation against the new standards set for the clean development mechanism. *Climate Law, 3*(2), 139–163.

Tolliver, C., Fujii, H., Keeley, A. R., & Managi, S. (2021). Green innovation and finance in Asia. *Asian Economic Policy Review, 16*(1), 67–87.

Ugwu, M. C., Adewusi, A. O., & Nwokolo, N. E. (2024). The role of public-private partnerships in building clean energy infrastructure in the United States and Nigeria. *International Journal of Management & Entrepreneurship Research, 6*(4), 1049–1068.

Van-Voss, B. H., & Rafaty, R. (2022). Sensitive intervention points in China's coal phaseout. *Energy Policy, 163*, 112797.

Vulturius, G., Maltais, A., & Forsbacka, K. (2024). Sustainability-linked bonds–their potential to promote issuers' transition to net-zero emissions and future research directions. *Journal of Sustainable Finance & Investment, 14*(1), 116–127.

Wang, J., Ju, Y., & Fujikawa, K. (2024). Climate policies in China: Renewable energy introduction and national emissions trading scheme. In *Empirical research on environmental policies in China: China towards decarbonization and recycle economy* (pp. 3–18). Springer Nature Singapore.

Wimbadi, R. W., & Djalante, R. (2020). From decarbonization to low carbon development and transition: A systematic literature review of the conceptualization of moving toward net-zero carbon dioxide emission (1995–2019). *Journal of Cleaner Production, 256*, 120307.

Xu, G., Wang, Y., & Rehman, H. (2023). The future trajectory of carbon emissions in the process of carbon neutrality in South Korea. *Journal of Environmental Management, 345*, 118588.

Yang, L., Wang, Y., Wang, R., Klemeš, J. J., Almeida, C. M. V. B. D., Jin, M., Zheng, X., & Qiao, Y. (2020). Environmental-social-economic footprints of consumption and trade in the Asia-Pacific region. *Nature Communications, 11*(1), 4490.

Zhang, S., Wang, K., Xu, W., Iyer-Raniga, U., Athienitis, A., Ge, H., woo Cho, D., Feng, W., Okumiya, M., Yoon, G., Mazria, E., & Lyu, Y. (2021). Policy recommendations for the zero energy building promotion towards carbon neutral in Asia-Pacific Region. *Energy Policy, 159*, 112661.

Zhong, Z., Chen, Y., Fu, M., Li, M., Yang, K., Zeng, L., Liang, J., Ma, R., & Xie, Q. (2023). Role of CO_2 geological storage in China's pledge to carbon peak by 2030 and carbon neutrality by 2060. *Energy, 272*, 127165.

Chapter 2
Harnessing Human Capital for a Sustainable Future: HR's Pivotal Role Toward Net Zero Policies

Rubee Singh, David Crowther, and Kamalesh Ravesangar

Abstract The urgency of tackling climate change requires global adaptation to capture any carbon dioxide emissions. To successfully navigate this transformation, companies need to effectively mobilize their human capital, as they play a vital role in the success of the organization. This study explores the role of HR in creating a sustainable focus in human capital and for organizations is moved to its value of zero. Through a comprehensive review of existing literature, this study aims to demonstrate the observable impact of effective HR practices in reducing an organization's carbon footprint. In addition, this study discusses the challenges of green HRM, limitations of the role of HR in net zero policy and future recommendations. The findings of this study provide valuable insights for HR practitioners and organizational leaders seeking to align their staffing policies with net zero policies. By harnessing the potential of human capital, companies can become powerful agents of positive change and contribute to a sustainable future. This chapter bridges the gap between theory and practice, and provides valuable insights for HR practitioners, business leaders, and practitioners for diverse audiences such as HR functions, knowledge management, and organizational innovation specialists.

Keywords Human capital · Sustainability · Sustainable future · HRM practices · Net zero policies

---

R. Singh (✉)
Institute of Business Management, GLA University, Mathura, Uttar Pradesh, India
e-mail: dr.rubeerajput@gmail.com

Department of Management Studies, Kumaun University, Nainital, Uttarakhand, India

D. Crowther
Social Responsibility Research Network, London, UK

K. Ravesangar
Department of Accounting and Business, Tunku Abdul Rahman University of Management and Technology, Kuala Lumpur, Malaysia

R. Singh and D. Crowther (eds.), *Transition Towards a Sustainable Future*, Approaches to Global Sustainability, Markets, and Governance,
https://doi.org/10.1007/978-981-97-5756-5_2

2.1 Introduction

It is an undeniable fact that the environment is changing, while ecologists and visionaries insist that the quality of our environment is deteriorating due to drastic changes in tides and in the environment, for being a clear symptom of the global warming problem. In today's world, changes in our lifestyles and collective efforts to address urgent environmental issues are needed. The significance of ecologically conscious progress is reflected in the fact that the United Nations has established very ambitious sustainable development benchmarks for all countries to achieve by 2030. There has been splendid human capital growth has led to serious threats to herbal assets and the surroundings (Aboramadan, 2020). Public efforts to deal with this dire situation were restrained. In Southeast Asia, relentless efforts to obtain a sturdy economic system through industrialization have raised serious issues approximately the transition of these economies to an environmentally friendly economic system, with wind sinless and spherical (Afshar Jahanshahi et al., 2021). This milestone has caused many international organizations to reprioritize their focus on things other than financial gains that are more about environment care and persons' well-being. In addition to the environmental and social causes, this shift also involves increased corporate social responsibility and promotion of sustainable business practices (Ajzen, 2020). As of now, the main concern is how businesses operating in less developed countries achieve their economic objectives as well as how can they contribute toward sustainable development actively.

There has been a significant number of scholars and researchers who have examined how human capital, environmental preservation and sustainable development influence the use of energy resources (Bali Swain & Yang-Wallentin, 2020; Fatimah et al., 2020). With social and environmental responsibility taking center stage in many organizations, their objectives go beyond financial gains to encompass environmental and social goals (Wang et al., 2020). The goals that were identified by the Millennium Development Goals (MDGs) focuses on closing the development gaps especially for Chams and Garca-Blandon documented in 2019. These 17 goals are under three umbrellas which are social, economic, and environmental sustainability which is embedded in the five P's: 'Planet, People, Peace, Prosperity, and Sharing'. From this perspective of job creation SDGs become the central point considering that their main agenda is building a new economy that is 'sustainable', 'job-rich' particularly for young people. Organizations need to ensure the well-being as well as skills training for their human resources. In addition to making wise energy decisions, it is also important to protect environment from harmful emissions such as CO_2. This way both individuals and institutions either private or public sector will manage to stop pollution spreading within our societies (Ilyas et al., 2020).

Sustainable development is a vital aspect in modern business as it attempts to satisfy the demands of the current organization and individuals without threatening the natural environment of the future generation (Nadeem et al., 2017). But the traditional market models are mainly concerned with the short-term efficient utilization of organizations' natural, social, and human capital as against a long-term complex view of knowledge-based economy that is dynamically evolving and fluctuating. Ehnert (2014) indicated that, sustainable development requires the exploration of innovative options and long-term improvements of economic systems, institutions and minimization of detrimental impacts on social and global welfare. It is pertinent to consider that sustainability concerns the organization internal HR base sustainability and thus shifts attention to organizational performance. There is increasing interest in sustainability and human resource management, according to numerous studies One can define the principles of strategic human resource management (SHRM) as those that establish the employment relationship and contribute to its continuous improvement (Ehnert, 2009a; Gollan, 2000; Zaugg et al., 2001a, 2001b). More SHRM practices are needed for sustainable development because it offers HRM a field of action, this is what Ehnert (2009b) says. Sustainable development is seen as an area that requires greater SHRM involvement by Ehnert (2009b) who provides the following definition: "Strategic Human Resource Management is a deliberate or emergent tactic which uses employment policies, systems, and activities to achieve certain objectives while at the same time ensuring organizational sustainability and preventing harmful side effects". Bush (2020) stressed the importance of adoption of sustainable HRM practices to avoid any unfavorable environmental impacts. Sustainable HRM practices assist the organizations to integrate their financial, social, and environmental aspirations while catering the internal and external stakeholder expectations, minimize unintended consequences and negative reactions, and attempt to manage the stakeholder expectations. The above findings may be interpreted differently by various organizations (Lopez-Cabrales & Valle-Cabrera, 2020). Human dimensions are at the heart of net-zero energy systems, hence the transition to net-zero systems is as much a technical challenge as it is a social challenge. HRM practices involve making employees to work effectively by investing in knowledge, skills and abilities of employees. Within this context, human resource function in organizational context may be considered critical in terms of improving capacity to attract and develop talents to meet current and future business needs, providing motivation (intrinsic and extrinsic) that will influence employees' behavior in terms of low-carbon emissions (Ravesangar et al., 2024).

Sustainable management of human resources can be successful if it creates a stable job atmosphere, supports welfare and flexibility among employees, champions all-round leadership, and preserves enduring competitive edges. It also supports cost-effectiveness, individual accountability, and a harmonious work-life equilibrium (Stahl et al., 2020). This approach acknowledges the necessity of creating value for all stakeholders and regards HR as a multifaceted obligation for sustainability. It has acknowledged that the success of an organization relies on employees' motivation, and HR can be impacted adversely. Addressing climate change demands thorough planning, and it is imperative for companies to assume a pivotal role. Despite the relevance of technological progress and environmentally conscious methodologies, the primary driver of reducing carbon emissions is sometimes disregarded and that impacts HR. Here, HR assumes significance by mobilizing employees toward this critical objective. While HR typically dedicates attention to recruitment, training, and talent retention to boost productivity, this study proposes an alternative approach.

Prioritizing sustainable employee development enables HR to play a critical part in the endeavor to achieve net zero emissions.

2.2 United Nations Sustainable Development Goals

The UN Sustainable Development Goals are designed to address the unique needs of each nation. However, the latter tactic failed taking into account socio-economic aspects as well as ecology because it only focused on financial progress. The notion of sustainable development can be traced back to December 28th, 1983, when the General Assembly adopted a resolution which defined goals for International Environmental Development Council's activities. As per UN—sustainable development involves protecting nature (1987; Arulrajah & Opatha, 2016; Paillé et al., 2017). In an empirical study of all the theoretical principles mentioned above in the development of the previous definition, we will show the central flaws in this definition, reveal shortcomings and contradictions in the current concept, and understand why the efforts of international organizations to achieve global sustainability have not yet succeeded; we seek to answer the following questions: In the end, a profound concern for the future of the human race has been shown to be the clearest message of this idea: the environment is increasingly ruinous under the pressure of economic activity. This is the reason why the concept of human capital development and preparation through training and education is fundamental. To make sure that Sustainable Development Goals (SDGs) succeeds, it's important that the meaning of this is advanced through fostering personal initiatives. Highlighting the interrelationship between the SDGs, it is clear that sharing of information on sustainable development should be rationally and selectively selected through active HRD emphasis, as this will bring a better environment result. The study explored the UN Sustainable Development Goals proposal and sought to identify issues in the proposal that can be used to improve organizational performance and build better internal and external stakeholder relationships. Organizations can improve performance by introducing some different principles to guarantee a successful environmental outcome. To achieve the UN's 17 sustainable development goals depicted in Table 2.1, it is crucial to promote alternative methods and HRD. Therefore, the administration must incorporate "green learning and development" within strategic space awareness in order to address these issues through HRD practice. According to UN (2013), the Millennium Development Goals as well as the 2030 Agenda sentence noble aims and relevance to human resource development in line with gender equality, minority rights and other rights spelled out in various international instruments which can be actively pursued by HRD working jointly with governments at national level.

Table 2.1 Sustainable development goals (SDG's)

Sustainable Development Goals (SDG'S)	
• Goal 1 →	No Poverty
• Goal 2 →	Zero Hunger
• Goal 3 →	Good Health and Well-being
• Goal 4 →	Quality Education
• Goal 5 →	Gender Equality
• Goal 6 →	Clean Water and Sanitation
• Goal 7 →	Affordable and Clean Energy
• Goal 8 →	Decent Work and Economic Growth
• Goal 9 →	Industry, Innovation and Infrastructure
• Goal 10 →	Reduced Inequality
• Goal 11 →	Sustainable Cities and Communities
• Goal 12 →	Responsible Consumption and Production
• Goal 13 →	Climate Action
• Goal 14 →	Life Below Water
• Gaol 15 →	Life on Land
• Gaol 16 →	Peace and Justice Strong Institutions
• Gaol 17 →	Partnerships to achieve the Goals

2.3 Sustainable Human Capital

Long-term development requires investing in the people that work for you because they provide great benefits when properly managed such as increased productivity due to their experience or knowledge gained over time. These include abilities acquired through training at school or other institutions prior to employment which could be technical secondary school diploma plus higher education degree depending on job description but all these needs financial support not only from organizations but also government what is crucial for social progress. It is necessary to remember that human resource is part of the environment thus sustainable growth demands taking care of different aspects of life like ecosystem however most importantly; there should be deliberate efforts toward investment in them more so through education and training. Enhancing employee education and skills within a company can yield numerous benefits, such as reducing adverse environmental impacts. All aspects of human capital positively impact life expectancy, a common trend influenced by the population policies of many countries. Acknowledging human capital is believed to significantly contribute to the advancement of new eco-friendly technologies (Zhang & Li 2023). Human capital and energy consumption are highly associated with economic growth. Nevertheless, there is a strong negative correlation between research and development (R&D) spending and energy use (Shahbaz et al., 2022). Sustainable Development Goals (SDGs) largely depend on understanding the link between economic growth and green growth which includes human capital and education expenditure (Ngo et al., 2022). Overall efficiency of industrial sectors for transition toward greening economies as well as sustainable development can be achieved through management of human capital under environmental regulations plus foreign direct investments (FDI). Additionally, it should be noted that education and health have constructive influences on economic growth in relation to human capital in green economies where higher growth is encouraged (Jayadevan, 2021). With regard to the level of a country's economic development health-care expenditure largely depends on its people knowledge or skilled labor force which is measured by their literacy rates since they are more productive when healthy than sick hence this creates differences in productivity between similarly endowed countries unless one takes into account such factors otherwise known as endowments hypothesis.

Investing in the development of human capital can significantly contribute to promoting environmental sustainability in various ways. Education plays a vital role in nurturing an individual's comprehension of environmental concerns, their origins, and potential resolutions. By offering programs that focus on the environment, businesses can heighten employees' awareness of the significance of environmental conservation and the adoption of sustainable methods. Additionally, knowledgeable staff members are likely to have understanding of sustainable technologies and methods. Businesses can advocate for the adoption of environmentally friendly practices, like reducing waste, improving energy efficiency, and conserving resources, through educational programs. Besides, encouraging creativity and discovery is crucial. Allocating resources to education and research can drive the development of eco-friendly technologies and methods. A well-informed workforce is more capable of creating original solutions to tackle environmental challenges and promote sustainable manufacturing, ultimately helping companies reduce their environmental footprint.

Also operationalizing consciousness with environmentally friendly laws is fundamental. Imparting knowledge and training will help staff understand and be able to adhere to ecological regulations. Enterprises can ensure their team meets the environmental standards by imparting knowledge to employees; additionally, this will help them assimilate sustainability practices as part of their day-to-day activities. Education equips individuals with problem-solving skills that are useful in critical thinking. Entrepreneurs who have received good formal education may think about how each decision affects our surroundings thus making more environmentally friendly choices while running their businesses which may include integrating principles of sustainability into company's operation hence making it more eco-friendly.

2.4 Sustainable HRM

Employees have a crucial role in the establishment and enhancement of an organization's human and social capital. They are not only valuable sources of knowledge and support, but also integral to sustainable human resource development and implementation. Recognizing the significance of employees, organizations should reorient their focus from mere human resource management to the comprehensive development of their workforce, particularly in an uncertain labor market. Sustainable HR has emerged to address societal changes, labor market dynamics, and employment relationships, with the goal of enhancing employee retention (Macke & Genari, 2019). It goes further than normal HRM by including a broader range of objectives such as economic, social, environmental, and organizational impacts. Organizations can improve their workforce management and foster rejuvenation, expansion as well as innovation through implementing sustainable HRM practices (Ravesangar et al., 2024). In line with this organizations need policies that promote sustainability regarding employee engagement, performance effectiveness at work tasks among other things may include opportunities for enhancing well-being or reducing stress levels while supporting work-life balance. Sustainable HRM integrates traditional human resource management concepts with those from corporate sustainability, social responsibility sustainable work systems but also draws on them from other disciplines too—it is therefore an interdisciplinary field (De Prins et al., 2014). Most HRM models emphasize the importance of workplace flexibility in managing personnel within organizations. Favorable working conditions, sustainable leadership, collaboration, and inclusiveness, ethical governance, value creation, health and safety, employee engagement, and environmental sustainability all fall within the purview of sustainable HRM. Sustainability becomes the guiding principle of HRM itself, encompassing the dual tasks of promoting the enduring sustainability of individual employees and the overall HRM system to attract, retain, and develop motivated and engaged employees.

Cohen et al. (2012) underscored three major aspects of sustainable human resource management (HRM), which are equity, well-being and development of staff. These elements are of great significance to our study, more so the aspect of employee growth. Going further, Zaugg (2001a) looked into more dimensions of sustainable HRM that include flexibility, staff engagement, principles, strategy competence and expertise among others stakeholder involvement as well nurturing trustworthy staff employer relationships. Additionally, six themes Park (2017) associated with sustainable HRM were connected to it such as diverse workforce; staff development; flexibility; voluntary work; employee welfare; environmentally friendly HRM. However, Nyameh (2013) came up with different model for sustainable hrm and organizational performance which focused on training, incentives and involvement as key drivers for achieving positive outcomes like job satisfaction motivation and retention of an employee.

Additionally, HRM tactics execute a significant function of attracting, inspiring, and retaining employees which in return ensures the organization steadfastness. Human capital's importance toward the achievement of the organizational objectives is widely recognized hence effective implementation of HRM practices increases staff commitment (Roscoe et al., 2019). Also, job satisfaction and organizational commitment are part of well-being among employees that comes as a result of high-performance management impacted on by organizational performance (Sharma et al., 2021). In addition to this point, training and development among other HRM practices lead to improved worker productivity thus promoting organizational efficiency.

In other words, through training they get equipped with necessary knowledge and skills needed for them perform their duties satisfactorily hence higher output. Moreover, sustainable HRMs characteristics have a significant effect on job satisfaction level of workers. So then for us to make sure we are able maintain these best practices within our organizations there should be provision favorable working environment which meets all needs for people to work comfortably (Macke & Genari, 2019).

2.5 HR's Pivotal Role Toward Net Zero Policies:

The term "net zero" refers to the state where the carbon dioxide emissions produced and removed by human activities in the atmosphere are in equilibrium. In other words, the emissions not absorbed are equal to the emissions removed, resulting in no net emissions within the atmosphere (Singh et al., 2024). Collaboration to fight against climate change is needed, including from businesses, and as businesses pursue net-zero emissions, or balancing what is produced with what is removed, Human Resource's (HR) role in the net zero journey is critical. HR can be a leader in promoting sustainability in the workplace, with HR professionals being at the forefront of this journey. There are three main ways HR can assist in progress toward net zero, by first creating a culture of sustainability, second by promoting eco-friendly office behaviors, and third by hiring a workforce that lives sustainability, shown here in Fig. 2.1.

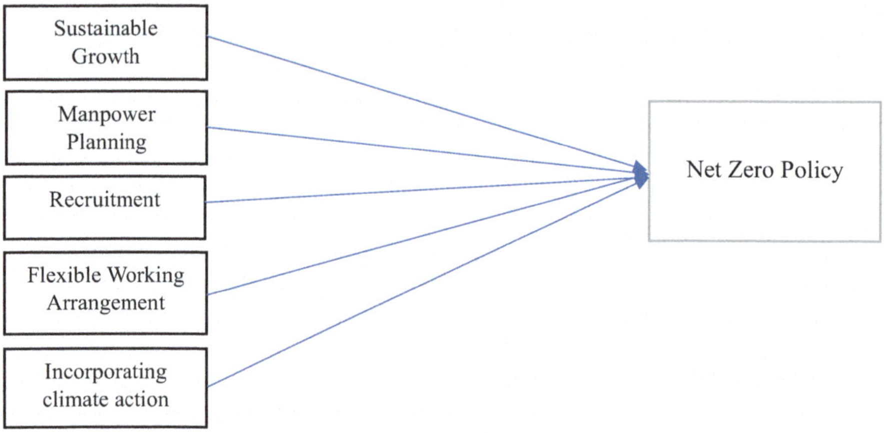

Fig. 2.1 Conceptual framework of HR's role toward net zero policies

2.5.1 Sustainable Growth

Despite the daunting task at hand, it is important that business and HR leaders should take responsibilities in addressing the pressing issue of climate change. As organizations consider ways to recover and stimulate growth in the aftermath of the pandemic, there is a unique opportunity to integrate sustainability into both their business processes and day-to-day operations, and they walk thus reducing carbon footprint. Although strategic human resource development (SHRD) is predominantly additive, concentrating on the economic and financial outcomes of human capital in organizations, human resource management (HRM) is more transformative, creating a new workplace that includes internal and external employees and diverse community stakeholders to enhance awareness and accountability for environmental protection (Chams & Garcia-Blandón, 2019). This new workplace will eventually lead to organizations and their employees increasing their resource allocation and consumption of resources within the organization for production, under conditions of competition. Ehnert et al. (2014) further argue that sustainable development requires the coexistence of three key components: human responsibility, corporate efficiency, and infrastructure. Furthermore, HRM plays an important role in contributing to sustainable performance supporting four factors (Esen & Süral, 2020). Human Resource Capability are fundamental in the organization's ability to maintain a cycle of learning, HRM and shared psychological agreements must include the need for planning and developing a commitment from all parties to identify an increasing need for understanding effective HR processes to suit the different perspectives of stakeholders.

2.5.2 Manpower Planning

At the strategic level, HR professionals have the ability to influence and affect the employee consequences of investing in innovative green technologies or technologies aimed at reducing the impact of goods and services on the ecosystem directly. For example, in the construction industry, the integration of heat pumps can contribute significantly to reducing the carbon footprint of UK construction (Shamoushaki & Koh, 2023) but there are significant skills development challenges, particularly through training involves sufficient engineers capable of installing these so planning It is important for employers in this sector to engage in strategic workforce planning to anticipate and address emerging skills gaps or deficiencies. This may involve partnerships with local educational institutions to nurture the skills gap. Across all companies, strategic employee management is a powerful tool for HR leaders to drive organizational strategies to track every aspect of carbon. A strategic personnel strategy begins with a comprehensive PESTLE analysis, which examines the broader political, environmental, technological, legal and social contexts in which organizations operate (Wu, 2020). This study seeks to highlight the implications for current and future professional skills development of changes in business models or paradigms.

2.5.3 Recruitment

Strengthening the organization's commitment to climate change in job ads or during interviews could better reach potential candidates and encourage them to want to work for the company. Additionally, there are steps that can be taken to lower the carbon footprint of the hiring process. These can include a shift in thinking toward digital workflows and more virtual interviews to cut down on unnecessary travel (which can be applied sooner due to the outbreak). In addition, data concerning the environmental goals of the organization and the expectations toward employees will be incorporated into the onboarding process. According to Ogbu Edeh and Okwurume (2019), Green Recruitment (GR) is defined as an approach that lays out an organization's environmental sustainability orientation as a consideration in the recruitment process. GR uses digital tools to reduce environmental impact. They use online channels such as email, web-based job applications and global talent pools to attract candidates. Some of the measures implemented in the GR program include the use of telecommunication gadgets such as video or telephone interviews in order to curb environmental costs that are associatively related to traveling known by Mwita and Kinemo (2018). This also identified that unlike the outdated methods for example through producing a brochure flyer or an advertisement in the newspapers, online recruitment allows for the posting of considerable amounts of information at one time which in more moderation minimizes damage to the environment effectively (Davidescu et al., 2020).

There is a heightened emphasis on the need to incorporate green initiatives as a result of the increasing awareness and emphasis on environmental sustainability by both employees and employers. Today, so many talent acquisition professionals want to work for a company, which supports such environmentally friendly recruitment measures as favorable consequences like increased efficiency, decreased expenses, and employees' turnover (Pham & Paillé, 2020).

2.5.4 Flexible Working Arrangement

Introducing more versatile and diverse business models can go a long way toward reducing an organization's carbon footprint by reducing emissions from outsourcing employers need to consider the impact of remote work on resources environment and support sustainable practices such as switching to renewable energy suppliers. HR professionals can help create a travel plan that doesn't provide an ecosystem for travel and work-related travel. In addition, training and development programs can be optimized to reduce the need for employees to go to work through a mixture of methods and technologies such as virtual training and collaborative online learning to reduce environmental impact.

One of the most important prerequisites for obtaining global competitive advantage by organizations is the ability to have flexible processes that enable their employees to arrange working hours and locations while discussing with clients, globally. Another aspect relates to the image commonly perceived by the target audience as a friendly and supportive family-oriented company interested in environmental issues. Based on Choi (2018), those organizations that support the values of corporate social responsibility and value environmental conservation should know that resilience is fundamental. A flexible workplace offers tremendous benefits to employees, not just in economic terms. Reduced business hours, flexible start, and finish times make it easier to use public transportation, bike, or walk to get between home and work. According to Herttuala et al. (2020), offering flexible work options such as telecommuting, leave, and adjustable schedules not only improves work-life balance and well-being but also leads to more environmentally friendly workplaces with reduced traffic, lower fuel use, and decreased emissions. This positive correlation between satisfied employees and pro-environmental practices extends beyond the workplace, impacting home life as well.

2.5.5 *Incorporating Climate Action Mindset*

Small adjustments to policies in human resources and personnel management can cumulatively result in a considerable decrease in an organization's carbon footprint. To demonstrate that HR professionals can be an example for the employees, eco-friendly practices should be incorporated through collaborating with various departments. When coming up with effective business growth and profitability plans, it is important to have an economic, social, and environmental considerations based on sustainable or net zero approach. This often involves actively involving employees in reshaping the organization's operations, highlighting the significance of sustainable HR management. For example, picture an organization constructing an environmentally conscious manufacturing facility by incorporating energy-efficient features and utilizing renewable energy sources like solar power (Wamsler et al., 2020). The entire process, from engaging architects to implementing recycling programs, requires skilled and knowledgeable employees who can design and execute sustainable practices. It is essential to align employees' mindset and values with organizational goals to embed the net zero policy into the organization's core operations, which the organization heavily depends on. Sustainability is deeply ingrained in key components of effective governance, ethical conduct, and compliance with legal requirements (Accenture, 2016). However, it is not just limited to complying to create more avenues in reducing business risks, improving business standing image, and seizing business advances. Through a sustainability and net zero strategy, as a company, one also helps to improve the state of the planet; at the same time, as a business, it becomes more future-proof. It is revolutionizing the pattern by which business organizations formulate their strategies, come to decisions, implement their operations, manage their human resources and customer and stakeholder relations, not to mention meeting the multifaceted expectations of these entities in the growing and openly networked age of commerce. For this, sustainable business development also needs the support of an organizational culture that integrates the strategic vision of the company.

2.6 Resource-Based View (RBV) Theory

Therefore, green HRM as used in the context of the current description pertains to the integration of green practices in the HRM with the need to manage the organization's resources efficiently based on RBV theory advanced by Barney (1991). Nonetheless, the critics argue that the RBV theory is static and does not explain how firms mobilize their resources and capabilities to create value premium, and does not take into consideration external environment in the management of resources (Bettis & Hitt, 1995). However, the RBV theory deals with the P and E conditions or environment of an organization, and although claiming on the exploitation of organizational resources and efficiency for gaining competitive advantage, it neglects the aspects related to the environment for the attainment of strategic advantage more acutely in dynamic market structures (Wiggins & Ruefli, 2002). Therefore, to fill the gaps started with environmental researches the new theory known as dynamic capabilities was proposed in addition to the RBV theory (Teece et al., 1997). Although many scholars have utilized RBV to examine organizational strategies concerning the environment, there are few studies on the dynamic capabilities' perspective of sustainability. This perspective suggests the firm's ability to develop as well as to acquire new capabilities and competencies to create new theories as well as strategies, stances, and resources that lead to long-run sustainable competitive edge (Amui et al., 2017; Bhupendra & Sangle, 2015). Amui et al. (2017) have also stressed on the positive link between dynamic capability and the performance contribution of eco-capability resources which is an important area of research that has not been explored much by scholars and researchers.

The theory of dynamic capabilities, originally developed by Teece et al. (1997), looks into how organizations learn and transform their resources internally and externally based on the changes that occur within the business environment. While organizational capabilities do not possess such formulations, dynamic capabilities are framed to tackle changes in the environment (Winter, 2003). March (1991) draws a clear line between organization capabilities and dynamic capabilities, the former is often explained as the excellent management of existing resources, whereas the latter is the ability to create new wealth by repositioning the firm. Dynamic capability relates to a firm's ability of gaining efficiencies in the identification of market signals besides changes to external context and the available resources, making the firm more productive and thus perform better (Barreto, 2010). So, researchers' discussions and analyses about the concept of dynamic capabilities have been marked by a lack of consensus on its core meaning. Teece et al. (1997) proposed three: The three basic activities described in the imperialism literature are sensing, seizing, and reconfiguring/transforming. Competencies involving sensing enable an organization to identify opportunities and threats in a context that includes competitors, Customer, Market Intelligence and being able to predict customers' needs before competitors in the operating environment. In other words, sensing capability assists manufacturing industries to obtain essential and pertinent information regarding internal and external settings and Milieus as a way of catering for the customers' needs and hence gaining competitive edge. Seizing capabilities or the ability of strategically capturing new opportunities is the opportunity for an organization focusing on acquiring new enterprise structures, procedures, designs, interests, and restraints that complement to the business strategy of the company (Jiao et al., 2011). Transformational capabilities refer to the process of reconfiguring tangible and intangible resources across the organization to assist the firm to mold itself to the operating environment. All in all, these capabilities were mirrored to the development of dynamic capabilities in an organization and assist the management to find the new value via resources integration.

2.7 Challenges of Implementing Green HRM Practices

Sustainable Human Resource Management (HRM), which deals with the integration of environmental issues into company's human resources department, is considered to be a major opportunity in achieving net-zero policies. However, several barriers may hinder its effective implementation such as the process of behavior change which will be time-consuming. It is essential to exercise patience and perseverance when endeavoring to redirect employee behavior toward environmentally sustainable practices (Wang et al., 2019). Cultivating eco-friendly habits, parallel to acquiring any new skill, necessitates consistent effort. Individuals require patience in order to internalize and integrate actions such as turning off lights or utilizing reusable containers into their daily routine. Some individuals may not immediately recognize the benefits of sustainability. Explaining the positive environmental impacts and the financial advantages of eco-friendly choices can promote greater employee engagement.

Furthermore, employee receptiveness to Green HRM initiatives will not be standardized. Some employees are enthusiastic environmental advocates and readily endorse green initiatives, while others may lack knowledge or interest in environmental matters (Roscoe et al., 2019). Workplace changes could potentially cause disruptions, as employees heavily reliant on paper documents may be hesitant to transition to paperless operations. Certain employees may demonstrate greater resistance to change compared to their peers. Individuals who prefer established routines may be more reluctant to embrace new environmentally friendly practices. Moreover, the personal values of certain employees may not entirely align with sustainability objectives. Offering alternative options and emphasizing the long-term benefits can facilitate acceptance. Prior investment in Green Human Resource Management (HRM) practices, including training, infrastructure modifications, or technological enhancements, may not result in immediate returns. Quantifying the environmental benefits in financial terms can be complex, leading to a delayed return on investment (ROI) (Xie et al., 2022). Several Green HRM practices offer long-term benefits that require accumulation over time. For instance, educating employees on energy conservation may not immediately generate cost savings on electricity bills, but it can lead to substantial savings in the long term. Some benefits of Green HRM practices are indirect and pose greater challenges in measurement. For example, prioritizing employee well-being through green initiatives such as improved air quality or access to natural light can enhance employee morale and productivity, although these gains may not be immediately evident in financial reports.

Identifying employees who possess both the necessary qualifications and a strong commitment to sustainability can be a challenging endeavor. The field of sustainability is still in its emerging stages compared to more traditional professions, resulting in a limited pool of candidates with the requisite skills and experience for sustainability-focused roles. Despite having enthusiastic candidates, there may be a discrepancy between their passion and the technical proficiency demanded by the position (Khan et al., 2020). Assessing a candidate's dedication to sustainability can be a subjective and challenging task during the interview process. Conventional interview techniques may not work well to identify candidates who truly are committed to the organization's environmental/ social responsibility agenda. In fact, assessing the effectiveness of Green HRM is one of the HRM challenges. Deciding whether and to what extent Green HRM strategies have impacted the behavior of employees can be quite challenging. There are some factors like increased environmental awareness or personal sense of responsibility for sustainability that cannot be easily captured by traditional HR metrics (Rashid & Alam, 2020). The effects of going green through Green HRM practices in changing the behavior of employees regarding wastage or energies use might not be easily noticed. It necessitates long-term data collection and analysis to establish a clear cause-and-effect relationship. There is no universally applicable method for assessing the effectiveness of Green HRM. The metrics employed will vary based on specific goals and initiatives. Establishing a comprehensive measurement framework can be a difficult undertaking.

2.8 Limitations of Study on HR's Role in Net Zero Policies

Due to the need to reduce net zero emission by 2050, there has been a significant change to all the spheres of the corporate activities (Sun et al., 2021). HR departments are becoming more and more important in helping companies make the move to remote work. HR strategies can have a big impact on an organizations environmental impact by shaping how employees behave, helping them develop skills, and promoting the efficient use of resources (Islam et al., 2021). Despite the increasing interest in the impact of HR on accomplishing net zero objectives, scholarly research in this domain remains initial, with prior reviews underscoring various limitations.

Prevailing these constraints is critical for attaining a comprehensive comprehension of how HR can effectively strengthen the pursuit of this ambitious target. HR practices in the context of net zero policies are continuously evolving, as evidenced in antecedent studies. Companies are presently investigating diverse approaches, encompassing guidelines for remote work, campaigns for paperless offices, and eco-friendly performance appraisals. This ongoing exploration creates a dynamic landscape where effective strategies are consistently evolving. While this fosters creativity, it stimulates a challenge for scholars seeking to establish broadly applicable best practices, as highlighted in preceding analyses. Moreover, evaluating the precise impact of HR strategies in diminishing emissions poses a challenge that has been stressed in prior research (Mahmood & Nasir, 2023). Various HR initiatives, such as promoting a sustainable ethos or engaging employees in environmental initiatives, produce complex secondary effects that are difficult to separate and accurately quantify. Traditional HR metrics may not adequately capture these indirect environmental benefits, as highlighted in antecedent research. The establishment of dependable metrics that can effectively assess the environmental ramifications of HR practices remains an ongoing undertaking.

Research has demonstrated that the most efficient HR strategies for achieving net zero emissions can significantly differ based on the industry (Obeidat et al., 2018). For example, a manufacturing company with a substantial physical presence and high energy consumption may necessitate different approaches in contrast to a service-oriented business with a primarily digital operation. Neglecting these industry-specific variations in resource utilization and environmental impact can constrain the applicability of research findings. Therefore, further inquiry is essential to comprehend how HR strategies can be tailored to suit the distinct attributes of various industries. For example, a study investigating HR's contribution to the goal of net zero emissions in the manufacturing industry may find that the primary areas may include the encouragement of energy saving in manufacturing and the adoption of measures to reduce waste. Conversely, studies in the service sector may examine the effectiveness of remote work policies and paperless office practices. Previous appraisals have highlighted that the prevailing research landscape may favor large corporations with dedicated sustainability teams and greater resources to execute extensive sustainability programs. In contrast, smaller businesses often encounter challenges due to their constrained personnel and financial capacity to invest in large-scale sustainability initiatives (Ben Amara et al., 2020). To address this inequality, it is imperative for research to concentrate on bridging the gap, as highlighted in prior appraisals, by investigating how HR strategies can be customized for smaller organizations with limited resources. Such pursuit might involve examining economical green HR practices, evaluating prospects of partnership with other companies as well as evaluating possibilities of introducing innovations into the process of implementing sustainability initiatives.

2.9 Conclusion and Future Directions

The pursuit of achieving net zero emissions has driven a fundamental shift across all domains of business. Traditionally focused on managing employee relations and talent, Human Resources (HR) departments have now assumed a pivotal role in catalyzing this environmental revolution. HR methodologies possess the potential to significantly influence an organization's environmental footprint by shaping employee conduct, facilitating skill enhancement, and optimizing resource consumption patterns. Nevertheless, the body of research in this domain is still in its emerging stages, and it is imperative to address certain limitations. By acknowledging these constraints and charting a course for the future, HR departments can fully harness their capability to strengthen organizational sustainability.

The challenges encountered in scrutinizing the role of HR in net zero policies indicate a path for the future despite their limitations. Through the adoption of a data-driven approach, nurturing a sustainable attitude, streamlining HR processes, engaging with key stakeholders, and leveraging technology, HR departments can emerge as influential champions in combatting climate change. This transformation necessitates a fundamental overhaul in HR practices, yet the potential rewards, such as a healthier planet and a sustainable future, are immeasurable.

To ensure a robust data foundation for assessing the environmental impact of HR practices, it is imperative to foster collaboration between HR professionals and researchers. There are some key measures are recommended. It is essential to engage in collaborative efforts with environmental researchers to structure studies aimed at gathering data on the environmental implications of HR practices. These endeavors will aid in establishing causal relationships between specific HR undertakings and quantifiable decreases in emissions. In addition, research can assist in comprehending the requirements, inclinations, and likely obstacles of employees concerning sustainable behaviors. Subsequently, HR personnel can formulate approaches that are not only efficient but also positively perceived by the workforce.

Besides, cooperate with sustainability experts and researchers to devise standardized metrics for evaluating the environmental impact of diverse HR practices will be one of the effective methods. These metrics should encompass both direct (e.g., reduction in paper usage resulting from a paperless office policy) and indirect (e.g., enhanced employee morale leading to reduced turnover and emissions related to recruitment) environmental benefits. The utilization of clear metrics to monitor progress fosters transparency and accountability within the organization. HR teams can effectively showcase the influence of their practices on reducing emissions. Additionally, using data analysis tools to identify patterns and best practices that support achieving net zero goals is beneficial. This data can then be used to develop flexible HR strategies for various sectors. Building a strong data foundation allows HR departments to move beyond anecdotal evidence and show the measurable environmental impact of their efforts. These data are essential for gaining support from leadership, using resources effectively, and constantly improving HR practices to maximize environmental advantages.

In order to effectively align HR practices with environmental impact, it is crucial to strategically coordinate different HR functions with the organization's net zero targets. There are several key principles to consider. One important principle is to implement remote work policies whenever possible to decrease employees' commuting emissions and office energy usage. This can be accomplished by providing employees with the necessary technology and infrastructure for efficient remote work. Remote work reduces the need for office space, leading to less heating, cooling, and lighting usage. Additionally, reduced physical presence of employees lowers the continuous operation of office equipment, such as printers and computers, resulting in significant energy savings. Set up green procurement policies that ensure that prospective suppliers and supplies are environmentally sound as well. Work together with your suppliers to find sustainable alternatives and minimize the carbon footprint of your organization's entire supply chain. Further collaboration with suppliers to discover the potential for reducing the environmental footprint of their products or processes is also essential for your organization. This may involve the exploration of alternative materials, improving the methods of production or establishing steps for waste reduction.

Moreover, it is crucial to support eco-friendly travel options like video conferencing, train travel, and carbon offsetting for necessary business trips. Also, encourage environmentally friendly practices in the company by discouraging unnecessary travel and promoting carpooling for crucial face-to-face meetings. One way to do this is by creating a corporate travel policy that demonstrates the company's commitment to sustainability. This policy can outline a preference for sustainable travel choices, urge trip consolidation, and discourage unnecessary travel. Furthermore, it is crucial to create career advancement initiatives aimed at providing workers with the skills and expertise to lead sustainability endeavors. This could include educational courses on environmental laws, eco-friendly business techniques, and the integration of environmentally friendly technology. It is also advisable to develop career paths within the human resources department that concentrate on sustainability, establishing specific positions to guide and supervise the company's environmental campaigns. By aligning HR strategies with net zero goals, HR departments can create a beneficial impact across the organization. This will help decrease emissions and appeal to and keep environmentally aware employees who are looking for companies dedicated to sustainability.

References

Aboramadan, M. (2020). The effect of green HRM on employee green behaviors in higher education: The mediating mechanism of green work engagement. *International Journal of Organizational Analysis, 30*, 7–23.

Accenture. (2016). *Great expectations: Redesigning the employee experience for the workforce of the future*. www.accenture.com/t00010101T000000__w__/gb-en/_acnmedia/PDF-26/Accenture-Strategy_UKGradsStudy_GreatExpectations_PoV.pdf

Afshar Jahanshahi, A., Maghsoudi, T., & Shafighi, N. (2021). Employees' environmentally responsible behavior: The critical role of environmental justice perception. *Sustainability: Science, Practice and Policy, 17*, 1–14.

Ajzen, I. (2020). The theory of planned behavior: Frequently asked questions. *Human Behaviour and Emerging Technologies, 2*, 314–324.

Amui, L. B. L., Jabbour, C. J. C., de Sousa Jabbour, A. B. L., & Kannan, D. (2017). Sustainability as a dynamic organizational capability: A systematic review and a future agenda toward a sustainable transition. *Journal of Cleaner Production, 142*, 308–322. https://doi.org/10.1016/j.jclepro.2016.07.103

Arulrajah, A. A., & Opatha, H. H. (2016). Analytical and theoretical perspectives on green human resource management: A simplified underpinning. *International Business Resource, 9*(12), 153. https://doi.org/10.5539/ibr.v9n12p153

Bali Swain, R., & Yang-Wallentin, F. (2020). Achieving sustainable development goals: Predicaments and strategies. *International Journal of Sustainable Development & World Ecology, 27*(2), 96–106. https://doi.org/10.1080/13504509.2019.1692316

Barreto, I. (2010). Dynamic capabilities: A review of past research and an agenda for the future. *Journal of Management, 36*(1), 256–280. https://doi.org/10.1177/0149206309350776

Barney, J. B. (1991). Firm resources and sustained competitive advantage. *Journal of Management, 17*, 99–120

Ben Amara, D., Chen, H., & Hafeez, M. (2020). Role of entrepreneurial opportunity identification factors in the eco-innovation of agribusiness. *Business Strategy & Development*, 1–14. https://doi.org/10.1002/bsd2.107

Bettis, R. A., & Hitt, M. A. (1995). The new competitive landscape. *Strategic Management Journal, 16*(S1), 7–19. https://doi.org/10.1002/smj.4250160915

Bhupendra, K. V., & Sangle, S. (2015). What drives successful implementation of pollution prevention and cleaner technology strategy? The role of innovative capability. *Journal of Environmental Management, 155*, 184–192. https://doi.org/10.1016/j.jenvman.2015.03.032

Bush, J. T. (2020). Win-Win-Lose? Sustainable HRM and the promotion of unsustainable employee outcomes. *Human Resource Management Review, 30*(3), 100676. https://doi.org/10.1016/j.hrmr.2018.11.004

Chams, N., & García-Blandón, J. (2019). On the importance of sustainable human resource management for the adoption of sustainable development goals. *Resources, Conservation and Recycling, 141*(6), 109–122.

Choi, S. (2018). Managing flexible work arrangements in government: Testing the effects of institutional and managerial support. *Public Personnel Management, 47*(1), 26–50.

Cohen, E., Taylor, S., & Muller-Camen, M. (2012). HRM's role in corporate social and environmental sustainability. *SHRM Representative, 1*, 1–16.

Davidescu, A. A., Apostu, S.-A., Paul, A., & Casuneanu, I. (2020). Work flexibility, job satisfaction, and job performance among Romanian employees—Implications for sustainable human resource management. *Sustainability, 12*, 6086.

De Prins, P., Van Beirendonck, L., De Vos, A., & Segers, J. (2014). Sustainable HRM: Bridging theory and practice through the 'Respect Openness Continuity (ROC)'-model. *Management Review, 25*, 263–284.

Ehnert, I., Harry, W., & Zink, K. (2014). Sustainability and HRM. In I. Ehnert, W. Harry, & K. Zing (Eds.), *Sustainability and human resource management: Developing sustainable business organizations* (pp. 3–32). Springer.

Ehnert, I. (2009a). *Sustainability and HRM. sustainability and human resource management* (pp. 3–32). Springer.

Ehnert, I. (2009b). *Sustainable human resource management. A conceptual and exploratory analysis from a paradox perspective* (pp. 33–78). Physica-Verlag.

Ehnert, R. C. (2014). *Corporate sustainability: A stakeholder management approach* (2nd ed.). Greenleaf Publishing.

Esen, D., & Süral, Ö. P. (2020). Sustainable human resource management (HRM): A study in Turkey context and developing a sustainable HRM questionnaire. *International Journal of Management Economics and Business, 16*(3), 550–580.

Fatimah, Y. A., Govindan, K., Murniningsih, R., & Setiawan, A. (2020). Industry 4.0 based sustainable circular economy approach for smart waste management system to achieve sustainable development goals: A case study of Indonesia. *Journal of Cleaner Production, 269*, 122–132. https://doi.org/10.1016/j.jclepro.2020.122263

Gollan, P. J. (2000). Human resources, capabilities and sustainability. In D. Dunphy, J. Benveniste, A. Griffiths, & P. Sutton (Eds.,) *Sustainability: The corporate challenge for the 21st century* (pp. 55–77). Allen and Unwin.

Herttuala, N., Kokkinen, L., & Konu, A. (2020). Social-and healthcare managers' work wellbeing—Literature review and key informant interviews. *International Journal of Workplace Health Management, 13*(6), 633–648. https://doi.org/10.1108/IJWHM-05-2019-0077

Islam, T., Khan, M. M., Ahmed, I., & Mahmood, K. (2021). Promoting in-role and extra-role green behavior through ethical leadership: Mediating role of green HRM and moderating role of individual green values. *International Journal of Manpower, 42*, 1102–1123. https://doi.org/10.1108/IJM-01-2020-0036

Ilyas, G., Rahmi, S., Tamsah, H., Munir, A., & Putra, A. H. P. K. (2020). Reflective model of brand awareness on repurchase intention and customer satisfaction. *Journal of Asian Finance, Economics and Business, 7*(9).

Jayadevan, C. M. (2021). Impacts of health on economic growth: Evidence from structural equation modelling. *Asia-Pacific Journal of Regional Science, 5*, 513–522. https://doi.org/10.1007/s41685-020-00182-4

Jiao, H., Alon, I., & Cui, Y. (2011). Environmental dynamism, innovation, and dynamic capabilities: The case of China. *Journal of Enterprising Communities: People and Places in the Global Economy, 5*(2), 131–144. https://doi.org/10.1108/17506201111131550

Khan, N. U., Bhatti, M. N., Obaid, A., Sami, A., & Ullah, A. (2020). Do green human resource management practices contribute to sustainable performance in manufacturing industry? *International Journal of Environment and Sustainable Development, 19*, 412–432.

Lopez-Cabrales, A., & Valle-Cabrera, R. (2020). Sustainable HRM strategies and employment relationships as drivers of the triple bottom line. *Human Resource Management Review, 30*(3), 100689.

Macke, J., & Genari, D. (2019). Systematic literature review on sustainable human resource management. *Journal of Cleaner Production, 208*, 806–815.

Mahmood, F., & Nasir, N. (2023). Impact of green human resource management practises on sustainable performance: Serial mediation of green intellectual capital and green behaviour. *Environmental Science and Pollution Research, 30*, 90875–90891. https://doi.org/10.1007/s11356-023-28541-6

March, J. G. (1991). Exploration and exploitation in organizational learning. *Organization Science*, 71–87. https://doi.org/10.1287/orsc.2.1.71

Mwita, K. M., & Kinemo, S. M. (2018). The role of green recruitment and selection on performance of processing industries in Tanzania: A case of Tanzania tobacco processors limited (TTPL). *International Journal of Human Resource Studies, 8*, 3546.

Nadeem, S., Khan, A. S., & Chishti, N. S. (2017). Sustainable development in the textile dyeing industry of Pakistan. *Journal of Cleaner Production, 162*, 1124–1135. https://doi.org/10.1007/978-3-030-38545-3_2

Ngo, T., Trinh, H., Haouas, I., & Ullah, S. (2022). Examining the bidirectional nexus between financial development and green growth: International evidence through the roles of human capital and education expenditure. *Resources Policy*.

Nyameh, J. (2013). Impact of sustainable human resource management and organizational performance. *International Journal of Asian Social Science, 3*, 1287–1292.

Obeidat, S. M., Al Bakri, A. A., & Elbanna, S. (2018). Leveraging "green" human resource practices to enable environmental and organizational performance: Evidence from the Qatari oil and gas industry. *Journal of Business Ethics*. https://doi.org/10.1007/s10551-018-4075-z

Ogbu Edeh, F., & Okwurume, C. N. (2019). Green human resource management and organisational sustainability of deposit money banks in Nigeria. *Journal of Management and Science, 3*, 224–235.

Paillé, P., Chen, Y., Boiral, O., & Jin, J. (2017). The impact of human Resource management on environmental performance: An employee-level study. *Journal of Business Ethics, 121*(3), 451–466. https://doi.org/10.1007/s10551-013-1732-0

Park, L. Y. (2017). EOR-based HRM system and organizational ambidexterity. *Academy of Management*.

Pham, D. D. T., & Paillé, P. (2020). Green recruitment and selection: An insight into green patterns. *International Journal of Manpower, 41*, 258–272.

Rashid, A., & Alam, W. (2020). The impact of green human resource management on organizational performance: A case of manufacturing industry. *City University Research Journal, 10*, 289–299.

Ravesangar, K., Ping, L. L., & Pachar, S. (2024). A review on the sustainable HRM practices in building net-zero transformation: An emerging trend in the workplace. In *Net zero economy, corporate social responsibility and sustainable value creation*. Springer. https://doi.org/10.1007/978-3-031-55779-8_8

Roscoe, S., Subramanian, N., Jabbour, C. J. C., & Chong, T. (2019). Green human resource management and the enablers of green organisational culture: Enhancing a firm's environmental performance for sustainable development. *Business Strategy and Environment*.

Shahbaz, M. M., Song, S., Ahmad, & Vinh, X. (2022). Does economic growth stimulate energy consumption? The role of human capital and R&D expenditures in China. *Energy Economics*.

Sharma, S., Prakash, G., Kumar, A., Mussada, E. K., Antony, J., & Luthra, S. (2021). Analysing the relationship of adaption of green culture, innovation, green performance for achieving sustainability: Mediating role of employee commitment. *Journal of Cleaner Production*.

Shamoushaki, M., & Koh, S. C. L. (2023). Heat pump supply chain environmental impact reduction to improve the UK energy sustainability, resiliency and security. Scientific Reports, 13, 20633. https://doi.org/10.1038/s41598-023- 47850-x

Singh, R., Ravesangar, K., Narayanan, S., & Himat Singh, K. K. (2024). HRM practices in creating net zero: Supporting organizational sustainability. In *Net zero economy, corporate social responsibility and sustainable value creation*. Springer. https://doi.org/10.1007/978-3-031-55779-8_1

Stahl, G. K., Brewster, C. J., Collings, D. G., & Hajro, A. (2020). Enhancing the role of human resource management in corporate sustainability and social responsibility: A multi-stakeholder, multidimensional approach to HRM. *Human Resource Management Review, 30*(3), 100708. https://doi.org/10.1016/j.hrmr.2019.100708

Sun, T., Ocko, I. B., Sturcken, E., & Hamburg, S. P. (2021). Path to net zero is critical to climate outcome. *Science and Reports, 11*(1), 22173. https://doi.org/10.1038/s41598-021-01639-y

Teece, D. J., Pisano, G., & Shuen, A. (1997). Dynamic capabilities and strategic management. *Strategic Management Journal, 17*(7), 509–533.

United Nations Documents. (1987). Report of the world commission on environment and development: Our common future.

United Nations. (2013). United Nations millennium development goals. http://www.un.org/millenniumgoals/bkgd.shtml

Wamsler, C., Schapke, N., Fraude, C., Stasiak, D., Bruhn, T., Lawrence, M., Schroeder, H., & Mundaca, L. (2020). Enabling new mindsets and transformative skills for negotiating and activating climate action: Lessons from UNFCCC conferences of the parties. *Environmental Science Policy, 112*, 227–235.

Wang, S., Wang, J., Ru, X., Li, J., & Zhao, D. (2019). Understanding employee's electricity conservation behavior in workplace: Do normative, emotional and habitual factors matter? *Journal of Cleaner Production*.

Wang, X., Yuen, K. F., Wong, Y. D., & Li, K. X. (2020). How can the maritime industry meet Sustainable Development Goals? An analysis of sustainability reports from the social entrepreneurship perspective. *Transportation Research Part D: Transport and Environment, 78*, 102–113. https://doi.org/10.1016/j.trd.2019.11.002

Wiggins, R. R., & Ruefli, T. W. (2002). Sustained competitive advantage: Temporal dynamics and the incidence and persistence of superior economic performance. *Organization Science, 13*(1), 81–105. https://doi.org/10.1287/orsc.13.1.81.542

Winter, S. G. (2003). Understanding dynamic capabilities. *Strategic Management Journal, 24*(10), 991–995. https://doi.org/10.1002/smj.318

Wu, Y. (2020). *The marketing strategies of IKEA in China using tools of PESTEL, five forces model and SWOT analysis* (pp. 348–355). Atlantis Press.

Xie, Y., Fang, Y., Zhang, D. (2022). How environmental performance affects financial performance in the food industry: A global outlook. *Sustainability, 14*(4), 2127. https://doi.org/10.3390/su14042127

Zaugg, R., Blum, A., & Norbert, T. (2001a). Sustainability in human resource management. In *Evaluation report* (Vol. 1). IOP Press.

Zaugg, R. J., Blum, A., & Thom, N. (2001b). Sustainability in human resource management. Evaluation report. Survey in European Companies and Institutions, Switzerland.

Zhang, J., & Sh, L. (2023). The impact of human capital on green technology innovation—Moderating role of environmental regulations. *International Journal of Environmental Research and Public Health, 20.*

Chapter 3
Embracing Environmental Sustainability by Resilient Net Zero Carbon Supply Chains: A Fuzzy AHP Approach

Sonal Khurana, Bisma Mannan, Vaibhav Jain, Abid Haleem, and Vanita Bhardwaj

Abstract In light of growing climate change and global warming, it is now imperative for organizations worldwide to implement net zero carbon supply chain (NZCSC) practices in order to achieve carbon neutrality. Therefore, in order to assist the stakeholders, it is critical to identify the obstacles to the implementation of resilient NZCSC. As climate change and global warming escalate, organizations worldwide are increasingly prioritizing the implementation of net-zero carbon supply chain (NZCSC) practices. Identifying the barriers to implementing resilient NJCSC is crucial to supporting stakeholders in achieving carbon neutrality. The primary barriers to implementing resilient net zero carbon supply chain initiatives were recognized by reviewing the literature and expert opinions. Fuzzy AHP approach is used for understanding the most significant barrier. From the results of the fuzzy AHP approach, "Reluctance of managing authorities" was identified as the main barrier having the maximum weight. Barrier "Unavailability of suppliers" obtained the minimum weight. The study's findings can assist managers in their implementation of strategies aimed at achieving a NZCSC.

Keywords Net zero · Carbon supply chain · Environmental sustainability · Net zero policies

S. Khurana (✉) · V. Bhardwaj
Department of Engineering, Vivekananda Institute of Professional Studies, New Delhi, India
e-mail: sonal.khurana@gmail.com

B. Mannan
Department of Mechanical and Marine Engineering, University of Plymouth, Plymouth, UK

V. Jain
Department of Mechanical Engineering, Maharaja Agrasen Institute of Technology, New Delhi, India

A. Haleem
Department of Mechanical Engineering, Jamia Millia Islamia, New Delhi, India

R. Singh and D. Crowther (eds.), *Transition Towards a Sustainable Future*, Approaches to Global Sustainability, Markets, and Governance,
https://doi.org/10.1007/978-981-97-5756-5_3

3.1 Introduction

In the last few years, it is becoming more obvious how the emission of greenhouse gases is having a pervasive impact on both human livelihoods and the natural world (Rogelj et al., 2018). The rise in greenhouse gas emissions is seen as a contributing factor to adverse impacts on vegetation health, water availability, the spread of various diseases (Mikhaylov et al., 2020). Yet, the primary concern remains global warming. It is estimated that average global surface temperatures will increase by approximately 3–4 °C by the year 2100 compared with the average recorded during the period from 1986 to 2005 (Javid et al., 2014).

At the global level, greenhouse gas emissions saw a notable increase at the start of the twentieth century, primarily driven by the substantial escalation in fuel usage across diverse industrial activities. CO_2 emissions rose in India from 615 metric tons (MT) in 1990 to 1897 MT in 2013, accounting for approximately 5% of the emissions globally (Rogelj et al., 2018). The largest contribution to these statistics comes from the power sector, accounting for approximately 50% (Garg et al., 2017). Additionally, the release of gases form supply chains are far higher than the emissions from other sources, both direct and indirect (Mcdowall et al., 2018).

Taking this into account, several countries have established strategies and aims to achieve zero emissions by 2050. However, achieving net zero emissions presents numerous technological, financial, and policy hurdles. Developing countries, in particular, have considerable challenges because of limited access to innovative technologies and financial resources. For example, the switch to renewable energy sources necessitates significant initial infrastructure investments, which many developing countries cannot afford without international support. Furthermore, policy structures in many countries may not be strong enough to sustain the massive reforms required to attain net zero emissions.

By the middle of the century, large financial investments will be required to keep global warming's temperature rise below 2 °C (UNFCCC, 2015). The Indian union budget for 2022 emphasizes the country's transition to a carbon–neutral economy. It also makes recommendations to reduce emissions in industries such as energy and to issue sovereign green bonds to fund environmentally friendly projects (Chakraborty, 2022). To ensure compliance and improvement, these rules will need to be implemented through effective governance, public–private partnerships, and ongoing monitoring and assessment.

According to the US Environmental Protection Agency, an organization's supply chain frequently makes up for more than 90% of the release of greenhouse gases. Businesses in India run the risk of losing out on roughly $274 billion in export revenue as an increasing number of global corporations insist that their suppliers use sustainable business practices. The main obstacle in creating a green supply chain is creating a model that satisfies an organization's objectives for profitability and sustainability.

Moreover, integrating sustainable practices into supply chains is a complicated problem that must overcome a variety of obstacles. These challenges include a lack of standardized sustainability measures, limited supply chain training and awareness, and the high costs of green technologies. Companies must also handle the intricacies of global supply chains, which include numerous stakeholders with various levels of commitment to sustainability. Overcoming these challenges will need coordinated efforts across the supply chain, backed by strong leadership and a clear vision for sustainability.

Furthermore, the notion of sustainable business operations is extensive, incorporating goals like zero emissions as well as zero plastics, trash, and so forth. Because of this, it may be challenging for companies to create a green policy that works and accomplishes several goals at once. Businesses need to have a comprehensive strategy that takes into account all facets of sustainability while maintaining the flexibility to adapt to shifting conditions and create best practices. This necessitates constant learning and creativity in addition to a readiness to try out novel technologies and practices that can initially seem costly or dangerous.

Additionally, a supply chain is composed of multiple nodes that aren't always centralized, which adds to the complexity and friction in building a sustainable supply chain. Achieving sustainability goals depends on these nodes' efficient coordination and communication. This could mean creating unambiguous guidelines and standards for sustainability in addition to fostering a climate of cooperation and shared accountability. It is important to motivate supply chain partners to work together toward shared objectives, utilizing one another's skills and assets to go beyond challenges and attain higher sustainability.

Furthermore, the concept of sustainable business practices is broad and can include goals like zero plastics, zero waste, etc. besides having zero emissions. For this reason, it can be a challenge for firms to create an efficient green policy that simultaneously accomplishes several goals (Vulturius et al., 2024). Furthermore, a supply chain is made up of numerous nodes that aren't always under central control, which leads to conflict because of this lack of centrality and raises issues with creating a sustainable supply chain.

After fulfilling the goal of becoming carbon neutral in a number of industries and facing pressure from global governments, it is now imperative to switch to net zero carbon supply chain initiatives. For this, it is necessary to investigate ways to increase the sustainability and resilience of supply chains for meeting net zero emission target and address disruptions while also reducing global emissions significantly. In order to do that, it is necessary to identify the barriers that must be overcome in order to convert an existing supply chain model into a resilient and sustainable one. Thus, the present study recognizes the important barriers to build resilience in a net zero carbon supply chain. This problem statement is being addressed by identifying the barriers to adopting net zero carbon supply chain by the review of literature and by the opinion of experts.

Thus, the current analysis acknowledges the significant challenges to developing resilience in net zero carbon supply chains. These obstacles include technological constraints, budgetary constraints, legislative gaps, and the need for increased collaboration and coordination among supply chain stakeholders. Addressing these impediments would necessitate a comprehensive approach involving stakeholders from all stages of the supply chain, as well as strong leadership and a dedication to sustainability.

The emphasis should also be on creating creative solutions to improve the sustainability and resilience of supply networks. This involves looking into innovative technology and techniques that can help reduce emissions, enhance efficiency, and increase the overall sustainability of supply chain operations. Furthermore, it is critical to foster a sustainable culture inside organizations, encouraging employees and partners to adopt sustainable practices and play an active role in meeting net zero emission targets.

The remaining of the paper proceeds as follows. Literature review is carried out in Sect. 3.2. The discussion of the methodology is done in Sect. 3.3. Section 3.4 contains the results and discussions. Section 3.5 concludes the present study with the future scope.

3.2 Literature Review

In this section, extensive review of literature is conducted based on the existing net zero carbon supply chain scenario. A net zero carbon supply chain balances the quantity of greenhouse gases released with an equal amount of emissions removed or offset in order to reduce carbon emissions to net zero at every stage of the supply chain. Reaching this lofty objective will require a variety of tactics, technologies, and cross-sector cooperation.

3.2.1 Net Zero Carbon Supply Chain

It is critical to monitor the current global greenhouse gas emissions due to growing concerns about global warming and the ensuing danger to nature brought about by activities of humans. Thus, it is essential to lay the foundation for a low-carbon transition in the supply chain infrastructure that links each individual node (Satola et al., 2021). The idea of Net Zero Carbon Emissions (NZCE) was born out of the need to reduce the release of greenhouse gases. The concept of net zero carbon emissions is to remove the same quantity of carbon that each activity adds to the atmosphere. Maintaining global temperatures within a tolerable range in line with sustainable development goals is the aim of setting NZCE targets (Fankhauser et al., 2022).

Since supply chain companies play a significant role in the world's carbon emissions, it is imperative to develop strategies to reduce release of gases by implementing net zero efforts. Stakeholders can use this research to strengthen their implementation of a net zero plan of action throughout the supply chain. Enhanced cooperation among supply chain participants also contributes to the development of NZCSC. This collaboration includes exchanging best practices, pooling resources to invest in environmentally friendly technologies, and developing uniform measures to track the achievement of net zero objectives. Accelerating the shift to a sustainable future can also be achieved by utilizing digital technology and data analytics to optimize supply chain operations, cut waste, and find new potential for emission reductions.

3.2.2 Barriers in Net Zero Carbon Supply Chain

The Indian government has taken the lead in developing regulations pertaining to the NZCSC and circular economy, according to the country's most recent Union Budget for 2022. Nevertheless, given the numerous barriers it must overcome, it may still take some time for these regulations to be put into effect.

One of the main concerns is the **lack of awareness among consumers and suppliers** about alternative materials and processes, as well as current policies (Van Sluisveld et al., 2021). Moreover, it is more difficult to meet the goal of a NZCSC when there is a **lack of robust government action and policy** regarding GHG emissions (Robati et al., 2019). It is noted that for technical or financial reasons, **there is a lack of transparency regarding emission data** throughout the supply chain, which results in emissions control falling short of the desired level of efficiency. The difficulties encountered during policy implementation can also be attributed to the vagueness and immaturity of carbon accounting. Also, the suppliers adopting sustainable practices are not available as it is challenging for organizations to determine and verify that their suppliers are implementing green practices (Pye et al., 2021). **The significant initial investment** needed to incorporate new technology makes the transition to the sustainable practices challenging for organizations (Sparrevik et al., 2018). Moreover, **there is lack of cooperation** among potential suppliers to use environmentally friendly materials (Wimbadi & Djalante, 2020). Also, without a **definite assurance** that the long-term, deep decarbonization processes in their supply chain will pay off, companies find it challenging to commit to them (Davis et al., 2018). It has been observed that **managing authorities are reluctant** to adopt net-zero carbon policies because doing so would increase their financial concerns (Garvey et al., 2022). The barriers are listed in Table 3.1 and are identified by review of literature and opinion of the experts.

3.2.2.1 Lack of Awareness Among Consumers and Suppliers

Implementing a net-zero carbon supply chain is significantly hampered by suppliers' and customers' ignorance. In the absence of a comprehensive comprehension of the ecological consequences of their decisions, buyers might persist in requesting goods without taking into account their carbon footprint, thus sustaining non-sustainable behaviors. In a similar vein, suppliers might put efficiency and cost over sustainability, passing up chances to use greener technology and procedures. The shift to

Table 3.1 Barriers identified with their references

S. No.	Notation	Barriers	References
1	B1	Lack of awareness among suppliers	Albuquerque et al. (2020), Ramandi and Bafruei (2020)
2	B2	Lack of robust government action and policy	Van Sluisveld et al. (2021), Pan and Pan (2019)
3	B3	There is a lack of transparency regarding emission data	Comello et al. (2021), Higgins et al. (2020), Kim et al. (2018)
4	B4	Unavailability of suppliers adopting sustainable practices	Pye et al. (2021), Huang et al. (2021)
5	B5	Significant initial investment required	Bertheau and Lindner (2022), Khurana et al. (2021a, 2021b), Sachin and Rajesh (2022)
6	B6	Lack of cooperation among potential suppliers	Adomako, (2020), Khan et al. (2020)
7	B7	No assurance of the benefits	Papadis and Tsatsaronis (2020), Pye et al. (2021)
8	B8	Reluctance of the managing authorities	Hafner et al. (2021), Kemp et al., (2021)

a net-zero carbon supply chain is hampered by this ignorance, which also encourages damaging environmental practices that exacerbate climate change and its effects. Thus, raising awareness among suppliers and customers is essential to bringing about significant change that will lead to a sustainable and carbon–neutral future.

3.2.2.2 Lack of Robust Government Action and Policy

One major obstacle to putting net-zero policies into practice is the lack of strong policy frameworks and government action. In the absence of well-defined policies and incentives encouraging the adoption of sustainable practices, companies might put immediate financial gain ahead of long-term environmental objectives. Inertia in the adoption of carbon–neutral technologies and practices can result from this lack of guidance, as businesses may be reluctant to make costly upgrades in the absence of market or regulatory demand. Furthermore, the absence of government leadership could result in disparities in industry-specific standards and practices, impeding cooperation and inventiveness in the pursuit of a common net-zero objective. In order to provide the necessary direction, incentives, and enforcement mechanisms to drive systemic change toward a net-zero carbon economy, effective government action and policies are imperative.

The two most important tactics in this regard are education and knowledge sharing. Companies who want to educate their suppliers about the value of sustainability and the advantages of implementing eco-friendly practices might provide training sessions, seminars, and workshops. By emphasizing the long-term financial benefits and competitive advantages associated with sustainability, such programs might motivate suppliers to make investments in more environmentally friendly products and procedures. Additionally, consumer awareness programs can be started to inform consumers about how their purchases affect the environment. Through informing customers on the carbon footprint linked to various products, companies can stimulate the market for more environmentally friendly choices. Another important function that sustainability indicators can play is in product labeling and marketing. Consumers are more inclined to select options that are consistent with their values when products prominently display their environmental credentials, thus encouraging greener choices in the marketplace. Businesses can also use social media and digital platforms to interact with a wider audience and raise awareness of the significance of cutting carbon emissions in supply chains. Working together and taking advantage of environmental organizations' knowledge can help raise awareness of campaigns in new ways (Carter & Rogers, 2008).

Through the cultivation of a sustainable culture and the active engagement of suppliers and consumers in the pursuit of net-zero carbon emissions, companies can establish a supply chain that is both sustainable and ecologically benign. In order to solve the urgent problems caused by climate change and guarantee a sustainable future for all, we must work together.

3.2.2.3 Lack of Transparency Regarding Emission Data

One major barrier to the implementation of net-zero policies is the lack of transparency surrounding emission data. Businesses and policymakers find it difficult to pinpoint areas for improvement and monitor the movement toward carbon neutrality in the absence of precise and easily available data on emissions across the supply chain. Because of this opacity, carbon footprints may be underestimated, which makes it challenging to set reasonable reduction goals and distribute resources wisely. Furthermore, in the absence of transparency, customers find it more difficult to hold businesses responsible for their environmental impact and to make educated decisions. Companies need to make investments in reliable data collection and reporting systems that offer precise and up-to-date information on emissions in order to address this problem. Standardized reporting frameworks can be put into place to assist guarantee data comparability and consistency across various industries and supply chains. Furthermore, by utilizing cutting-edge technologies like blockchain, emissions data may be made more transparent and traceable, facilitating information access and verification for all parties involved. By increasing openness, companies may gain the trust of stakeholders and customers by showcasing their dedication to environmental responsibility and paving the way for more successful climate action (Bai & Sarkis, 2010).

3.2.2.4 Unavailability of Suppliers Adopting Sustainable Practices

One major obstacle to putting net-zero policies into effect is the unavailability of suppliers who are embracing sustainable practices. Businesses find it difficult to source goods and services that support their net-zero objectives without access to sustainable suppliers, which impedes their attempts to lower carbon emissions across the supply chain. This restriction may result in more waste, more emissions, and damage to the environment. It also hinders opportunities for collaboration and innovation because it becomes difficult to find partners who share your values. Businesses can take proactive measures to engage with suppliers, increase awareness, and provide incentives for sustainable behavior to address this. In order to encourage the adoption of eco-friendly practices, promote sustainability throughout the supply chain, and ease the transition to a net-zero economy, government regulations and industry standards can be extremely important.

3.2.2.5 The Significant Initial Investment

One of the main obstacles to implementing net-zero policies is the large initial investment that is needed. Making the shift to carbon–neutral technologies and practices frequently requires large upfront expenditures for workforce training, research and development, and infrastructure upgrades. Furthermore, there may be a delay in the return on investment for green practices and technologies, which makes financial planning and decision-making even more difficult. Furthermore, organizations may find it more difficult to make the required investments to shift to net-zero operations if they have restricted access to capital or financing options. Innovative financing strategies, encouraging government incentives, and a long-term outlook that acknowledges the financial advantages of sustainability in the form of cost savings, increased competitiveness, and resilience to upcoming regulatory changes and market shifts are all necessary to overcome these financial obstacles.

3.2.2.6 Lack of Cooperation Among Potential Suppliers

The implementation of net-zero policies is significantly impeded by the potential suppliers' lack of cooperation. Throughout the supply chain, supplier cooperation and coordination are crucial to meeting sustainability targets. But without collaboration, it becomes difficult to coordinate tactics, pool resources, and put uniform guidelines and policies in place to cut carbon emissions. This misalignment may lead to dispersed efforts, redundant use of resources, and lost chances to maximize efficacy and efficiency in lowering the environmental impact of operations. Fostering a culture of cooperation, increasing stakeholder trust, providing incentives for common objectives, and setting up open channels of communication and information sharing are all necessary to remove obstacles to cooperation. Businesses can only effectively address the complex challenges associated with achieving net-zero carbon emissions through cooperation and collective action.

Government incentives and creative financing techniques are essential to addressing these financial obstacles. For instance, green bonds provide a feasible financing solution by contributing the required funds designated for environmentally friendly projects (Flammer, 2021). It has been demonstrated that green bonds draw investors who are more concerned with ethical and sustainable investing, which lowers the cost of money for businesses ready to take on green projects. Another important factor in lowering the financial load on businesses is government incentives. Grants, tax exemptions, and subsidies can greatly reduce the upfront expenses of adopting green practices and technologies. To assist sectors in making the transition to more sustainable operations, the European Union's Green Deal, for example, provides significant money and incentives (European Commission, 2019). These kinds of programs not only offer direct cash assistance but also foster a supportive regulatory framework that promotes sustainability investment. A long-term perspective that acknowledges the financial benefits of sustainability is also crucial. Businesses must realize that although there may be significant upfront costs, there may be significant long-term rewards. According to Porter and van der Linde (1995), these advantages include lower costs from increased energy efficiency, increased competitiveness from a better environmental reputation, and more resistance to future changes in regulations and market dynamics. Businesses that make sustainability investments are frequently better equipped to handle harsher environmental laws and can steer clear of the possible expenses connected with non-compliance.

3.2.2.7 Lack of Definite Assurance of the Benefits

Businesses are impeded in their efforts to adopt net-zero policies by the absence of a clear guarantee of the benefits. If the firms are unsure of the possible returns on investment, they might be reluctant to commit resources. This uncertainty can make it difficult to take action, especially for risk-averse businesses that put short-term profits ahead of long-term sustainability objectives. Strong data and case studies outlining the economic, environmental, and social advantages of enacting net-zero policies are necessary to overcome this barrier. Clear proof of the possible returns on investment can allay worries and encourage companies to take steps toward becoming carbon neutral (Mikhaylov et al., 2020).

Comprehensive cost–benefit evaluations that show the direct and indirect advantages of implementing net-zero practices are one technique to offer assurance. Along with the immediate financial savings from increased waste reduction and energy efficiency, these evaluations should also account for the long-term benefits from increased customer loyalty, improved brand reputation, and regulatory compliance. Businesses that have successfully made the switch to net-zero operations, for instance, frequently report improved operational efficiencies, lower energy-related expenses, and higher resistance to changes in the market and in regulations (Satola et al., 2022). Furthermore, case studies and real-world examples of companies that have benefited greatly from net-zero projects can be extremely motivating. These kinds of cases can show how profitable and feasible sustainable practices can be, giving other businesses a model to follow. Moreover, businesses can gain important insights and support from working with industry leaders and joining sustainable networks, which will enable them to comprehend and manage the shift more skillfully. Companies are encouraged to take action toward being carbon neutral. Businesses may invest in sustainability more confidently by providing strong proof of the observable advantages of net-zero policies, which will hasten the transition of the world economy to a low-carbon one (Fankhauser et al., 2022).

3.2.2.8 Reluctance of Managing Authorities

Businesses' confidence in implementing net-zero policies can be greatly impacted by managing authorities' unwillingness to prioritize and enforce them. Businesses are left in the dark when regulatory bodies are reluctant or slow to establish explicit rules, guidelines, and rewards for carbon neutrality. Businesses may view the switch to net-zero policies as risky or unnecessary in the absence of a strong regulatory framework, particularly if they believe there won't be many consequences for non-compliance. In the end, businesses lose faith in managing authorities' ability to successfully navigate the shift to net-zero policies when they are unwilling to act decisively. Managing authorities must show strong leadership, offer precise direction, and establish regulatory frameworks that encourage and support the adoption of sustainable practices in order to effectively address this challenge.

3.3 Methodology Used

The Fuzzy Analytical Hierarchy Process (AHP) is an effective methodology that provides an organized way to assess complex problems while evaluating and prioritizing factors in decision-making scenarios. AHP offers unique benefits that set it apart from other Multi-Criteria Decision-Making (MCDM) tools such as ANP, ISM, TOPSIS, ELECTRE, and DEMATEL, making it a better option in many situations (Khurana et al., 2021a, 2021b). For example, although ISM makes it easier to create hierarchies, it is unable to explore the relative significance of variables, which creates a gap in thorough analysis. In a similar manner, although being alternatives, TOPSIS and ELECTRE are not well-known or well-accepted by experts and scholars alike.

To effectively utilize AHP in problem-solving, practitioners typically follow a systematic three-step process. First, the decision problem is broken down into its component parts and organized into a hierarchy. In this step, the problem is divided into its component parts, which are then arranged in a hierarchy according to their dependencies and relationships with one another. By doing this, it gets simpler to understand the complexities of the issue and identify the key factors involved. To create judgmental matrices, information from domain experts is gathered in the second stage. Quantifying the relative importance of factors and their interrelationships is based on these matrices. By means of discussions with specialists in the field, information is gathered on the factors and how they stack up against other factors that are associated. These insights are then converted into structured correlation matrices that represent the experts' opinions. Linguistic terms are used to describe the relative importance of a criterion over the other. Similar to conventional AHP, the decision problem is organized hierarchically, with the primary goal at the top and criteria and other options at lower levels. Pairwise comparisons are used to compare each criteria and alternative with regard to the higher-level element.

Pairwise Comparison: Using fuzzy numbers (e.g., triangular or trapezoidal fuzzy numbers) or linguistic terms (e.g., "very important," "somewhat important," "equally important," etc.), decision-makers present pairwise comparisons of criteria and alternatives. Fuzzy matrices are frequently used to illustrate these comparisons.

Fuzzy Weight Determination: Fuzzy AHP uses fuzzy arithmetic operations to determine the weights of criteria and alternatives. The performance of alternatives in relation to each criterion and the relative importance of each criterion are represented by fuzzy weights.

Similar to conventional AHP, the decision problem is organized hierarchically, with the primary goal at the top and criteria and other options at lower levels. Pairwise comparisons are used to compare each criteria and alternative with regard to the higher-level element.

Pairwise Comparison: Using fuzzy numbers (e.g., triangular or trapezoidal fuzzy numbers) or linguistic terms (e.g., "very important," "somewhat important," "equally important," etc.), decision-makers present pairwise comparisons of criteria and alternatives. Fuzzy matrices are frequently used to illustrate these comparisons.

Fuzzy Weight Determination: Fuzzy AHP uses fuzzy arithmetic operations to determine the weights of criteria and alternatives. The performance of alternatives in relation to each criterion and the relative importance of each criterion are represented by fuzzy weights.

Aggregation and Ranking: To ascertain the overall rankings of the alternatives, fuzzy AHP aggregates the fuzzy weights. Depending on the type of decision problem, different aggregation techniques like fuzzy weighted geometric mean or fuzzy weighted averaging can be used.

Sensitivity analysis is a technique used to evaluate how resilient the outcomes are to shifts in the preferences of decision-makers or fluctuations in the input data. This facilitates comprehension of the decision outcomes' stability in various scenarios.

When making decisions in situations where uncertainty, ambiguity, or subjective judgments are important factors, fuzzy AHP is especially helpful. It facilitates more realistic modeling of decision problems and offers an organized framework for managing imprecise information. Furthermore, it makes it possible for decision-makers to integrate their qualitative evaluations into the process, resulting in decisions that are stronger and better informed.

3.3.1 Data Collection

The panel of experts consisted of experts from various fields/organizations. To ensure thorough and accurate data collection, a diverse group of experts from various fields and organizations must be assembled for the purpose of gathering data for a net-zero carbon supply chain. A diverse group of experts was assembled. The objectives of data collection were made clear to them. The questionnaires were distributed to all the experts using the e-Delphi approach. The e-Delphi method improves accessibility and efficiency by facilitating the traditional Delphi process through the use of digital communication tools (Khurana et al., 2019). This approach entails conducting several rounds of anonymous surveys to get input and opinions from experts on a given subject. Every round builds on the one before it, with participants editing their own responses after reading through an overview of the group's responses. Until a consensus is reached or diminishing returns are noticed, this iterative process is continued. Experts contribute in an anonymous manner, which lessens the power of powerful people and promotes sincere and objective criticism. Surveys are conducted in multiple rounds, with the information refined after each round based on input from experts. After every round, participants receive a summary of the group's responses, giving them the opportunity to revise and amend their opinions. The responses undergo statistical analysis in order to pinpoint trends, central tendencies, and expert consensus.

When applying the e-Delphi method to net-zero carbon supply chains, a variety of specialists, such as data analysts, supply chain managers, and environmental scientists, can provide valuable insights. This approach guarantees a thorough and cooperative comprehension of optimal methodologies, obstacles, and inventive resolutions for accomplishing net-zero objectives. When it comes to complicated, multidisciplinary problems like creating net-zero carbon supply chains, where a variety of expert viewpoints are essential for thorough decision-making, the e-Delphi approach is especially helpful.

3.4 Results and Discussions

The fuzzy Analytic Hierarchy Process (AHP) approach was utilized to identify and prioritize the barriers preventing organizations from achieving a net zero carbon supply chain.

The results of fuzzy AHP approach depicts "Reluctance of Managing Authorities" as the barrier having maximum significance for resilience toward net zero supply chains based on the weights obtained by fuzzy AHP approach. The hesitation or resistance from top management, often due to concerns about cost, lack of awareness, or the perceived complexity of implementing net zero carbon strategies, emerged as the primary issue. Table 3.2 depicts the weight of the barriers obtained by applying a fuzzy AHP approach. Barrier "Reluctance of Managing Authorities" obtained the normalized weight of 0.87769 thereby emphasizing it as the most essential barrier to develop resilience toward net zero supply chain. The study by Zhang et al. (2022) also discusses the similarity in result that inadequate management interest as the significant reason for not shifting toward net zero carbon supply chain. Barrier "Lack of transparency regarding emission data" obtained the next higher weight. This factor emphasizes the difficulties in getting transparent and accurate emission data. Organizations find it more difficult to accurately measure, report, and manage their carbon footprints when there is a lack of trustworthy data. The barrier "Lack of cooperation among potential suppliers" obtained the next highest weight. This obstacle highlights the challenges in encouraging supplier cooperation and coordination. Coherent and effective implementation of sustainable supply chain practices requires effective cooperation. Barrier "Unavailability of suppliers" obtained the least weight depicting its dependency on other parameters which have got more weight. According to the study, creation of effective Government policies is paramount for the success of any organization toward net zero supply chain (Chang et al., 2021). Similar results are shown by the present research which strengthen the results obtained by the present research. The most significant obstacles found offer guidance on where focus should be directed. As the primary barrier, managing authorities' reluctance implies that increasing leadership awareness and commitment is essential. Educating top management about the long-term advantages of net zero carbon strategies through the implementation of focused training programs and workshops is one way to accomplish this. Presenting case studies that are successful and best practices to show the benefits and viability of sustainable supply chains is also promoted. Encouraging an organizational culture that integrates environmental goals into the main business plan in order to foster sustainability. The absence of transparency in relation to emission data emphasizes the necessity of creating uniform techniques for reporting and measuring emissions, putting money into innovations that offer precise, up-to-date emissions data and encouraging open data sharing among interested parties in order to promote accountability and trust. The lack of cooperation among potential buyers can be mitigated by establishing reliable channels of communication and connections with suppliers, promoting cooperative projects and joint ventures with a sustainability focus and laying out explicit guidelines and rewards for suppliers who use sustainable practices. The lack of suppliers implementing sustainable practices suggests that the following are necessary. Providing grants, subsidies, or preferential treatment in contracts to assist and encourage suppliers to switch to sustainable practices. Extending the scope of the search and selection process to encompass international vendors who fulfill sustainability requirements. And supplying tools and training to local suppliers so they can create and apply sustainable practices.

In conclusion, the challenges associated with implementing net zero carbon supply chains have been effectively prioritized by the fuzzy AHP approach. It is imperative to tackle the aforementioned obstacles, specifically the hesitancy of regulatory bodies, absence of clarity in emissions data, and inadequate collaboration among providers. Organizations can create more effective strategies to meet their sustainability objectives and make the shift to a net zero carbon supply chain by concentrating on these areas.

Table 3.2 Pairwise comparison matrix of barriers

	B1	B2	B3	B4	B5	B6	B7	B8	Fuzzy geometric means	Fuzzy geometric weights	Normalized weights	Rank
B1	(1,1,1)	(4,5,6)	(1/4,1/3,1/2)	(5,6,7)	4,5,6	(1/4,1/3,1/2)	3,4,5	(1/6,1/5,1/4)	(2.5, 13.33,78.75)	(0.0016, 0.00088, 0.00098)	0.001162375	4
B2		1,1,1	(1/5,1/4,1/3)	2,3,4	1,2,3	(1/4,1/3,1/2)	(1/5,1/4,1/3)	(1/6,1/5,1/4)	(0.00083, 0.0083, 0.0833)	(5.379E−07,5.5229E−07, 1.0473E-06)	7.12509E−07	6
B3			1,1,1	4,5,6	3,4,5	4,5,6	4,5,6	(1/6,1/5,1/4)	(256, 1500, 6480)	(0.1652, 0.0994, 0.0814)	0.115366445	2
B4				1,1,1	(1/6,1/5,1/4)	(1/5,1/4,1/3)	(1/5,1/4,1/3)	(1/6,1/5,1/4)	(7.71605E−06, 3.333E−05, 0.0002170)	(4.980E−09, 2.209E−09, 2.727E−09)	3.30575E−09	8
B5					1,1,1	(1/4,1/3,1/2)	(1/5,1/4,1/3)	(1/7,1/6,1/5)	(0.00014, 0.00092, 0.0083)	(9.6055E−08, 6.1366E−08, 1.0473E−07)	8.73846E−08	7
B6						1,1,1	2,3,4	(1/6,1/5,1/4)	(10.66, 75,432)	(0.0068, 0.0049, 0.0054)	0.005761747	3
B7							1,1,1	(1/6,1/5,1/4)	(0.033, 0.2, 1.25)	(2.1516E−05, 1.3255E−05, 1.5709E−05)	1.68271E−05	5
B8								1,1,1	(1280, 13,500, 72,576)	(0.8262, 0.8947, 0.9121)	0.877691803	1

Maximum Eigen Value = 9.57559; C.I. = 0.225084; B1: lack of awareness among suppliers; B2: Lack of robust government action and policy; B3: lack of transparency regarding the emission data; B4: unavailability of green suppliers adopting sustainable practices; B5: high initial investment required; B6: lack of cooperation among potential suppliers; B7: no assurance of benefits; B8: reluctance of managing authorities

3.5 Implications

In recent years, the idea of sustainable supply chains and its advancement have generated a lot of discussion. The goal of the current study is to determine how different industries' implementation of NZCSC has been hindered by related issues. In order to reduce global emissions, the research findings are focused on overcoming the obstacles and developing robust NZCSCs in a variety of industries. The study points out the significant obstacles to creating a robust net-zero carbon supply chain.

After that, they are ranked in order of importance to determine the essential elements of creating a NZCSC. The difficulties found can provide a framework for further research on the creation of sustainable supply chains. Experts from the industry have contributed to the fuzzy AHP approach that has been used to weigh the various challenges. The present study's methodology holds potential for identifying obstacles and promoting the advancement of net zero carbon objectives in various industries, including manufacturing, energy production, infrastructure development, and so on. Additionally, by combining the results of numerous studies on the creation of sustainable supply chains across a range of industries, this study adds to the body of knowledge already available on supply chain management and sustainable operations.

The main goal is to get over the obstacles in the way of creating a NZCSC. The practitioners and policymakers involved in supply chain development and implementation must consider a wide range of factors in order to achieve the same goal.

The results of the current study offer information that can help those in charge of supply chain planning and execution comprehend how the issues under investigation may affect the emissions in a chain. Building a roadmap toward a resilient NZCSC will be made easier for supply chain development practitioners if they have a clear understanding of the interrelationships between the various challenges explored and their relative importance.

The recommendations made in this study aim to help supply chain development professionals create a robust and sustainable NZCSC network. There are several advantages to putting the suggested solutions into practice.

- Climate Impact: Organizations can lower their emissions and have a positive climate impact by tackling the issues covered in the study. Consequently, this can assist organizations in fulfilling their sustainability objectives and the environmental, social, and governance requirements.
- Efficiency: By putting in place a NZCSC, organizations will automatically use less water and energy, which will result in less waste being produced. Consequently, there will be a decrease in incidental expenses. Additionally, this would stress getting the most output out of the least amount of input.
- Appeal: Using a NZCSC can be a great way to appeal to buyers who are concerned about the environment and are increasingly demanding goods that are sourced responsibly.

- Transparency: Information about how activities are conducted at each supply chain node must be known in order to establish a NZCSC.

This may facilitate information exchange among the chain's numerous nodes, which may streamline the chain overall.

3.5.1 Societal Implications

There are several advantages to implementing a NZCSC for human lifestyles. Previous studies highlight the significance of social factors as a crucial component and stress the need to take them into account in addition to environmental and economic factors when creating sustainable supply chains. Therefore, the study's conclusions can be used to create an ethical supply chain that has numerous positive effects on society. Among the advantages mentioned above are as follows:

- Employment and Skill Development: Converting supply chain operations from a traditional model to a net-zero carbon model calls for a change in operational thinking. This includes reevaluating different supply chain nodes and hiring staff members who are capable of carrying out sustainable supply chain practices. Such a significant change will necessitate the hiring of qualified staff members for the NZCSC building.
- Cleanliness of the Environment: A large reduction in GHG emissions would result from the implementation of a NZCSC. Furthermore, this would also result in less waste being produced. The quality of the land, water, and air would all benefit from this. This consequently suggests improved health, hygienic food, and clean water. By implementing a NZCSC, greenhouse gas emissions would be significantly reduced, which would reduce waste generation and improve the quality of the air, water, and land. Consequently, this would lead to improved health, purer water, and more hygienic food selections.
- Ethics in Labor: Creating a NZCSC requires openness in all of its processes. As a result, it makes it possible to keep an eye on supply chain procedures and assists in highlighting disparities in labor practices, which can then be appropriately addressed to guarantee socially conscious sourcing throughout the chain. To create a NZCSC, all processes must be transparent. As a result, this makes it easier to keep an eye on supply chain activities and detect differences in labor practices. These differences may then be appropriately addressed to guarantee socially conscious sourcing throughout the entire chain. Moreover, by ensuring fair treatment of employees and promoting transparency in labor norms, a NZCSC fosters confidence and collaboration among stakeholders. Prioritizing ethical labor standards enhances a business's standing and strengthens relationships with vendors, employees, and the community.

3.6 Future Scope

Innovative approaches to overcoming these difficulties are becoming more and more the focus of industry practices and research. Future research topics include improving supply chain resilience against climate impacts, developing more effective carbon capture and storage technologies, and developing thorough frameworks for net-zero transitions in various sectors. All things considered, the shift to net-zero carbon supply chains is a complex process that calls for cooperation, technological innovation, and significant financial outlay. Businesses can achieve significant progress toward net-zero emissions by utilizing digitalization, renewable energy, circular economy principles, and sustainable finance.

It is necessary to use proactive measures to reduce the risks associated with extreme weather, resource scarcity, and other climate-related disruptions in order to address supply chain resilience against climate impacts. This entails incorporating adaptable sourcing practices, utilizing data analytics for in-the-moment monitoring and reaction, and including climate resilience considerations into supply chain design. To achieve net-zero emissions, technological advancements in carbon capture and storage are crucial. The goal of research is to create scalable and reasonably priced methods of absorbing and storing carbon dioxide emissions from power plants, industrial processes, and other sources. Direct air capture and carbon mineralization are two examples of innovative carbon capture technologies that hold great potential for cutting emissions and slowing down global warming. In-depth net-zero transition frameworks are essential for directing industry and government officials toward reaching carbon neutrality in a range of industries. These models cover all-inclusive approaches to cutting emissions, switching to renewable energy, improving energy economy, and encouraging sustainable behaviors all the way through the supply chain. They make sure that no communities or stakeholders are left behind by addressing the transition's social and economic ramifications as well. In general, the transition to net-zero carbon supply chains necessitates a multimodal strategy that incorporates financial, policy, and technological solutions. Companies that embrace digitalization, invest in renewable energy infrastructure, embrace the circular economy, and have access to sustainable financing choices can achieve significant strides toward net-zero emissions. To spur innovation, expand sustainable practices, and quicken the shift to a low-carbon economy, cooperation between business, academia, government, and civil society is crucial.

3.7 Conclusion

Following the review of literature and by taking the opinion of the experts, the present study recognizes and analyzes 8 barriers to implementing a resilient net zero carbon supply chain. Through a thorough investigation, this study seeks to give stakeholders the information they need to put the required steps into place in order to reach the goal of net zero carbon. The present study focuses on the mathematical results for providing the stakeholders with the necessary information for implementing resilience toward net zero carbon supply chains. Based on the results of the present study, "Reluctance of managing authorities" is identified as the most significant barrier and "Unavailability of suppliers" is found to be the least significant factor. The analysis of the barriers is performed using a fuzzy AHP approach. This study identifies the influential factors for implementation of a resilient net zero carbon supply chain. But, this study is based on qualitative analysis. The further research can be performed by quantitative analysis which will help to validate the results of present study.

The roadmap functions as a thorough guide for policymakers to concentrate their implementation of NZCSC policies and adheres to a bottoms-up solution procedure. It was created to address the identified challenges based on their criticality as per Fuzzy AHP. These steps can hasten the process of achieving the various SDGs and, as a result, increase an institution's positive impact on sustainability and the climate. The study's implications provide insight into earlier research in this area and how the current work aims to advance the body of knowledge on creating resilient and sustainable NZCSCs. A change in management strategy toward net zero carbon policies could lead to additional chances for social and ecological advancement. The study examines decision-making factors that impact the creation of a resilient NZCSC; however, the paper unavoidably has limitations. The dataset obtained for fuzzy AHP is based on the opinion of the experts. The technique might be further optimized with a panel that is far more diverse. Furthermore, the suggested solution roadmap is restricted to a wide range of problems. These policy recommendations can be further simplified to meet the requirements of particular industries. Therefore, it should be the goal of future research to include a global expert panel that is both larger and more diverse. This will give a much clearer picture of the difficulties that NZCSC faces. Along with fuzzy AHP, emphasis could also be placed on combining other MCDM approaches.

References

Adomako, S. (2020). Environmental collaboration, sustainable innovation, and small and medium-sized enterprise growth in sub-Saharan Africa: Evidence from Ghana. *Sustainable Development, 28*(6), 1609–1619.

Albuquerque, F. D., Maraqa, M. A., Chowdhury, R., Mauga, T., & Alzard, M. (2020). Greenhouse gas emissions associated with road transport projects: Current status, benchmarking, and assessment tools. *Transportation Research Procedia, 48*, 2018–2030.

Bai, C., & Sarkis, J. (2010). Integrating sustainability into supplier selection with grey system and rough set methodologies. *International Journal of Production Economics, 124*(1), 252–264.

Bertheau, P., & Lindner, R. (2022). Financing sustainable development? The role of foreign aid in Southeast Asia's energy transition. *Sustainable Development, 30*(1), 96–109.

Carter, C. R., & Rogers, D. S. (2008). A framework of sustainable supply chain management: Moving toward new theory. *International Journal of Physical Distribution & Logistics Management, 38*(5), 360–387.

Chakraborty, L. (2022). Union budget 2022–23: Fiscal-monetary interface (No. 22/378). https://www.nipfp.org.in/publications/working-papers/1977/.

Chang, Y., Ji, Q., & Zhang, D. (2021). Green finance and energy policy: Obstacles, opportunities, and options. *Energy Policy, 157*, 112497.

Comello, S., Reichelstein, J., & Reichelstein, S. (2021). Corporate carbon reduction pledges: An effective tool to mitigate climate change? *ZEW-Centre for European Economic Research Discussion Paper*, (21–052).

Davis, S. J., Lewis, N. S., Shaner, M., Aggarwal, S., Arent, D., Azevedo, I. L., Benson, S. M., Bradley, T., Brouwer, J., Chiang, Y. M., Clack, C. T., & Caldeira, K. (2018). Net-zero emissions energy systems. *Science, 360*(6396), eaas9793.

European Commission. (2019). The European green deal. Retrieved from European Commission Green Deal

Fankhauser, S., Smith, S. M., Allen, M., Axelsson, K., Hale, T., Hepburn, C., Kendall, J. M., Khosla, R., Lezaun, J., Mitchell-Larson, E., Obersteiner, M., & Wetzer, T. (2022). The meaning of net zero and how to get it right. *Nature Climate Change, 12*(1), 15–21.

Flammer, C. (2021). Corporate green bonds. *Journal of Financial Economics, 142*(2), 499–516.

Garg, A., Shukla, P. R., Kankal, B., & Mahapatra, D. (2017). CO_2 emission in India: Trends and management at sectoral, sub-regional and plant levels. *Carbon Management, 8*(2), 111–123.

Garvey, A., Norman, J. B., & Barrett, J. (2022). Technology and material efficiency scenarios for net zero emissions in the UK steel sector. *Journal of Cleaner Production, 333*, 130216.

Hafner, S., Jones, A., & Anger-Kraavi, A. (2021). Economic impacts of achieving a net-zero emissions target in the power sector. *Journal of Cleaner Production, 312*, 127610.

Higgins, C., Tang, S., & Stubbs, W. (2020). On managing hypocrisy: The transparency of sustainability reports. *Journal of Business Research, 114*, 395–407.

Huang, Y., Xue, L., & Khan, Z. (2021). What abates carbon emissions in China: Examining the impact of renewable energy and green investment. *Sustainable Development, 29*(5), 823–834.

Javid, R. J., Nejat, A., & Hayhoe, K. (2014). Selection of CO_2 mitigation strategies for road transportation in the United States using a multi-criteria approach. *Renewable and Sustainable Energy Reviews, 38*, 960–972.

Kemp, J., McCowage, M., & Wang, F. (2021). Towards net zero: Implications for Australia of energy policies in East Asia| Bulletin–September 2021.

Khan, Z., Ali, M., Kirikkaleli, D., Wahab, S., & Jiao, Z. (2020). The impact of technological innovation and public-private partnership investment on sustainable environment in China: Consumption-based carbon emissions analysis. *Sustainable Development, 28*(5), 1317–1330.

Khurana, S., Haleem, A., & Mannan, B. (2019). Determinants for integration of sustainability with innovation for Indian manufacturing enterprises: Empirical evidence in MSMEs. *Journal of Cleaner Production, 229*, 374–386.

Khurana, S., Haleem, A., Luthra, S., & Mannan, B. (2021a). Evaluating critical factors to implement sustainable oriented innovation practices: An analysis of micro, small, and medium manufacturing enterprises. *Journal of Cleaner Production, 285*, 125377.

Khurana, S., Haleem, A., Luthra, S., Huisingh, D., & Mannan, B. (2021b). Now is the time to press the reset button: Helping India's companies to become more resilient and effective in overcoming the impacts of COVID-19, climate changes and other crises. *Journal of Cleaner Production, 280*, 124466.

Kim, J. H., Kim, H. J., & Yoo, S. H. (2018). Consumers' willingness to pay for net-zero energy apartment in South Korea. *Sustainability, 10*(5), 1564.

McDowall, W., Rodriguez, B. S., Usubiaga, A., & Fernández, J. A. (2018). Is the optimal decarbonization pathway influenced by indirect emissions? Incorporating indirect life-cycle carbon dioxide emissions into a European TIMES model. *Journal of Cleaner Production, 170*, 260–268.

Mikhaylov, A., Moiseev, N., Aleshin, K., & Burkhardt, T. (2020). Global climate change and greenhouse effect. *Entrepreneurship and Sustainability Issues, 7*(4), 2897.

Pan, W., & Pan, M. (2019). Opportunities and risks of implementing zero-carbon building policy for cities: Hong Kong case. *Applied Energy, 256*, 113835.

Papadis, E., & Tsatsaronis, G. (2020). Challenges in the decarbonization of the energy sector. *Energy, 205*, 118025.

Porter, M. E., & Linde, C. V. D. (1995). Toward a new conception of the environment-competitiveness relationship. *Journal of Economic Perspectives, 9*(4), 97–118.

Pye, S., Broad, O., Bataille, C., Brockway, P., Daly, H. E., Freeman, R., Gambhir, A., Geden, O., Rogan, F., Sanghvi, S., Tomei, J., & Watson, J. (2021). Modelling net-zero emissions energy systems requires a change in approach. *Climate Policy, 21*(2), 222–231.

Ramandi, M. D., & Bafruei, M. K. (2020). Effects of government's policy on supply chain coordination with a periodic review inventory system to reduce greenhouse gas emissions. *Computers & Industrial Engineering, 148*, 106756.

Robati, M., Daly, D., & Kokogiannakis, G. (2019). A method of uncertainty analysis for whole-life embodied carbon emissions (CO_2-e) of building materials of a net-zero energy building in Australia. *Journal of Cleaner Production, 225*, 541–553.

Rogelj, J., Shindell, D., Jiang, K., Fifita, S., Forster, P., Ginzburg, V., Handa, C., Kheshgi, H., Kobayashi, S., Kriegler, E., Mundaca, L., & Zickfeld, K. (2018). Mitigation pathways compatible with 1.5 C in the context of sustainable development. In *Global warming of 1.5 C* (pp. 93–174). Intergovernmental Panel on Climate Change.

Sachin, N., & Rajesh, R. (2022). An empirical study of supply chain sustainability with financial performances of Indian firms. *Environment, Development and Sustainability, 24*(5), 6577–6601.

Satola, D., Balouktsi, M., Lützkendorf, T., Wiberg, A. H., & Gustavsen, A. (2021). How to define (net) zero greenhouse gas emissions buildings: The results of an international survey as part of IEA EBC annex 72. *Building and Environment, 192*, 107619.

Satola, D., Wiberg, A. H., Singh, M., Babu, S., James, B., Dixit, M., Sharston, R., & Gustavsen, A. (2022). Comparative review of international approaches to net-zero buildings: Knowledge-sharing initiative to develop design strategies for greenhouse gas emissions reduction. *Energy for Sustainable Development, 71*, 291–306.

Sparrevik, M., Wangen, H. F., Fet, A. M., & De Boer, L. (2018). Green public procurement–A case study of an innovative building project in Norway. *Journal of Cleaner Production, 188*, 879–887.

UNFCCC. (2015). Key aspects of the Paris agreement. https://unfccc.int/mostrequested/key-aspects-of-the-paris-agreement

Van Sluisveld, M. A., de Boer, H. S., Daioglou, V., Hof, A. F., & van Vuuren, D. P. (2021). A race to zero-assessing the position of heavy industry in a global net-zero CO_2 emissions context. *Energy and Climate Change, 2*, 100051.

Vulturius, G., Maltais, A., & Forsbacka, K. (2024). Sustainability-linked bonds–their potential to promote issuers' transition to net-zero emissions and future research directions. *Journal of Sustainable Finance & Investment, 14*(1), 116–127.

Wimbadi, R. W., & Djalante, R. (2020). From decarbonization to low carbon development and transition: A systematic literature review of the conceptualization of moving toward net-zero carbon dioxide emission (1995–2019). *Journal of Cleaner Production, 256*, 120307.

Zhang, A., Alvi, M. F., Gong, Y., & Wang, J. X. (2022). Overcoming barriers to supply chain decarbonization: Case studies of first movers. *Resources, Conservation and Recycling, 186*, 106536.

Chapter 4
Financing the Transformation to Net Zero Goals: IREDA Investment for Sustainable Initiatives

Daravath Vikas Nayak, A. Arun Kumar, Rakesh Suryadevara, and Shahbaz Khan

Abstract This study focused on the trends and fluctuations of renewable energy projects for net zero goals, concentrating on sector-wise funding sanctions, funding distribution, and capacity additions in India. To understand the investment patterns and outcomes in various renewable energy projects of Wind power, Hydropower, Biomass, and Solar, 2007–2023 data were assessed. The study found that IREDA is pumping extreme investments in solar renewable energy projects instead of hydro and biomass projects. Capacity addition in solar and wind projects establishes momentous growth in employing renewable energy infrastructure. This emphasizes the potential of the renewable energy sector to combat environmental change and initiate net zero goals. The study needs strategic investment distribution, cognizant policy preparation, effective risk management approaches, and extensive project preparation, embracing technology implementation, encouraging partnerships, collaborations, and stakeholders' engagement. These inferences guide management involvement in renewable energy policy development and financial decision-making to attain environmentally sustainable initiatives.

Keywords Renewable energy · Net zero · IREDA · Financial · Sustainable initiatives

D. V. Nayak
Loyola Academy (Degree and PG College), Old Alwal, Secunderabad, Telangana 500010, India

A. Arun Kumar (✉) · R. Suryadevara
ICFAI Law School, ICFAI Foundation for Higher Education (Deemed to be University Under Sec 3 of UGC Act, 1956), Hyderabad, Telangana 501203, India
e-mail: arunkumar@ifheindia.org

R. Suryadevara
e-mail: rakeshsuryadevara@ifheindia.org

S. Khan
Department of Industrial Engineering, University of Tabuk, Tabuk, Saudi Arabia

R. Singh and D. Crowther (eds.), *Transition Towards a Sustainable Future*, Approaches to Global Sustainability, Markets, and Governance,
https://doi.org/10.1007/978-981-97-5756-5_4

4.1 Introduction

With the existing and growing concerns about climate change and the global need for sustainable development options, there is a rising push towards various renewable energy options to cut back on greenhouse gas emissions and realize net zero (Lin et al., 2023). India is one of the leading countries in the world's renewable energy sector. India has achieved this status through its aggressive renewable energy targets and the country's massive renewable energy resource capacity (Hoang et al., 2023). It makes a fascinating case study regarding the role that renewable energy funding plays in promoting environmental sustainability and achieving net zero. India, one of the world's most rapidly developing economies with an expanding population and increasing energy needs struggles to balance its energy demands and concerns about the environmental implications of energy generation (Das et al., 2023).

India has taken various policy measures supported by the government to promote the widespread use of renewable energy sources and reduce fossil fuel consumption. Current Indian renewable energy capacities show considerable changes that have been possible due to the government's policies, technological development, and cost reduction of renewable energy systems (Tang et al., 2023). India has the world's third-largest renewable energy capacity, with wide-ranging sources, including solar, wind, biomass, and hydroelectric power. The development and deployment of solar and wind energy have received a major impetus from the National Solar Mission and National Wind Energy Mission introduced by the Government of India. On the other hand, the Green Energy Corridors Project and the KUSUM Scheme are geared toward ensuring proper grid integration and encouraging decentralized renewable energy generation processes (Ali et al., 2023).

The importance of finance in the transition and achieving net-zero emissions cannot be overlooked. Renewable energy projects require significant capital investment to deploy the technology, build the required infrastructure, and scale up the capacity (Vulturius et al., 2022). In this context, the importance of available and low-cost sources of finance is essential to finance the scaling of renewable energy projects and promote environmental sustainability. First, IREDA is the main provider of financial support and subsidized investment in developing renewable energy projects. Secondly, the introduction of new types of financing, such as green bonds, venture capital investment, and carbon financing, has generated additional sources of fund allocation for the expansion of renewable resources (Musah et al., 2024). Moreover, international partnerships with such organizations as the World Bank, Green Climate Fund, Asian Development Bank, and others provide technical assistance grants, a concession for everything and everyone loans to mobilize funding for renewable energy projects in India (Rawat et al., 2024).

Renewable energy projects are provided with a number of subsidies and financial incentives by the Indian Government. Capital and interest subsidies, as well as tax breaks and accelerated depreciation, are part of this. The Pradhan Mantri Kisan Urja Suraksha evam Utthan Mahabhiyan (PM-KUSUM) project funds the installation of solar pumps and grid-connected solar power plants in rural regions. In order to secure longer term funding for Renewable Energy Projects, public sector banks and finance institutions play an important role. Key players in this area are the Indian Renewable Energy Development Agency IREDA and the Power Financing Corporation PFC. Renewable energy has also been designated as a priority lending sector by the Reserve Bank of India, which requires banks to allocate an amount for this industry. In order to raise funds for renewable energy projects, green bonds have become a valuable instrument (Rathnasiri et al., 2024). India is one of the first countries that issued green bonds, which were intended to finance photovoltaic and wind projects. The Securities and Exchange Board of India (SEBI) has issued rules for green bonds to ensure openness and accountability in their use.

There are certain hurdles in this process of financing renewable projects for India.

Significant initial investment, in particular for Solar and Wind energy projects, is needed. Private investors, in particular in a market characterized by policy uncertainty and fluctuating energy prices, may be deterred by high capital costs. A significant risk is the financial instability of state owned DISCOMs. A number of DISCOMs are in a poor state of finances, which lead to delays in payments for renewable energy producers and affect the investment environment as a whole.

Investor confidence can be jeopardized by incompatibility of policy frameworks and regulatory uncertainties. The complexity of financing renewable energy projects is further aggravated by frequent changes in tariffs, delays in project approvals and land acquisition issues. Access to affordable and long term funding is still a challenge, although there are many sources of financing. In India, interest rates are very low and there is a need for more innovative financial instruments to cut the cost of capital (Singh et al., 2024).

Further to achieve its Net Zero 2070 targets, India has to follow certain approaches to improve the funding scenario of these renewable projects. It is essential that there be stability and transparency in the policy environment. Consistency in policies and clear long term signals for investors should be ensured by the government. Faster implementation of projects will be facilitated by streamlining the legislative approval process and land acquisition procedures. Public–private partnerships (PPPs) can use private sector's efficacy and expertise. The risks linked to investment in renewables can be mitigated by establishing risk sharing mechanisms, e.g., guarantees and insurance policies. The Green Bond Market is getting more and more developed. Investor confidence will be strengthened by the development of clearer guidance and standards for green bonds. Diversification of sources of funding can be facilitated by promoting the issuance of Green Bonds by State and municipal governments. Another source of capital can be provided by the development of new financial instruments such as green banks, yields, and renewable energy investment trusts. The cost of capital and de-risking investments may be reduced by blended finance models, combining PPAs. There is a need for reforms to improve the financial health of the DISCOMs. Initiatives such as the Ujwal DISCOM Assurance Yojana UDAY are intended to deal with financially distressed companies, but further efforts need to be made in ensuring their long term viability. Significant financial resources can be generated by actively participating in international climate finance mechanisms. Grants, concessional loans and technical assistance from the Global Climate Fund should continue to be sought by India. Technological transfer and capacity building can be facilitated through enhanced international partnerships.

It is a monumental task for India to achieve its goal of reducing carbon emissions by 2070 and requires an integrated and innovation approach in the financing of renewables. Significant progress has been achieved by the government in creating a climate for investment in renewables, but there are still many obstacles to overcome.

It is essential to address high capital costs, improve the fiscal position of DISCOMs, ensure policy stability, and increase access to funding at reasonable rates. By strengthening policy frameworks, enhancing Public Private Partnerships, increasing the Green Bond Market and using International Climate Finance, India can build a robust funding ecosystem that supports its transition toward Renewable Energy.

The impacts of climate change will be mitigated and energy security strengthened, jobs created and sustainable growth stimulated if we achieve net zero emissions. India is in a position to lead the world's Renewable Energy sector with coordinated efforts and strategy investments. In addition, the shift to renewable energy can lead to other multiple environmental benefits. These include curbing air and water pollution, protecting and sustaining natural resources, and reducing the negative influences of climate change (Raihan & Tuspekova, 2023). Solar photovoltaic and wind energy technologies are conducive to a clean, secure, and abundant energy source with fewer environmental implications than fossil fuels (Li et al., 2024). In this regard, action was also required to balance greenhouse gas emissions and removals to address the climate emergency and maintain the global temperature well below 2° above pre-industrial levels. By investing in renewable energy infrastructure and implementing energy efficiency measures, India can significantly contribute to global efforts towards achieving net-zero goals by 2070 (Du et al., 2023).

4.2 Review of Literature

India's potential for renewable energy is large and diverse, which is essential for meeting its Net Zero Emissions (NZE) targets. Renewable energy sources found in the nation include geothermal, hydro, biomass, solar, and wind energy. Of them, solar energy is the most promising, with estimates pointing to a theoretical capacity of up to 749 GW because of the country's vast sunny regions. Bhattacharya et al. (2023), emphasized that although solar energy capacity has grown quickly, it still makes up a small portion of India's total energy mix. According to a map of India's solar energy potential created by the National Institute of Solar Energy (NISE), areas with favorable climates like Rajasthan, Gujarat, and Maharashtra have a disproportionately high potential for producing solar electricity. Significant potential exists for wind energy as well, particularly in coastal areas like Tamil Nadu, Gujarat, and Maharashtra. At 100 m above sea level, the Ministry of New and Renewable Energy (MNRE) projects a potential capacity of 302 GW. Norman et al. (2024), mentioned how, despite advancements in wind energy production, there are still a lot of unexplored places, highlighting the need for stronger regulations and financial support to properly utilize this resource. International partnerships have been very beneficial to India's renewable energy industry since they have brought in cutting-edge technologies, capital, and experience. These collaborations are essential to overcoming obstacles at home and quickening the switch to renewable energy.

One such partnership that aims to advance solar energy through technology transfer and joint partnerships is the Indo-German Energy Forum. Germany and India have partnered to set up solar projects and train Indian engineers using Germany's cutting-edge solar technology. Panicker et al. (2023), emphasized how this collaboration has significantly advanced India's use of solar photovoltaic (PV) technology. Large-scale solar parks, like the Rewa Ultra Mega Solar Park in Madhya Pradesh, which has established standards for solar electricity pricing and efficiency, have been made possible by German investment and expertise. The International Solar Alliance (ISA) is a coalition of solar resource-rich nations that was started in 2015 by France and India. Mehmood et al. (2023), described how the ISA wants to make it easier for more than 1000 GW of solar energy to be deployed by 2030. Significant foreign investment has been drawn to this program, which has also encouraged cooperation on solar energy projects. Through financial assistance, technical support, and capacity-building initiatives, the ISA assists its member nations, assisting India in quickening the deployment of solar energy and the development of its infrastructure. With an emphasis on the development, deployment, and study of renewable energy technology, the United States and India have collaborated extensively in the clean energy space. Chaturvedi (2021), has taken note of important programs like the Partnership to Advance Clean Energy (PACE) and the US-India Clean Energy Finance Task Force. These partnerships have helped India achieve its goals for renewable energy by fostering the growth of bioenergy, solar, and wind energy plants. Notably, the introduction of the Solar Decathlon India competition has stimulated professional and student creativity in sustainable building ideas.

The India–Denmark Green Strategic Partnership focuses on renewable energy development, including wind energy. Denmark, a global pioneer in wind energy technology, has teamed with India to improve wind turbine efficiency and integrate wind energy into the country's grid. Singh et al. (2024), highlighted that the goal of this partnership is to use Danish experience to maximize India's wind energy potential, allowing for significant expansion in this sector. Joint partnerships and technology transfer agreements have resulted in the development of high-efficiency wind turbines and more effective grid integration approaches. Ordonez et al. (2023), discussed India and Japan's renewable energy collaboration, with an emphasis on technology exchange and investments in solar and hydrogen energy. Japan's excellent renewable energy technology capabilities, combined with India's immense resource potential, form a synergistic partnership that drives innovation and growth in the Indian renewable energy scene. This partnership has resulted in numerous key initiatives, including the development of improved solar cells and hydrogen fuel cells, which are critical to India's move to greener energy. According to Midha and Tomar (2024), the Australia-India Energy Dialogue intends to strengthen collaboration in renewable energy, notably solar and wind. Australia's expertise with large-scale solar and wind installations offers vital ideas and technological help to India's renewable energy aspirations. Joint research initiatives and financing agreements have been developed to promote sustainable energy alternatives, such as the creation of high-efficiency solar panels and wind turbines that are suited to Indian climate conditions.

The National Solar Mission, launched in 2010, aspires to make India a global leader in solar energy. The mission's aim of achieving 100 GW of solar capacity by 2022 was nearly fulfilled, with installed capacity reaching roughly 98 GW by the end of 2021. Singh (2022), highlighted the mission's accomplishment in lowering solar prices through competitive bidding and large-scale solar park development. This program has also encouraged decentralized solar power generation, which benefits rural and distant communities with limited grid access. The government has implemented a number of initiatives to promote wind energy, including the Generation-Based Incentives (GBI) and the wind-solar hybrid policy. Jörgensen et al. (2015), noted that these regulations promote the establishment of wind farms in high-potential states, as well as the combination of wind and solar electricity to improve land use and grid stability. The government's focus on replacing aging wind turbines with newer, more efficient versions is another important step towards increasing wind energy generation.

Biomass energy, obtained from agricultural and forest leftovers, has great potential in India. The Ministry of New and Renewable Energy (MNRE) helps biomass projects with financial incentives and technical assistance. Sokołowski (2019), discussed the importance of small hydro projects in rural electrification and providing stable energy sources in remote places. These efforts are critical to expanding India's renewable energy portfolio and guaranteeing energy security. The Punjab Biomass Power Plant generates power from agricultural leftovers including rice husks and wheat straw. Chaturvedi and Malyan (2022), discussed how this initiative not only provides a sustainable energy source but also solves the problem of agricultural waste combustion, which adds to air pollution. The biomass power station shows how biomass energy may contribute to India's renewable energy mix while also supporting rural development. The Muppandal Wind Farm in Tamil Nadu is one of the world's largest onshore wind farms, having a capacity of more than 1500 MW. Noh et al. (2024), noted that this project has greatly contributed to Tamil Nadu's position as a leader in wind energy generation. The Muppandal Wind Farm's performance demonstrates the potential for wind energy growth in India, as well as the relevance of ideal wind conditions and supporting regulations.

Rewa Ultra Mega Solar Park, located in Madhya Pradesh, is one of India's largest solar parks, with a capacity of 750 MW. Dafnomilis et al. (2023), noted that this project has established new norms for solar pricing, making solar electricity competitive with traditional energy sources. The park generates power for the Delhi Metro and other commercial organizations, demonstrating the potential of large-scale solar projects in India. Integrating renewable energy into the national grid presents substantial hurdles. The intermittent nature of solar and wind electricity necessitates effective grid management and storage technologies. Saini et al. (2022), emphasized the need of modern grid infrastructure, such as smart grids and energy storage systems, in accommodating renewable energy sources' changing output. Investment in grid upgrading and the development of battery storage technologies are critical to providing a stable and dependable electricity supply. Land acquisition for large-scale renewable energy projects can be difficult, resulting in delays and higher prices. Iyer-Raniga (2019), emphasized the need for faster regulatory processes and clear land-use rules to promote the development of renewable energy infrastructure. Implementing clear and efficient land acquisition methods, as well as community participation and benefit-sharing structures, can help to reduce conflict and speed up project completion. Continuous innovation in renewable energy technology is critical to increasing efficiency and lowering prices. Li et al. (2022), highlighted the importance of research and development (R&D) in expanding solar and wind technology, such as the creation of more efficient photovoltaic cells and wind turbines. Collaboration with foreign partners, as well as investment in local R&D skills, may accelerate technical improvements and facilitate the expansion of renewable energy projects.

A stable and supportive policy environment is critical to the expansion of the renewable energy sector. Ray (2019), emphasized the need of stable, long-term policy frameworks that give certainty and trust to investors and developers. Policies that encourage renewable energy adoption, such as tax credits, subsidies, and renewable purchase obligations (RPOs), can boost market growth and accelerate the transition to clean energy. Obtaining funding for renewable energy projects remains difficult, especially for small and medium-sized businesses (SMEs). Ko et al. (2022), discussed the need for novel financing arrangements, such as green bonds and public–private partnerships, to raise the funds required for renewable energy initiatives. Government policies that give financial incentives and lower risks for investors can increase the appeal of renewable energy projects.

Strengthening international collaborations and using global experience will help India expedite its renewable energy transition. Upadhyay (2024), considered extending bilateral and multinational cooperation to get access to innovative technology, attract investment, and exchange best practices.

This study tries to fill these gaps by offering a complete analysis of India's renewable energy environment, with an emphasis on the strategic use of solar and wind energy to reach net zero targets. By investigating the precise patterns and variations of renewable energy projects for net zero targets, with a focus on sector-specific financing sanctions, funding distribution, and capacity increases in India. This analysis provides innovative insights into how India might efficiently exploit its renewable energy potential to reach its net zero emission targets.

4.3 Objectives of the Study

1. To study the Indian renewable energy financing policy for Net Zero goals.
2. To analyze trends of Net Zero goals financing for sustainable initiatives.

4.4 Results and Discussions

The Indian Renewable Energy Development Agency (IREDA) is a government owned non-banking financial institution recognized in 1987. It plays an essential role in financing renewable energy projects in India contributing considerably to the country's changeover towards clean and sustainable energy sources (Supriya, 2023).

It operates under the Ministry of New and Renewable Energy (MNRE) and serves as the country's major financial institution for renewable energy initiatives. IREDA's finance policy landscape reveals an inclusive approach to promoting, developing, and financing renewable energy projects across India to achieve Net Zero initiatives (Ulpiani et al., 2023).

IREDA pooled financial assistance from international financial sources, domestic borrowings, equity, and reserves to promote renewable energy infrastructure, enhance energy access and affordability, and ensure renewable energy projects' financial viability and sustainability (Xu et al., 2023). It proposes different kinds of financial services to the requirements of renewable energy projects (See Fig. 4.1). It provides term loans for the growth, installation, and operation of renewable energy projects (Jin et al., 2023). Term loan favorable in terms and conditions to subsidize renewable energy, project developers. It provides short-term financing to bridge funding gaps to eligible projects during the project development phase (Alharbi et al., 2023). It provides refinance assistance to banks and other financial institutions to provide financial assistance for renewable energy projects. In addition to debt financing, IREDA provides equity financing in select projects that involve substantial strategic importance (Bożena Gajdzik, 2023).

The funding policy landscape contains detailed eligibility criteria for accessing its financing services based on renewable energy sources such as wind, hydro, biomass, and solar. It evaluates project developers' track record and credibility, technical feasibility, and compliance with relevant standards and regulations to minimize default risks (Ma et al., 2024). It offers competitive interest rates and flexible repayment terms to make financing accessible to many renewable energy project developers and ease financial burdens. It conducts thorough risk management practices to moderate market risks affecting project viability (Gao et al., 2024).

It provides a holistic approach to financing renewable energy projects in India. Ensures compliance with relevant regulations, policies, and guidelines issued by regulatory authorities such as the Central Electricity Regulatory Commission, Ministry of New and Renewable Energy, and state electricity regulatory commissions, plays a vital role in the growth of the renewable energy sector and advancing environmental sustainability in the country (Behera et al., 2023). It tracks key performance indicators (KPI) such as reducing carbon emissions, installed capacity, job creation, and enhancing energy access. IREDA finance policy is aligned with national renewable energy goals, energy policy objectives, and international commitments on climate change mitigation and sustainable development (Bie et al., 2024).

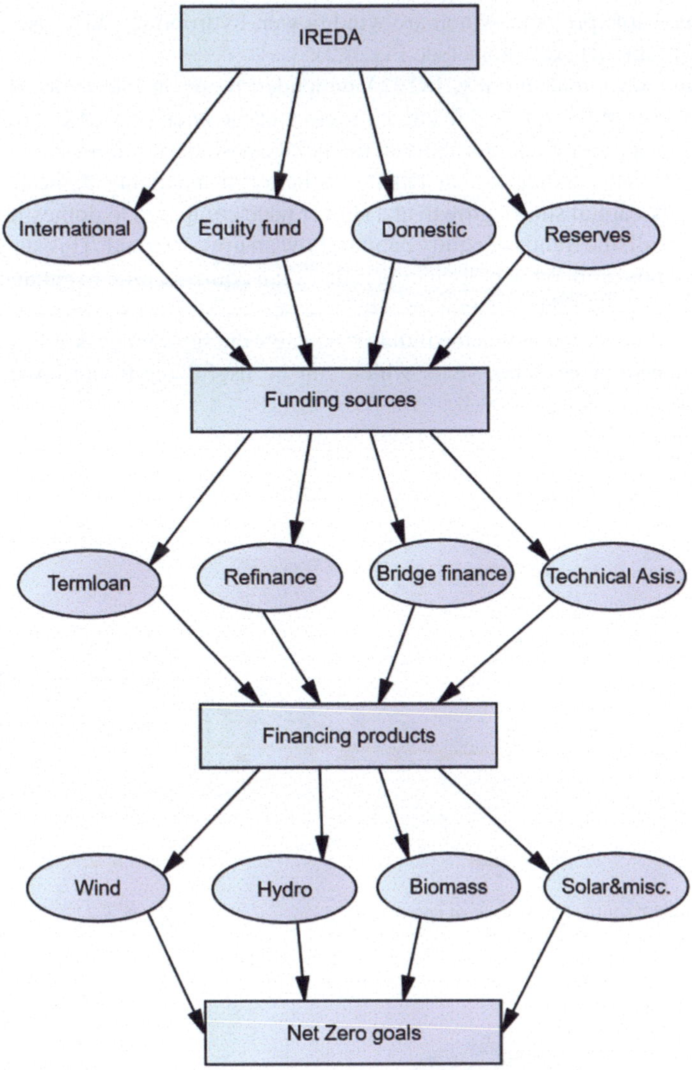

Fig. 4.1 IREDA financing strategies flowchart to attain net zero goals initiatives

4.4.1 Trends of Net Zero Goals Financing for Sustainable Initiatives

Renewable Energy Funding sources

During the period of 2007–2023, IREDA provided financial assistance from International assistance, equity capital, domestic borrowings, and internal reserves to support renewable projects, which are wind power, hydropower, biomass, and solar power, to initiate net zero goals (see Fig. 4.2).

The data shows that initially, IREDA depended more on international financial assistance to boost the renewable energy sector; in the later years, IREDA seems to shift from relying more on international financial assistance to domestic borrowings (see Fig. 4.3). This indicates that IREDA is better at managing domestic funding tactics. Equity capital shows growth at a slower pace compared to domestic financial sources. Overall, the trends of equity capital show an upward trend. This suggests that IREDA depends more on debt financing rather than equity capital to enhance renewable energy projects. The constant growth in Reserves/Surplus shows a strong financial position. This suggests that the Indian renewable energy agency is able to generate and retain earnings over the years which can be useful for future environmental investments.

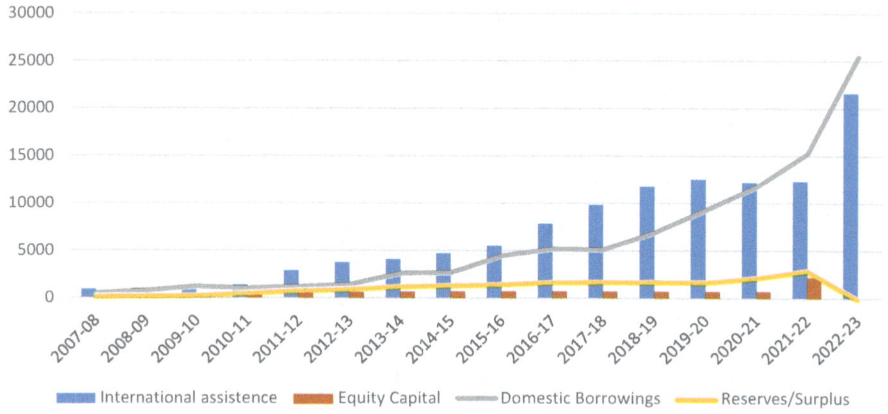

Fig. 4.2 Funding sources year wise in crores

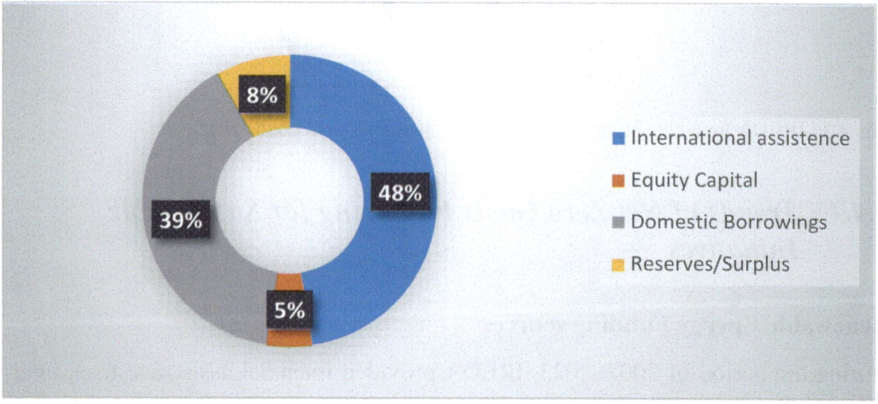

Fig. 4.3 Financing sources

The total investment from each individual source over the period from 2007 to 2023 was covered. IREDA majorly borrowed financial assistance for renewable energy with 48% from international financial assistance; later, domestic borrowing provided a contribution of 39%. The equity capital and reserves/surplus provide less amount of contribution, with 5% and 8%, respectively, towards renewable energy to achieve Net Zero goals.

Renewable Energy Capacity Addition

During 2007–2023, IREDA provided financial assistance to support renewable energy projects capacity and boost the energy sector with non-fossil fuels.

Indian renewable energy development agency provides capacity addition in wind power showing some significant variations over the years (See Fig. 4.4). There is a sudden upward trend in capacity addition in 2022–23, which indicates an extensive investment in technological advancement in wind power infrastructure. At the same time, hydro energy capacity addition shows significant variations from year to year. There was a sudden upward growth in 2013–14 followed by a decline in the subsequent years and then a slight upward trend again in 2019–2020. Capacity additions in biomass energy projects indicate a comparatively stable trend with some variations but no significant downward trend. Capacity addition in solar and miscellaneous projects demonstrates a momentous growth over the years. Particularly from 2015 to 2016 onwards, solar and miscellaneous projects capacity addition shows a tremendous upward trend in 2020–21, suggesting IREDA concentrated more on solar energy projects during that period.

Renewable Energy Sector-Wise Sanctions

During the period of 2007–2023, IREDA provided financial assistance to support renewable energy projects, which are wind power, hydropower, biomass, and solar power, to support power generation capacity (See Fig. 4.5).

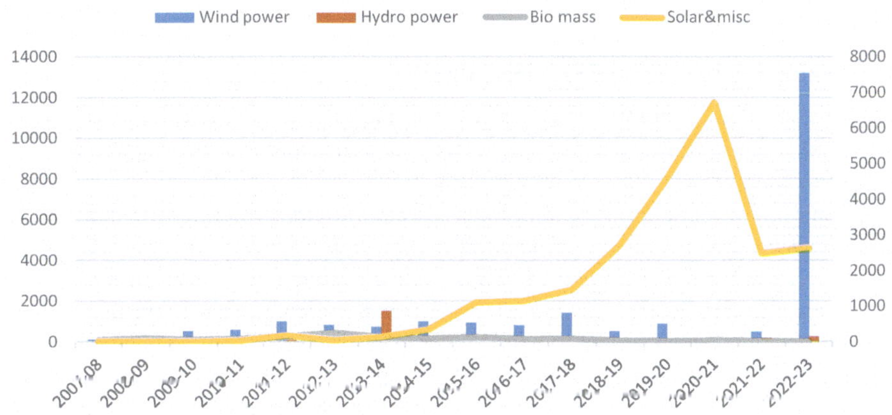

Fig. 4.4 Capacity addition year wise (MW)

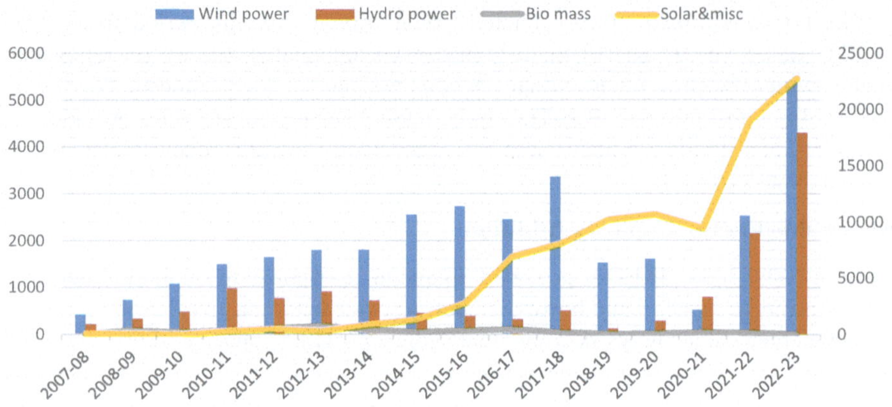

Fig. 4.5 Sector-wise sanctions in crore

Indian renewable energy development agency provides funding sanctions in wind power showing some significant upward trend over the years. A sudden upward trend in financial sanctions in 2021–22 and 2022–23 indicates an extensive investment in technological advancement in wind power infrastructure. At the same time, hydro energy funding sanctions show a significant upward trend from year to year. There is a sudden upward growth in 2021–22 followed by a subsequent upward trend in 2022–23, suggesting a growing trend in hydro energy projects. Financial sanctions in biomass energy projects indicate comparatively low wind and hydropower projects. There is a slight upward trend in 2021–22, but biomass trends remain uncertain compared to other renewable projects. Financial sanctions in solar and miscellaneous projects have exhibited momentous growth over the years. Particularly, from 2016 to 2017 onwards, solar and miscellaneous projects financial sanctions show a tremendous upward trend in 2020–21 and 2022–23, suggesting that the Indian renewable energy development agency concentrated more on solar energy projects during that period.

Renewable Energy Sector-Wise Distribution

During the period of 2007–2023, IREDA distributed financial assistance to support renewable energy projects, which are wind power, hydropower, biomass, and solar power, to support power generation capacity.

Indian renewable energy development agency provides funding distributions in wind power showing some significant upward trend over the years (See Fig. 4.6). There is a sudden upward trend in financial distribution, particularly from 2016 to 2017 onwards, and in 2021–22 and 2022–23, trends indicate an extensive investment on technological advancement in wind power infrastructure. At the same time, hydro energy funding distributions show a significant upward trend from year to year. There is a sudden upward growth in 2021–22 followed by a subsequent upward trend in 2022–23, suggesting a growing trend in hydro energy projects. Financial distributions in biomass energy projects indicate comparatively low wind and hydropower

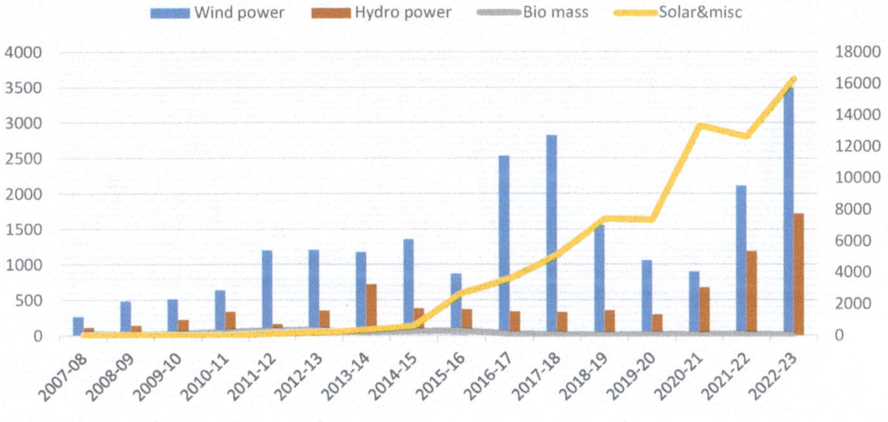

Fig. 4.6 Sector-wise distribution in crore

projects. There is a slight upward trend in 2021–22, but biomass trends remain uncertain compared to other renewable projects. Financial distributions in solar and miscellaneous projects have exhibited momentous growth over the years. Particularly from 2016 to 2017 onwards, solar and miscellaneous projects' financial distributions show a tremendous upward trend in 2020–21 and 2022–23, suggesting that the Indian renewable energy development agencies concentrated more on solar energy projects during that period.

Renewable Energy State-Wise Sanctions and Distributions in India

During the period of 2007–2023, IREDA provided financial assistance to different states in India to support renewable energy projects, which are wind power, hydropower, biomass, and solar power, to support power generation capacity.

The Indian renewable energy development agency provides financial assistance to different states in India (See Fig. 4.7). Among them, Andhra Pradesh state topped in loan sanction amount of Rs. 18,781.83 crore and distributed amount of Rs. 15,436.02 crore for renewable energy project development. A significant amount of financial assistance of Rs. 12,673.92 crore was sanctioned for Gujarat with a distribution of amount Rs. 5879.66 crore. Like Gujarat, Maharashtra allotted a significant sanction of Rs. 15,644.71 crore, with a distribution of Rs. 11,887.77 crore. Tamil Nadu provided a significant sanction of amount Rs. 13,282.8 crores, with a distribution of amount Rs. 10,526.56 crore, Uttar Pradesh had a sanction of amount Rs. 4062.96 crores, with a distribution of amount Rs. 2223.5 crore, Rajasthan established a considerable sanction of amount Rs. 21,196.81 crore, with a distribution of amount Rs. 13,129.31 crore, Karnataka had a considerable sanction of amount Rs. 19,316.18 crore, with a distribution of amount Rs. 13,096.41 crore. It indicates that a significant amount of the sanctioned funds has to be distributed.

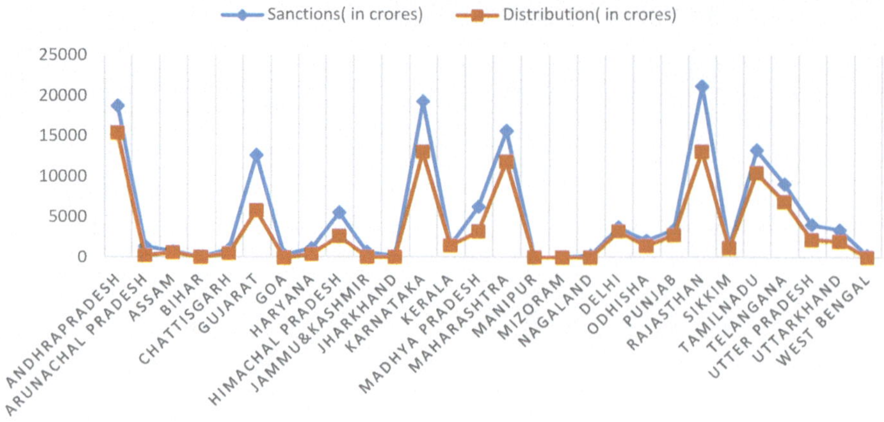

Fig. 4.7 State-wise sanctions and distributions in India

Renewable Energy Sanctions, Distributions, and Capacity Addition

During the period of 2007–2023, IREDA provided financial sanctions, financial distribution, and total capacity additions in Mega Watts to support renewable energy projects, which are wind power, hydropower, biomass, and solar power to support power generation capacity.

Indian renewable energy development agency financial sanctions and distributions show upward trends over the years, exhibiting a tremendous growth in renewable energy investment. The trends show some variations, but overall, there is an upward trend. Capacity additions show an upward trend over the years, reflecting the growth and development of the renewable energy sector (See Fig. 4.8). However, some variations in capacity addition are due to infrastructure development, demand variations, and policy changes. The financial distribution lower than the financial sanctions year by year indicates that IREDA does not distribute all funds immediately because of delays in the administrative process and project timelines, which affects financial allocation and capacity addition in India. Overall, the financial sanctions, distributions, and capacity addition in renewable energy show upward trends from 2017–18 to 2021–22, indicating growth in the renewable energy sector.

Projects Wise Sanctions, Distribution, and Capacity Addition in Renewable Energy

During the period of 2007–2023, IREDA provided financial sanctions, financial distribution, and capacity addition in renewable energy. The insights into the financial sanctions, distributions, and capacity addition in various types of renewable energy projects (See Fig. 4.9).

Solar and miscellaneous projects received more financial sanctions and distributions in these renewable energy projects, followed by wind energy projects. The capacity additions in wind power and solar energy projects are notably high compared

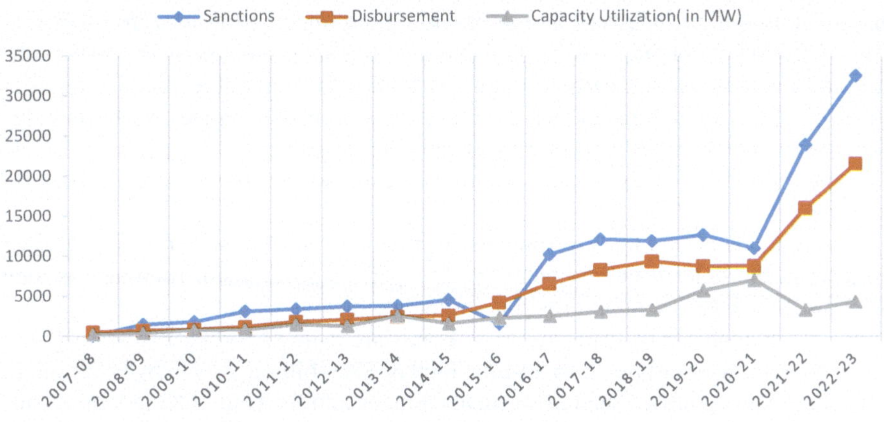

Fig. 4.8 Sanctions, distributions, and capacity addition year wise

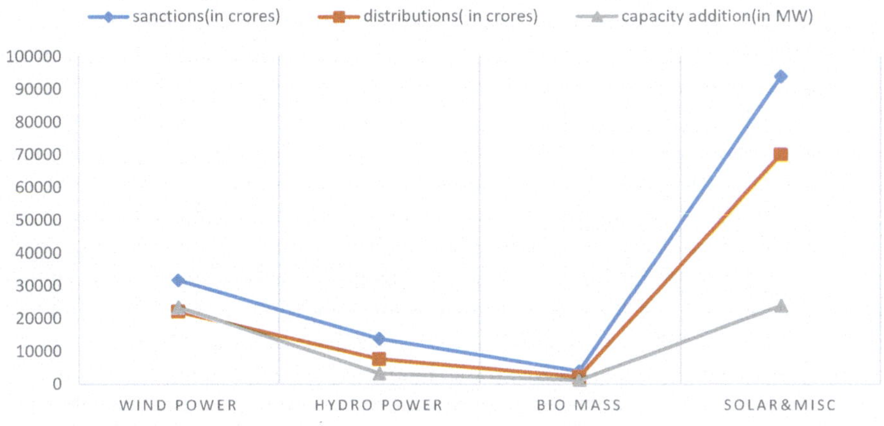

Fig. 4.9 Projects wise sanctions, distribution, and capacity addition

to hydropower and biomass energy projects. The upward trends in financial sanctions and distributions emphasize on capacity additions in all renewable energy projects.

Achieving net zero emissions has become a critical target for countries, businesses, and institutions as the world community increasingly acknowledges the urgency of addressing climate change. The equilibrium between the amount of greenhouse gas emissions generated and their elimination from the atmosphere is referred to as 'net zero'. A significant amount of funding is needed to support sustainable initiatives aimed at reducing the carbon footprint, improving energy efficiency and promoting renewable energy sources in order to meet these ambitious objectives. An analysis of the role of government policies, private sector investment, innovative financial instruments and international cooperation is presented in this article, which examines the emerging trends in financing these net zero goals. Governments have a key role to play in setting the framework and in providing the impetus for net zero financing. The environment conducive to sustainable investments is created by domestic policies and regulations, e.g. carbon pricing, subsidies for renewable energies or strict emission reduction targets. In fact, the European Union's Green Deal seeks to mobilize approximately EUR 1 trillion in sustainable investments over the next decade by using public funds to promote internal investment in environmentally friendly projects.

Green bonds issued by governments for the purpose of raising funds to finance projects in the field of environment are also part of government funding. As they offer a secure investment and funding initiatives such as wind farms, solar installations or sustainable infrastructure, these bonds are gaining popularity. In 2020, the green bond market will reach an all-time high of $280 billion, driven by government issuances from countries such as Germany, France and the United Kingdom. Financial viability and the need for sustainable investments are becoming more apparent to the private sector. Corporate undertakings are making commitments to achieving zero emissions due to shareholder pressure, consumer demand and the real risks of climate change. Renewable energy, energy efficiency and carbon capture technologies are being invested by businesses. In the case of Google and Microsoft, they have both committed to net zero emissions as well as investing heavily in renewable energy (Bhushan & Singh, 2024).

Pension funds, insurance companies, and such institutional investors are also aligning their portfolios with net zero goals. Investment strategies have been influenced by the rise of environmental, social, and governance indicators, and asset managers are integrating sustainability into their decision-making processes. New evidence of the potential for sustained investments to deliver competitive returns while reducing risks in the longer term is supporting that shift. To mobilize the required funds for net zero initiatives, it is essential to innovate in finance instruments. In addition to green bonds, various innovative mechanisms are gaining traction. These loans are linked to the performance of an entity in line with established sustainability objectives. If a company meets certain carbon reduction objectives, it may be able to take advantage of reduced interest rates. Sustainability loan guarantees encourage borrowers to improve environmental performance and align financial interests with sustainability results. These bonds are used to finance the transition of companies from high carbon into low-carbon operations. In particular, where the transition to net zero is complicated and costly in terms of capital investments, they are particularly important for sectors such as oil and gas, steel or cement.

More and more equity investments are being made specifically dedicated to green projects. Start-ups and companies offering innovations in sustainability, e.g., solar energy technology, sustainable agriculture or hybrid vehicles, are being supported by venture capital as well as public equity funds. In sustainable projects, the blended finance approach brings together public and private capital to reduce the risk of investments. Some of the risk is covered by state funds, which makes it more attractive for private investors to be involved. In developing countries where perceived risks are higher, blended finance is an important tool to mobilize funding for projects.

In particular, in developing countries that are facing significant financial and technological challenges, international cooperation is essential to fund global net zero goals. Multilateral development banks (MDBs) like Asian Development Bank, African Development Bank and even the World Bank, play a critical role in providing financial and technical assistance to sustainable development projects. Climate considerations are increasingly integrated into the funding criteria and project evaluation of MDBs. In order to help countries, develop and implement low-carbon strategies, they offer concessional loans, grants, and technical assistance. Indeed, more than $8 billion has been mobilized to combat climate change in developing countries through the Climate Investment Funds operated by the World Bank Group and financed jointly with another $58 billion.

Another efficient means of financing net zero goals is PPPs. Public private partnerships combine the strengths of both sectors by combining public sector oversight and regulation with effective performance in the private sector as well as innovation. The success of projects such as large-scale wind farms, development of urban infrastructure and energy efficiency programs in cities can be attributed to these examples. For instance, the Morocco Ouarzazate Solar Power Station, one of the largest solar power complexes in the world, was developed through a PPP. This project, which has been supported by International Donors and Private Investors, shows that cooperation can lead to significant progress toward the Net Zero targets.

However, there are still several challenges in funding net zero initiatives despite the good progress. Clear and consistent policies, the risk of greenwashing as well as a need for substantial upfront investment in emerging technologies are all part of this. Governments and financial institutions need to increase transparency and accountability for green financing in order to deal with these challenges. In order to ensure that funds contribute in a meaningful way towards the sustainability objectives, it is possible to create robust standards and verification mechanisms. In addition, it will be crucial to foster innovation in financial instruments and expand successful models such as blended finance and public private partnerships.

There are promising ways to advance net zero financing through the role of digital technologies in financial markets. For example, the use of block-chain technology to increase transparency and traceability in green finance may lead to a more accurate tracking and verification of investments. In the same way, a wider range of investors can benefit from financial technology solutions by making access to sustainable investments more accessible. A multidisciplinary approach involving governments, investors from the private sector, Innovative Financial Instruments and International Cooperation is needed to finance a net zero goal. The scale of the challenge calls for continued efforts and collaboration in all areas, although significant progress has been achieved. The world community can make significant progress towards the achievement of net zero emissions and against climate change through a combination of different types of funding mechanisms, as well as promoting an environment conducive to sustainable investment

4.5 Managerial Implications

The study has managerial implications. Indian Renewable Energy Development Agency strategically allocates resources and funds to renewable energy projects, understands the trends of financing different renewable energy projects, and prioritizes the energy sector with high potential for net zero goals. Management must identify the risk factors associated with renewable energy financing and understand the project timeline, policy dynamics, and technological uncertainties of projects to develop risk mitigation strategies. Policymakers can evaluate the impact of existing renewable energy policies on financing, capacity addition, and project outcomes. Policymakers can leverage existing investment strategies more effectively. Management can use insights to identify potential partnerships and collaboration among the renewable energy sector development stakeholders. Management must adopt a proactive approach and technological advancement to enhance its competitiveness and establish itself in an evolving market. Transparent communication can help to engage stakeholders of renewable energy projects, policymakers, the common public, and non-government organizations effectively. This leads to greater success in Net Zero initiatives.

4.6 Societal Implications

The shift to net zero emissions represents a fundamental shift in the way societies operate, requiring significant investment in sustainable initiatives. In this transition, the India Renewable Energy Development Agency IREDA plays an important role by providing funding to solar energy projects. This paper examines the social implications of financing the transformation to net zero goals, focusing on IREDA investment strategies and their impact on various societal dimensions.

The most significant social impact arising from the financing of Renewable Energy Projects under IREDA is economic growth and job creation. New sectors and industries are emerging as India expands its renewable energy capacities, creating new opportunities for employment. A significant manpower is required during the construction, operation or in maintenance of renewable energy infrastructure such as Solar Power Plants and Wind Energy Installations. By 2030, Council on Energy, Environment and Water (CEEW) estimates that India's renewable energy sector could employ more than 330,000 people.

Investment in IREDA also has a major impact on energy access and equity. Renewable energy projects, particularly those focused on decentralized solutions like home solar can provide electricity to isolated or hard through areas. This will improve the quality of life for residents and support social equity through a solution to the energy access gap between towns and villages. For instance, the Pradhan Mantri Kisan Urja Suraksha evam Utthan Mahabhiyan (PM-KUSUM) scheme, supported by IREDA, aims to install solar pumps and grid-connected solar power plants in rural regions, directly benefiting farmers and rural communities.

The transition to renewable energy offers considerable environmental benefits. Greenhouse gas emissions, climate change mitigation and its associated impacts are reduced by reducing the reliance on fossil fuels. There are significant public health benefits to cleaner air as a result of reduced pollution levels. Air pollution is the most significant risk factor for many chronic pulmonary or cardio vascular issues, according to World Health Organization WHO. IREDA is helping to reduce air pollution by investing in renewable energy, which could reduce the incidence of these diseases and reduce healthcare costs. Technological innovation and infrastructure development are supported by the financing of IREDA sustainable initiatives. R&D is driven by investment in renewable energy technologies that lead to progress which can be applied across different sectors. The development of smart grids, energy storage solutions and energy-efficiency technologies is supporting the renewable energy sector with more widespread applications as well as increasing total energy efficiency and reliability.

Social and cultural change is also necessary for the transition to renewable energy sources. A cultural shift toward sustainability is taking place as communities become more conscious of issues relating to the environment and its benefits from renewables. In this transformation, education and community engagement initiatives funded by IREDA investments are essential. Promoting renewable energy literacy helps to build public support for sustainable practices, bringing behavioral change, and contributing to environmental goals.

Financial inclusion and resilience to the economy can also be promoted by investing in green energy projects. IREDA promotes local entrepreneurship and diversification of economies through support to SMEs in the renewables industry. In this way, the vulnerability of communities to economic shocks can be reduced and the economic landscape can be more resilient. The key driver of other economic activities, such as support to SMEs and the promotion of regional economies, is also access to energy that is cost effective and secure. Despite the positive societal implications, financing the transformation to net-zero goals through IREDA's investments also presents challenges. Ensuring that the benefits of RE projects are distributed and exhibit careful planning. If proper management is not taken into account, there will be a risk of an excessive expansion of wind and solar energy infrastructure leading to deterioration in the environment. For example, large-scale solar installations or wind farms may require significant land that could potentially have an impact on local communities and ecosystems.

Moreover, stability of policy environments and strong legal frameworks are frequently decisive for the financial viability of renewables projects. Investment and the implementation of projects may be hampered by policy uncertainty, which may undermine potential social benefits. Therefore, in order to maintain the momentum of investments in renewables, it is necessary to provide continuous policy support and clearly establish legal guidance. Several policy recommendations can be considered to maximize the social benefits of IREDA investment in renewables. In order to ensure that their needs and concerns are met, local communities will be engaged in the planning and implementation of renewable energy projects. This can help to build local support and minimize social disruption. Ensure that, by creating employment opportunities and supporting local businesses, renewable energy projects contribute to inclusive economic growth. In order to maintain equity, special care should be paid to the disadvantaged groups. In order to encourage long term investment in renewables, ensure that the policy environment is stable and favorable. Clear regulatory guidelines, consistent subsidies and incentives for renewable energy projects are also part of this. In order to stimulate innovation in renewable energy technologies, investment is needed in research and development. This can improve the efficiency and effectiveness of renewable energy projects, as well as support more widespread applications in different sectors. In order to ensure that renewable energy projects do not adversely affect local ecosystems and communities, implement robust environmental impact assessment, and monitoring mechanisms.

4.7 Academic Implications

It is a technological and economic challenge as well as an enormous academic endeavor to transition towards net-zero emissions. There are important academic implications for funding this transformation, in particular by means of agencies such as the Indian Renewable Energy Development Agency IREDA. The impact of the shift to sustainable initiatives on research, education, interdisciplinary studies, policy analysis, and the development of new academic paradigms is explored in this article. One of the primary academic implications of financing renewable energy projects through IREDA is advancing research in this field. Extensive R&D is needed in order to improve efficiency, lower costs and enhance the reliability of renewables as a result of significant investments in them. The financing of IREDA can encourage cutting-edge research into solar, wind, hydro and biomass energy technologies. Funding opportunities for projects to develop new materials, improve energy storage solutions and optimize grid integration are likely to be increased for universities and research institutions. There are many opportunities to apply for research in the deployment of major renewables projects. In order to carry out field studies, collect empirical data and develop best practices for the implementation of renewable energy sources, academic institutions may cooperate with industrial sectors.

A workforce with specialized skills and knowledge is needed for the transition to renewable energies. This will have a significant impact on educational establishments. The curricula for renewable energy technologies, sustainable development and environmental sciences should be developed and extended by academic institutions. There will be a greater number of courses in solar and wind energy systems, energy policy, sustainable design as well as assessment of impacts on the environment. In order to prepare technicians, engineers and project managers in the renewable energy sector, there will be a growing need for vocational training programs. Practical training and certification schemes in line with industry standards must be made available by institutions. There is a need for knowledge from several disciplines regarding the complex nature of renewable energy systems. This not only provides comprehensive knowledge but also acknowledges the challenges and opportunities in renewables, universities may be increasingly offering integrated programs combining engineering, environmental sciences, economics or government policy.

Multifaceted challenges, which include different disciplines, will need to be addressed in order to transform the energy system into a sustainability one. This creates a conducive academic environment for interdisciplinary studies. In order to deal with the overall challenges of energy transition, researchers from various fields such as engineering, economics, environment, and social sciences will have to work together. Integrated research projects on the basis of technical, Economic, Social, and Environmental dimensions may thus be carried out. Detailed economics and policy analysis is needed for the financing aspects of Renewable Energy Projects funded under IREDA. In order to assess the effectiveness of different financial mechanisms, subsidies and regulatory frameworks that will contribute to more informed policy making, researchers will have to be required. The social impact of Renewable Energy Projects have to be explored in order to understand their effects on society. How these projects affect communities, contribute to social equity and improve public health outcomes can be investigated by sociologists, anthropologists and public health professionals.

The development of new academic paradigms reflecting the changing landscape of energy production and consumption can result from the completion of sustainable initiatives and the transition to net zero goals. The development of sustainability science as a separate discipline could result from an increasing emphasis on sustainable development. In order to address the various sustainability challenges and help sustain development, this field brings together knowledge of natural and Social Sciences. A systematic thinking approach is needed for the complexity of renewable energy systems. Frameworks that take into account the interdependence of energy systems, including technology, economics, environment, and social factors, need to be adopted and developed by researchers. Resilient and adaptation studies will become increasingly important in view of the increasing effects of climate change. In order to ensure reliable and sustainable access to energy, academic research will focus on how renewable energy systems can be designed for resilience against climatic or economic shocks.

The support of IREDA to fund renewable energy projects leads to increased collaboration among the academia and industry, which will lead to mutual benefits. Joint research initiatives, using industry funding and academic expertise, can be coordinated between academic institutions and industry players. Innovation and the implementation of research findings can be accelerated through this collaboration. There are many internships and job opportunities for students as a result of the expansion of the renewable energy sector. In order to give students hands on experience and career paths, educational institutions can form partnerships with industry. Collaboration between academia and industry facilitates the transfer of knowledge and technology. Academic research and innovation could help the industry, while giving academia a better understanding of practical challenges and needs in this sector that would lead to more effective and significant research.

4.8 Limitations and Scope of Future Research

The provided data may not cover all aspects of renewable energy projects, and additional data can provide a comprehensive understanding of financial distribution, project timeline, and technological specifications. There is limited data period available to understand renewable energy project developments. The data is limited to renewable energy projects, such as wind power, hydropower, biomass, and solar power, in India only. The reliability and accuracy of the secondary data may compromise the quality and accuracy of drawing meaningful conclusions.

Cross-country comparison of renewable energy projects on various investment strategies, technological adoption, and net zero goal policies can be considered a prospective research domain. A holistic approach to understanding renewable energy projects' socio-economic and environmental impacts on employment, community development, and carbon emission reduction can be studied. The effectiveness of renewable energy policies in investments and innovations to attain environmentally sustainable goals can be evaluated. The stakeholders of the Indian Renewable Energy Association and collective collaboration towards renewable energy transition can be studied.

4.9 Conclusion

The availability of data on renewable energy projects in India provides significant insights into financing sources, investment trends, and capacity additions. The data highlights the growing importance of the trends and fluctuations of wind power, hydropower, biomass, and solar renewable energy projects for net zero goals in India. The study emphasizes that Indian renewable energy development agency promotes net zero goals by providing huge investments in solar and wind projects. In capacity addition, solar projects grew momentously by strategically understanding the trends of financing different renewable energy projects and strategically prioritizing the energy sector with a high potential for net zero goals. Developing risk mitigation strategies for project timelines, policy dynamics, and technological uncertainties is required. Policymakers can leverage existing investment by evaluating the impact of present renewable energy policies on funding allocation, capacity addition, and project outcomes. Sustainable development is promoted by the IREDA's investment initiatives in financing the transformation to net zero goals. These initiatives are fostering economic growth. A transition to a low-carbon economy is accelerated by IREDA. IREDA's efforts harmonize financial viability with environmental protection, paving the way for a sustainable future. Projects should adopt a proactive approach and technological advancement to enhance their competitiveness and establish themselves in the competition. Implementation of proper promotion strategies leads to engaging stakeholders, policymakers, and local communities on economic growth and environmentally sustainable development.

References

Alharbi, S. S., et al. (2023). Green finance and renewable energy: A worldwide evidence. *Energy Economics*. https://doi.org/10.1016/j.eneco.2022.106499

Ali, K., et al. (2023). Testing the role of digital financial inclusion in energy transition and diversification towards COP26 targets and sustainable development goals. *Gondwana Research*. https://doi.org/10.1016/j.gr.2023.05.006

Behera, B., et al. (2023). Decoupling the role of renewable energy, green finance, and political stability in achieving the sustainable development goal 13: Empirical insight from emerging economies. *Sustainable Development*.https://doi.org/10.1002/sd.2657

Bhattacharya, S., et al. (2023). Bending the emission curve—The role of renewables and nuclear power in achieving a net-zero power system in India. *Renewable and Sustainable Energy Reviews*.https://doi.org/10.1016/j.rser.2023.113954

Bhushan, M., & Singh, R. (2024). Talent management facilitates net-zero transition through employee green behavior. In R. Singh, S. Khan, A. Kumar, S. Luthra, & H. Chokshi (Eds.), *Net zero economy, corporate social responsibility and sustainable value creation. CSR, sustainability, ethics & governance*. Springer. https://doi.org/10.1007/978-3-031-55779-8_7

Bie, F., et al. (2024). Transitioning to a zero-emission energy system towards environmental sustainability. *Gondwana Research*. https://doi.org/10.1016/j.gr.2023.03.022

Chaturvedi, V. (2021). A vision for a net-zero energy system for India. *Energy and Climate Change*. https://doi.org/10.1016/j.egycc.2021.100056

Chaturvedi, V., & Malyan, A. (2022). Implications of a net-zero target for India's sectoral energy transitions and climate policy. *Oxford Open Climate Change*.https://doi.org/10.1093/oxfclm/kgac001

Dafnomilis, I., et al. (2023). Achieving net-zero emissions targets: An analysis of long-term scenarios using an integrated assessment model. *Annals of The New York Academy of Sciences*.https://doi.org/10.1111/nyas.14970

Das, A., et al. (2023). Pathways to net zero emissions for the Indian power sector. *Energy Strategy Reviews*. https://doi.org/10.1016/j.esr.2022.101042

Du, J., et al. (2023). Modeling the green logistics and financial innovation on carbon neutrality goal, fresh insight for BRICS-T. *Geological Journal*.https://doi.org/10.1002/gj.4732

Gajdzik, B., et al. (2023). Process of transformation to net zero steelmaking: Decarbonisation scenarios based on the analysis of the Polish steel industry. *Energies*. https://doi.org/10.3390/en16083384

Gao, S., et al. (2024). Financial inclusion empowering sustainable technologies: Insights into the E-7 economies from COP28 perspectives. *Technological Forecasting and Social Change*. https://doi.org/10.1016/j.techfore.2023.123177

Hoang, A. T., et al. (2023). Technological solutions for boosting hydrogen role in decarbonization strategies and net-zero goals of world shipping: Challenges and perspectives. *Renewable and Sustainable Energy Reviews*.https://doi.org/10.1016/j.rser.2023.113790

Iyer-Raniga, U. (2019). Zero energy in the built environment: A holistic understanding. *Applied Sciences*. https://doi.org/10.3390/app9163375

Jin, C., et al. (2023). Green finance, renewable energy, and carbon neutrality in OECD countries. *Renewable Energy*. https://doi.org/10.1016/j.renene.2023.04.105

Jörgensen, K., et al. (2015). Multi-level climate governance in India: The role of the states in climate action planning and renewable energies. *Journal of Integrative Environmental Sciences*. https://doi.org/10.1080/1943815X.2015.1093507

Ko, I., et al. (2022). Have renewable energy leaders announced aggressive emission reduction goals? Examining variations in the stringency of country-level net-zero emission pledges. *Plos Climate*. https://doi.org/10.1371/journal.pclm.0000094.

Li, L., et al. (2022). Review and outlook on the international renewable energy development. *Energy and Built Environment*. https://doi.org/10.1016/j.enbenv.2020.12.002

Li, T., et al. (2024). Towards Paris Climate Agreement goals: The essential role of green finance and green technology. *Energy Economics*.https://doi.org/10.1016/j.eneco.2023.107273

Lin, Y., et al. (2023). Assessment of renewable energy, financial growth, and accomplishing targets of China's cities carbon neutrality. *Renewable Energy*.https://doi.org/10.1016/j.renene.2022.11.026

Ma, Z., et al. (2024). Financing net-zero energy integration in smart cities with green bonds and public-private partnerships. *Sustainable Energy Technologies and Assessments*. https://doi.org/10.1016/j.seta.2024.103708

Mehmood, A., et al. (2023). Achieving energy sustainability by using solar PV: System modelling and comprehensive techno-economic-environmental analysis. *Energy Strategy Reviews*. https://doi.org/10.1016/j.esr.2023.101126

Midha, A., & Tomar, A. (2024). Towards achieving net zero emissions in India by 2070. *Renewable Power for Sustainable Growth*. https://doi.org/10.1007/978-981-99-6749-0_66

Musah, M., et al. (2024). Achieving net-zero emission target in Africa: Are sustainable energy innovations and financialization crucial for environmental sustainability of sub-Saharan African state? *Applied Energy*.https://doi.org/10.1016/j.apenergy.2024.123120

Noh, Y., et al. (2024). A review on harnessing renewable energy synergies for achieving urban net-zero energy buildings: Technologies, performance evaluation, policies, challenges, and future direction. *Sustainability*. https://doi.org/10.3390/su16083444

Norman, J., et al. (2024). The role of repowering India's ageing wind farms in achieving net-zero ambitions. *Environment Research Letters*. https://doi.org/10.1088/1748-9326/ad28db

Ordonez, J. A., et al. (2023). India's just energy transition: Political economy challenges across states and regions. *Energy Policy*. https://doi.org/10.1016/j.enpol.2023.113621

Panicker, K., et al. (2023). Assessment of building energy performance integrated with solar PV: Towards a net zero energy residential campus in India. *Energy and Buildings*.https://doi.org/10.1016/j.enbuild.2022.112736

Raihan, A., & Tuspekova, A. (2023). Towards net zero emissions by 2050: The role of renewable energy, technological innovations, and forests in New Zealand. *Journal of Environmental Science and Economics*. https://doi.org/10.56556/jescae.v2i1.422.

Rathnasiri, M. S. H., Dewasiri, N. J., & Kumar, A. A. (2024). Policy framework and implementation strategies for Sri Lanka's transition to a net-zero economy. In R. Singh, S. Khan, A. Kumar, S. Luthra, & H. Chokshi (Eds.), *Net zero economy, corporate social responsibility and sustainable value creation. CSR, sustainability, ethics & governance*. Springer. https://doi.org/10.1007/978-3-031-55779-8_3

Rawat, A., et al. (2024). Analysis of the key hydrogen fuel vehicles adoption barriers to reduce carbon emissions under net zero target in emerging markets. *Energy Policy*.https://doi.org/10.1016/j.enpol.2023.113847

Ray, P. (2019). Renewable energy and sustainability. *Clean Technologies and Environmental Policy, 21*, 1517–1533.

Saini, L., et al. (2022). Net Zero Energy Consumption building in India: An overview and initiative toward sustainable future. *International Journal of Green Energy*.https://doi.org/10.1080/15435075.2021.1948417

Singh, N. (2022). India's strategy for achieving net zero. *Energies*. https://doi.org/10.3390/en15165852

Singh, R., Ravesangar, K., Narayanan, S., & Singh, K. K. H. (2024). HRM practices in creating net zero: supporting organizational sustainability. In R. Singh, S. Khan, A. Kumar, S. Luthra, & H. Chokshi, (Eds.), *Net zero economy, corporate social responsibility and sustainable value creation. CSR, sustainability, ethics & governance*. Springer. https://doi.org/10.1007/978-3-031-55779-8_1

Singh, U., et al. (2024). Sustainable development goals as means to motivate CO_2 capture and storage in Indian geologic formations. *Marine and Petroleum Geology*. https://doi.org/10.1016/j.marpetgeo.2023.106668

Sokołowski, M. M. (2019). When black meets green: A review of the four pillars of India's energy policy. *Energy Policy*. https://doi.org/10.1016/j.enpol.2019.03.051

Supriya, R. C. (2023). Low-CO_2 emission strategies to achieve net zero targets in the cement sector. *Journal of Cleaner Production*.https://doi.org/10.1016/j.jclepro.2023.137466

Tang, W., et al. (2023). Green innovation and resource efficiency to meet net-zero emission. *Resources Policy*.https://doi.org/10.1016/j.resourpol.2023.104231

Ulpiani, G., et al. (2023). Towards the first cohort of climate-neutral cities: Expected impact, current gaps, and next steps to take to establish evidence-based zero-emission urban futures. *Sustainable Cities and Society*.https://doi.org/10.1016/j.scs.2023.104572

Upadhyay, K. (2024). Fueling India's net-zero transition. In *Sustainable built environment* (pp. 245–259).

Vulturius, G., et al. (2022). Sustainability-linked bonds—Their potential to promote issuers' transition to net-zero emissions and future research directions. *Journal of Sustainable Finance & Investment*. https://doi.org/10.1080/20430795.2022.2040943

Xu, D., et al. (2023). The race to net-zero emissions: Can green technological innovation and environmental regulation be the potential pathway to net-zero emissions? *Technology in Society*.https://doi.org/10.1016/j.techsoc.2023.102364

Chapter 5
The Way Forward for Net Zero: Aligning Nationally Determined Contributions (NDCs) and Holistic Sustainable Development in South Asia

Narayanage Jayantha Dewasiri, W. K. A. C. Gnanapala, Mananage Shanika Hansini Rathnasiri, Mohit Yadav, Lalit Mohan Tewari, and Udgam Mishra

Abstract This chapter explores the intersection of Nationally Determined Contributions (NDCs) and sustainable development goals in the context of South Asia, with an emphasis on net zero emissions. It provides an overview of NDCs and their role in global climate agreements like the Paris Agreement, emphasizing the significance of net zero emissions for mitigating climate change. Following Preferred Reporting Items for Systematic Reviews and Meta-Analyses (PRISMA) guidelines, the study uses a systematic literature review method to analyze fifty-six (56) articles and identify critical insights. The research investigates the current status of NDCs

N. J. Dewasiri (✉)
Department of Accountancy and Finance, Faculty of Management Studies, Sabaragamuwa University of Sri Lanka, Belihuloya, Sri Lanka
e-mail: jayantha@mgt.sab.ac.lk

Adjunct Faculty, University Center for Research and Development (UCRD), Chandigarh University, Punjab, India

W. K. A. C. Gnanapala
Department of Tourism Management, Faculty of Management Studies, Sabaragamuwa University of Sri Lanka, Belihuloya, Sri Lanka
e-mail: gnanapala@mgt.sab.ac.lk

M. S. H. Rathnasiri
Department of Marketing Management, Faculty of Management Studies, Sabaragamuwa University of Sri Lanka, Belihuloya, Sri Lanka
e-mail: shanika@mgt.sab.ac.lk

M. Yadav
Jindal Global Business School, Jindal Global University, Haryana, India

L. M. Tewari
Department of Botany, DSB Campus, Kumaun University, Nainital, Uttarakhand, India

U. Mishra
Tribhuvan University, Kathmandu, Nepal

R. Singh and D. Crowther (eds.), *Transition Towards a Sustainable Future*, Approaches to Global Sustainability, Markets, and Governance,
https://doi.org/10.1007/978-981-97-5756-5_5

in South Asia, including country profiles, commitments, progress, and challenges. It also explores strategies for integrating Sustainable Development Goals (SDGs) with NDCs, sectoral approaches to net zero, policy frameworks, financial mechanisms, technology and innovation, monitoring and verification systems, and case studies of successful initiatives. We propose strategic policy recommendations for policymakers to enhance the alignment of NDCs with sustainable development goals. Additionally, areas for future research are identified, including impact assessment, technology innovation, policy integration, behavioral change, and climate justice. The paper concludes by emphasizing the importance of collective action and collaboration in advancing South Asia toward a sustainable and resilient future.

Keywords Net zero · Nationally determined contributions · NDCs · Holistic sustainable development · South Asia

5.1 Introduction

The quest for a sustainable future has become a defining challenge of our time, with climate change at its core. Since reducing greenhouse gas emissions is a significant challenge for the world community, achieving net zero has emerged as a critical goal. This chapter explores the Nationally Determined Contributions (NDCs) and holistic sustainable development, focusing on South Asia—a region marked by its vibrant diversity and significant climate vulnerabilities. In exploring the pathways to net zero, it is essential to understand the foundational role of NDCs within global climate agreements, the imperative of achieving net zero emissions (Akpan & Olanrewaju, 2023; Rathnasiri et al., 2024), and the unique challenges and opportunities that South Asian countries face in this endeavor.

5.1.1 Overview of NDCs

As the Paris Agreement states, Nationally Determined Contributions (NDCs) are essential to the global effort to combat climate change (Khan et al., 2021). The Nationally Determined Contributions (NDCs) that each nation that has ratified the Paris Agreement must disclose and explain the specifics of its post-2020 climate measures (Held & Roger, 2018). These pledges demonstrate a nation's commitment to cutting national emissions and mitigating the effects of climate change. NDCs are crucial because they provide a concise overview of each country's approach to assisting in the attainment of the global goal of limiting temperature increases to 1.5 °C over pre-industrial levels, with a focus on staying well below 2 °C (Lidskog & Sundqvist, 2022). NDCs are designed to be periodically updated, reflecting increased ambition and more robust climate actions over time. This dynamic approach encourages countries to enhance their commitments, progressively fostering global cooperation and a shared responsibility environment.

5.1.2 Importance of Net Zero

Reaching net zero emissions is critical to halting global warming. When the amount of greenhouse gases removed from the atmosphere equals the amount released, the situation is called "net zero" (Fankhauser et al., 2022). In order to prevent the most severe effects of climate change, such as extreme weather, rising sea levels, and biodiversity loss, this balance is essential for preserving stable global temperatures. For countries, the journey to net zero involves significant transformations in energy production, industrial processes, transportation, and land use (Van-der-Spek et al., 2022). It requires widespread adoption of renewable energy, increased energy efficiency, technological innovation in carbon capture and storage, and substantial changes in consumption patterns. In addition to addressing climate change, achieving net zero emissions has the potential to promote job creation, economic expansion, and the development of sustainable communities.

5.1.3 Focus on South Asia

In order to connect Nationally Determined Contributions (NDCs) with sustainable development goals, South Asia—which includes nations including India, Pakistan, Bangladesh, Sri Lanka, Nepal, Bhutan, and the Maldives—faces both unique opportunities and problems (Qudrat-Ullah, 2023). The region is characterized by diverse economies, ranging from rapidly industrializing nations to smaller, agrarian-based economies. Millions of people's livelihoods are at serious risk due to the effects of climate change, which are particularly likely to affect the South Asian region. These implications include intense heat waves, widespread flooding, and rising sea levels. The urgency of addressing these vulnerabilities is compounded by socio-economic challenges such as poverty, urbanization, and resource constraints.

Despite these challenges, South Asia also holds significant potential for climate action and sustainable development (Jat et al., 2022). Numerous renewable energy resources, including hydropower, wind, and solar power, are abundant in South Asia (Mitra et al., 2023). The necessity of integrating sustainable practices into economic planning and development plans is becoming more widely acknowledged (Jayarathne et al., 2021). Aligning NDCs with broader sustainable development goals (SDGs) can create synergies that enhance resilience, reduce emissions, and promote social and economic well-being. For South Asian countries, a holistic approach to sustainable development involves reducing emissions and addressing equity, justice, and inclusive growth issues.

This chapter will explore the pathways for South Asian countries to align their NDCs with holistic, sustainable development. We will examine the current status of NDCs, the integration of Sustainable Development Goals, sector-specific strategies, policy frameworks, financial mechanisms, technological innovations, community engagement, and effective monitoring and reporting systems. Through case studies and best practices, we will highlight successful initiatives and provide recommendations for policymakers to enhance the alignment of NDCs with sustainable development objectives in South Asia.

5.2 Methodology

To investigate the alignment of NDCs with holistic sustainable development in South Asia and the pathways to achieving net zero emissions, we employed a Systematic Literature Review (SLR) approach. This methodology allows for a comprehensive and unbiased synthesis of existing research, ensuring that the findings are based on thoroughly examining relevant literature. The PRISMA approach ensured transparency and replicability in the studies' selection and analysis, enhancing our findings' credibility as effectively used by Dewasiri et al. (2022), Jayawardena and Dewasiri (2023), Dewasiri et al. (2023), Dewasiri et al. (2024), Karunarthna et al. (2024), and Rathnasiri and De Silva (2023).

We began by developing a comprehensive search strategy to identify relevant literature. This involved selecting multiple academic databases, including Scopus and Google Scholar, to capture various studies. We used keywords connected to NDCs, net zero, sustainable development, and South Asia. Examples of search terms included "Nationally Determined Contributions," "net zero emissions," "sustainable development," "South Asia," and specific country names within the region. Inclusion criteria included peer-reviewed articles, studies focused on South Asia, and those addressing the intersection of NDCs and sustainable development. Exclusion criteria involved non-English publications, articles not peer-reviewed, and studies unrelated to the primary research questions.

The selection process consisted of several stages. Our initial search yielded 157 articles. We screened these papers based on topics and abstracts to remove duplicates and irrelevant articles. Accordingly, we retrieved the remaining articles and assessed them against our inclusion and exclusion criteria, reducing the pool to seventy-two (72). After a detailed evaluation, we selected 56 articles as the final sample for our review. This selection was based on the studies' relevance, quality, and comprehensiveness.

We extracted critical data from each selected article, focusing on study characteristics, research focus, and findings. Study characteristics included information about the study's authors, publication year, journal, and geographical focus. The research focus encompassed the study's main themes and objectives, including how it addresses NDCs, net zero, and sustainable development. We also noted key findings, contributions to the field, and any identified gaps or future research directions.

We conducted a thematic analysis to synthesize the findings, grouping studies into categories based on their focus areas, such as policy frameworks, technological innovations, financial mechanisms, and community engagement. We performed a comparative analysis to identify consistencies and divergences across studies' approaches and outcomes. By integrating insights from different studies, we comprehensively understood the alignment between NDCs and holistic, sustainable development in South Asia.

Our systematic literature review, guided by PRISMA guidelines, provides a robust foundation for understanding the sustainable road to net zero in South Asia. By synthesizing findings from 56 carefully selected studies, we offer a detailed analysis of the challenges, opportunities, and strategies for aligning NDCs with holistic, sustainable development in the region.

5.3 Findings and Discussion

5.3.1 Current Status of NDCs in South Asia

South Asian nations have articulated their climate action plans through Nationally Determined Contributions (NDCs), each reflecting their unique circumstances and ambitions. India's targets are to attain 40% of its installed electric capacity from non-fossil fuel sources by 2030, reduce its emissions intensity by 33–35% from 2005 levels, and create an additional carbon sink of 2.5–3 billion tonnes of CO_2 equivalent through forest and tree cover (United Nations India, 2018; Waghmare & Puranik, 2021). Pakistan planned to reduce its projected emissions by up to 50% by 2030, conditional on international support of 35 and 15% with domestic resources, with a strong focus on adaptation due to its high climate vulnerability (UNDP, 2021a). Bangladesh pledges to reduce its greenhouse gas emissions by 89.47 million tonnes by 2030 and up to 15% with international support while enhancing resilience to climate impacts (UNDP, 2021b).

Sri Lanka targets a 14.5% reduction in greenhouse gas emissions (GGEs) in the energy segment, including an unconditional reduction of 4% by 2030, conditional on international aid and emphasis on renewable energy and energy efficiency (UNDP, 2021c). Nepal commits to maintaining forest cover of at least 40% of the total area and achieving 15% of its energy demand from renewable sources by 2030, aiming to enhance climate resilience across various sectors (Mishra et al., 2024). Bhutan remains steadfast in its carbon neutrality, focusing on increasing forest cover, enhancing carbon sinks, and developing renewable energy, mainly hydropower (Suberi, 2020). Maldives has planned to reduce its GGEs by 26% by 2030 compared to business-as-usual levels, with a conditional target of 29% reduction with international support, focusing on renewable energy, energy efficiency, and adaptation to rising sea levels (UNDP, 2021d).

Progress toward these commitments varies across the region. India has significantly increased its renewable energy capacity, especially solar and wind power (Dawn et al., 2019). Still, it faces challenges balancing economic growth with emissions reductions and addressing air pollution and energy access issues. Pakistan has initiated several renewable energy and reforestation projects but needs help with political instability, financial constraints, and international support. Bangladesh has made strides in integrating renewable energy and enhancing climate resilience, yet faces high vulnerability to natural disasters, limited financial resources, and technological needs.

Sri Lanka has advanced in renewable energy development and efficiency improvements but relies heavily on international support and faces economic constraints and the effects of climate change, specifically on agrarian and water resources. Nepal has achieved progress in forest conservation and promoting renewable energy, yet financial constraints, political instability, and infrastructure development must pose significant challenges. Bhutan maintains its carbon neutrality and expands its renewable energy sector, mainly hydropower, but must balance development needs with conservation and manage climate impacts on its mountain ecosystems. Maldives focuses on renewable energy projects and adaptation measures to address rising sea levels. Still, the high infrastructure costs, reliance on international support, and existential threat from rising sea levels remain significant challenges.

Hence, South Asian countries are progressing toward their NDC commitments, but they face substantial challenges requiring enhanced international cooperation, financial resources, and technological support. These efforts are crucial for achieving their climate targets and ensuring sustainable regional development.

5.3.2 Integrating Sustainable Development Goals (SDGs) with NDCs

5.3.2.1 Synergies Between NDCs and SDGs

NDCs and SDGs aim to foster sustainable and resilient societies (United Nations, 2017). One of the most significant synergies is between climate action (SDG 13) and other vital SDGs. For instance, reducing poverty (SDG1) can be closely linked to climate action, as climate-focused agricultural initiatives could increase food security and livelihoods. Similarly, promoting clean energy (SDG7) through renewable energy projects directly supports climate mitigation efforts while providing affordable and sustainable energy to underserved populations. Sustainable cities (SDG11) benefit from climate initiatives that reduce urban heat islands, enhance public transportation, and improve air quality, thus making urban living more sustainable and healthier (United Nations, 2017).

5.3.2.2 Policy Integration

Integrating NDCs with national sustainable development policies requires a strategic approach that ensures coherence and mutual reinforcement (Khalid, 2023). Countries can align climate goals with broader development plans by embedding climate action into national policies, strategies, and regulatory frameworks. This includes integrating NDC targets into economic planning, infrastructure development, and social policies. For example, governments can promote renewable energy projects through supportive policies and incentives to reduce poverty and improve health outcomes. Effective policy integration also involves multi-sectoral coordination, ensuring that climate and development policies are developed and implemented collaboratively across different sectors and levels of government.

5.3.2.3 Co-benefits

Several initiatives demonstrate the co-benefits of aligning NDCs with SDGs, showcasing how climate action can simultaneously advance sustainable development. For instance, renewable energy projects in rural areas reduce GGEs, produce jobs, empower local economies, and improve energy access. Installing solar panels in off-grid communities in India (Akter & Bagchi, 2021) and Bangladesh (Held & Roger, 2018) has provided reliable electricity, enhancing education and healthcare services while reducing reliance on fossil fuels. Similarly, urban greening projects, such as creating parks and green spaces, contribute to climate mitigation by sequestering carbon and cooling urban areas while providing recreational spaces and improving mental health for city dwellers.

Another example is sustainable agriculture practices that enhance food security and resilience against climate change. In Nepal, climate-smart agricultural techniques, such as using drought-resistant crops and efficient water management systems, have increased farm productivity and resilience, helping to reduce poverty and hunger (Srivastav et al., 2021). These initiatives highlight the potential for climate action to deliver broad-based benefits, reinforcing the importance of integrating NDCs with SDGs to achieve holistic, sustainable development.

Hence, aligning NDCs with SDGs offers a strategic pathway to address climate change while advancing broader development objectives. By leveraging synergies, integrating policies, and focusing on co-benefits, South Asian countries can effectively enhance their climate resilience and achieve sustainable development goals, ensuring a more resilient and prosperous future for their populations.

5.3.3 Sectoral Approaches to Net Zero

Achieving net zero emissions requires targeted strategies across various sectors. Each sector presents unique challenges and opportunities, and adopting a tailored approach is essential for effective climate action.

5.3.3.1 Energy

Reducing greenhouse gas emissions in the energy industry requires a move toward renewable energy sources. To diversify their energy sources and reduce their reliance on fossil fuels, South Asian countries are gradually making investments in solar, wind, and hydropower (Rathnasiri et al., 2024). Enhancing energy efficiency measures, such as upgrading infrastructure and promoting energy-saving technologies, can significantly lower emissions. The role of decentralized energy systems, including off-grid solar and microgrids, is critical in rural areas, where they can provide reliable and sustainable electricity, thus supporting local economies and improving quality of life.

5.3.3.2 Agriculture

Sustainable agricultural practices are crucial for reducing emissions and enhancing resilience to climate change. Precision farming, organic agriculture, and agroforestry can improve productivity while minimizing environmental impact (Muhie, 2022). Reducing methane emissions from livestock and rice paddies by implementing improved management techniques and cutting-edge technologies is another crucial tactic. Enhancing carbon sequestration in soils through cover cropping, conservation tillage, and biochar application can also combat climate change, refining soil health and agricultural productivity.

5.3.3.3 Transportation

The transportation industry is another source of emissions, and transitioning to a low-carbon transport system is essential. The electrification of Transportation, including adopting electric vehicles (EVs) and developing necessary charging infrastructure, is a significant step toward reducing emissions (Muratori & Mai, 2020). Enhancing public transit systems to make them more efficient, accessible, and affordable can reduce reliance on private vehicles. Policies promoting non-motorized transport, such as cycling and walking, cut emissions, improve public health, and reduce congestion.

5.3.3.4 Industry

Innovations in industrial processes can lead to substantial emissions reductions in the industrial sector. Adopting circular economy principles, which emphasize reducing waste, reusing materials, and recycling, can significantly lower the environmental footprint of industrial activities. Technologies that capture and utilize carbon emissions from industrial processes, such as carbon capture and storage (CCS), are also crucial. Reducing emissions in heavy sectors, such as steel, cement, and chemicals, through energy efficiency improvements, alternative materials, and cleaner production methods is vital for achieving net zero targets.

Accordingly, sectoral approaches to net zero involve transitioning to renewable energy, adopting sustainable practices, and leveraging technological innovations. By addressing each sector's specific needs and challenges, South Asian countries can effectively reduce emissions and move toward a sustainable and resilient future. Integrating these sectoral strategies with national policies and international support will be vital in achieving comprehensive climate goals.

5.3.4 Policy Frameworks and Governance

5.3.4.1 National Policies

A robust policy framework is essential for effectively implementing NDCs and achieving sustainable development goals. South Asian nations developed various policies to face climate change. For instance, India's National Action Plan on Climate Change (NAPCC) focuses on solar energy, energy efficiency, sustainable agriculture, and water resources, providing a comprehensive approach to climate mitigation and adaptation (Chandel et al., 2016). Bangladesh's Climate Change Strategy and Action Plan (BCCSAP) focuses on managing climate impacts and integrating climate considerations into development planning (Islam et al., 2009). Pakistan's National Climate Change Policy aims to promote low-carbon growth and enhance climate resilience through energy, agriculture, and disaster risk reduction initiatives (Waheed et al., 2021). These national policies highlight the commitment of South Asian nations to exaggerate climate action into their broader net zero agendas.

5.3.4.2 Regional Cooperation

Regional cooperation is crucial for addressing climate change effectively, as many climate-related issues transcend national borders. The South Asian Association for Regional Cooperation (SAARC) is pivotal in fostering collaboration among its member countries. Initiatives such as the SAARC Action Plan on Climate Change aim to enhance regional cooperation in disaster management, water resource management, and renewable energy development (Ahmed et al., 2019). Collaborative efforts can facilitate the sharing of knowledge, technology, and best practices and leverage regional financial mechanisms to support climate initiatives. Strengthening regional cooperation can lead to more coordinated and effective responses to climate challenges, benefiting all countries in the region.

5.3.4.3 Governance Mechanisms

Effective governance structures are essential for the successful implementation of climate policies. This includes establishing clear and transparent policy formulation, implementation, and monitoring mechanisms. Engagement with all the stakeholders is critical to ensure that policies are inclusive and address the needs of all sectors of society (Rathnasiri et al., 2024). Transparency in policy implementation enhances accountability and builds public trust, which is vital for the long-term success of climate initiatives. Regular reporting, independent audits, and public consultations can help maintain transparency and implement policies effectively.

Governance mechanisms should also focus on building institutional capacity at all levels of government. This includes training and equipping local authorities to manage and implement climate initiatives and ensuring adequate financial and human resources to support these efforts. Strengthening legal and regulatory frameworks to enforce climate policies and standards is also crucial. By establishing robust governance structures, South Asian countries can enhance the effectiveness of their climate policies and ensure sustainable and equitable development.

A comprehensive policy framework and effective governance mechanisms are fundamental for achieving South Asia's NDC targets and sustainable development goals. National policies provide the foundation for climate action, while regional cooperation enhances collective efforts and resource sharing. Transparent and inclusive governance structures ensure policies are effectively implemented and all stakeholders are engaged. These elements create a strong foundation for South Asian countries to address climate issues and achieve net zero.

5.3.5 Financial Mechanisms and Investments

Achieving the ambitious targets in NDCs and sustainable development goals requires substantial financial resources. South Asian countries rely on domestic and international funding for climate and development projects. Domestic funding includes government budget allocations, national climate funds, and public–private partnerships. For instance, India implemented the National Adaptation Fund for Climate Change (NAFCC) to finance adaptation projects nationwide (Kaur et al., 2022). International funding sources include grants, concessional loans, and technical assistance from multilateral development banks, bilateral donors, and international organizations. These external funds supplement domestic efforts, especially in resource-constrained countries.

Further, climate finance mechanisms are vital in mobilizing resources for climate action. The Green Climate Fund (GCF) is a significant international climate finance mechanism that offers financial support to emerging nations for climate-related projects (Kalinowski, 2024). South Asian countries have accessed GCF funding for various projects, ranging from renewable energy installations to climate-resilient agricultural practices. The Global Environment Facility (GEF) also offers special financial concessions to support environmental projects contributing to global environmental benefits (Rahman & Ahmad, 2015). The Adaptation Fund also provides funding for initiatives and plans that help vulnerable communities in developing nations adjust to climate change. These systems are essential for bridging the funding gap and making it easier for large-scale development and climate efforts.

Moreover, the private sector is crucial in financing and implementing sustainable and climate-resilient projects. Encouraging private sector investment is critical to scaling climate action and achieving NDC targets. Governments can create an enabling environment for private investment through policy incentives, risk mitigation instruments, and public–private partnerships. For example, offering tax incentives and subsidies for renewable energy projects can attract private investors. Green bonds and climate bonds are innovative financial instruments that can mobilize private capital for climate-related projects. Furthermore, adding environmental, social, and governance (ESG) considerations to investment choices might increase the money the private sector is willing to give to sustainable projects.

Public–private partnerships (PPPs) invest in large-scale projects like renewable energy plants, sustainable transportation systems, and resilient urban infrastructure (Ugwu et al., 2024). By sharing risks and rewards, PPPs can attract significant private sector participation while ensuring that projects align with national climate and development goals. Furthermore, encouraging cooperation among governments, corporations, and financial institutions can help develop new financial services and products that promote sustainability and climate resilience.

Hence, mobilizing financial resources from diverse sources is critical for achieving South Asia's NDCs and sustainable development goals. Domestic and international funding, complemented by effective climate finance mechanisms, provide the necessary financial support for climate and development initiatives. Encouraging private sector involvement through policy incentives, innovative financial instruments, and public–private partnerships can significantly enhance investment in sustainable and climate-resilient projects. By leveraging these financial mechanisms, South Asian countries can make substantial progress toward their climate commitments and development aspirations.

5.3.6 Technology and Innovation

In order to achieve net zero emissions and advance sustainable development in South Asia, clean technology implementation is crucial (Tiwari et al., 2024). Clean technologies such as hydropower, wind, and solar offer sustainable substitutes for fossil fuels in the energy sector. Modern innovations like smart grids and enhanced battery storage increase the dependability and efficiency of renewable energy systems. In agriculture, clean technologies encompass precision farming, which utilizes data and sensors to optimize resource use, and bioenergy, which converts agricultural waste into energy. The industrial sector benefits from clean technologies through processes that reduce emissions, such as CCS, and by adopting energy-efficient machinery and practices.

Moreover, research and development (R&D) are critical for driving innovation and developing new technologies to address climate challenges. Investing in R&D enables countries to create solutions tailored to their needs and circumstances (Bataineh et al., 2024). South Asian countries increasingly recognize the importance of R&D in areas such as renewable energy, sustainable agriculture, and green manufacturing. For example, India's Department of Science & Technology (DST) supports numerous R&D initiatives to develop low-carbon technologies. Collaborative research programs involving universities, government agencies, and private sector entities pave the way toward developing and deploying novel solutions. Enhancing R&D capacity also helps countries stay at the forefront of technological advancements and fosters a culture of innovation.

Further, technology transfer from developed to developing countries is essential for bridging the technology gap and accelerating climate action (Kim et al., 2023). Effective mechanisms for technology transfer include international cooperation agreements, capacity-building programs, and financial incentives. The United Nations Framework Convention on Climate Change (UNFCCC) facilitates technology transfer through initiatives such as the Technology Mechanism, which supports countries in identifying technology needs and implementing technology projects (Puig et al., 2018). India's International Solar Alliance (ISA) promotes solar energy deployment across member countries by facilitating technology transfer and capacity building.

Therefore, technology and innovation are key drivers of climate action and sustainable development in South Asia. Adopting clean energy, agriculture, and industry technologies can significantly reduce emissions and enhance resilience. Investing in R&D is vital for new technologies and improving existing ones, while effective technology transfer mechanisms ensure that developing countries can access and benefit from these advancements. By fostering innovation and facilitating technology transfer, South Asian nations can accelerate their progress toward net zero emissions and sustainable development.

5.3.7 Monitoring, Reporting, and Verification (MRV)

Establishing robust Monitoring, Reporting, and Verification (MRV) systems is critical for tracking progress on NDCs and SDGs (de-Assis et al., 2022). MRV systems enable countries to systematically monitor their greenhouse gas emissions, evaluate the effectiveness of mitigation and adaptation measures, and report their findings transparently. These systems involve regular data collection, reporting protocols, and verification processes to ensure accuracy and reliability. South Asian countries increasingly recognize the importance of MRV systems and are working toward enhancing their capacity to implement them effectively. For instance, India has developed comprehensive MRV frameworks to track its climate actions and progress toward its NDC targets.

Moreover, accurate data collection and analysis are foundational to effective MRV systems. Reliable data provides the basis for informed decision-making and helps identify areas where interventions are needed. This involves collecting data on emissions from various sectors, such as energy, agriculture, transportation, and industry, and tracking progress on sustainable development indicators. Advanced technologies, like satellite monitoring, remote sensing, and geographic information systems (GIS), can enhance the accuracy and effectiveness of data gathering. Additionally, capacity building is crucial to ensure stakeholders have the necessary skills and resources to gather and analyze data effectively.

Further, ensuring transparency in reporting and accountability in policy implementation is essential for building trust and credibility in climate actions (Gupta & Mason, 2016). Transparent MRV systems allow for the clear communication of progress and challenges, both domestically and internationally. This involves publishing regular reports that detail emissions levels, policy impacts, and progress toward NDC and SDG targets. Engaging stakeholders in the MRV process can enhance transparency and accountability. Mechanisms such as public consultations, independent audits, and third-party verifications provide additional layers of oversight and credibility.

Accountability is further reinforced by establishing clear responsibilities and timelines for policy implementation and setting up mechanisms to address non-compliance or underperformance. By fostering a culture of transparency and accountability, South Asian countries can enhance the effectiveness of their climate policies and ensure sustained progress toward their climate and development goals.

Hence, robust MRV systems are crucial for tracking progress on NDCs and SDGs and rely on accurate data collection and analysis. Ensuring transparency and accountability in reporting and policy implementation builds trust and enhances the credibility of climate actions. By strengthening MRV frameworks, South Asian countries can improve the effectiveness of their climate strategies and demonstrate their commitment to achieving their climate and sustainable development objectives.

5.3.8 Case Studies and Best Practices

Several successful initiatives across South Asia demonstrate the effective alignment of NDCs with sustainable development goals, providing valuable insights for future efforts.

India's Solar Mission: Bhadla Solar Park (Kumar, 2022), Pavagada Solar Park (Meena et al., 2021), and Jawaharlal Nehru National Solar Mission (Yenneti, 2012) are a few flagship initiatives to increase the country's solar power capacity, which significantly contributed to India's renewable energy targets under its NDC. By fostering investments in solar power infrastructure, incentivizing solar projects, and promoting solar technology development, the mission has reduced greenhouse gas emissions, created jobs and improved energy access in rural areas. This initiative exemplifies how renewable energy projects can drive economic growth while advancing climate goals.

Bangladesh's Climate-Resilient Agriculture: Bangladesh has implemented various climate-focused agricultural initiatives and practices to enhance food safety and adapt to climate change. Programs such as the Climate-Smart Agriculture (CSA) initiative focus on integrating sustainable farming activities, like drought-resistant crops, effective water management initiatives, and improved soil health techniques (Riyadh et al., 2021). These practices have led to increased agricultural productivity, reduced vulnerability to climate impacts, and improved livelihoods for farmers. This case highlights the importance of integrating climate adaptation into agrarian development to achieve environmental and socio-economic benefits.

Bhutan's Forest Conservation: Bhutan has maintained its carbon–neutral status mainly due to its robust forest conservation policies (Yangka et al., 2019). The country's extensive forest cover, which absorbs more carbon dioxide than the nation emits, is protected through stringent conservation laws, community forestry programs, and sustainable forest management practices. This commitment to preserving natural ecosystems helps Bhutan meet its NDC targets and supports biodiversity, water resources, and the livelihoods of forest-dependent communities. Bhutan's approach underscores the significance of protecting natural carbon sinks in climate mitigation strategies.

Sri Lanka's Renewable Energy Expansion: Sri Lanka aims to enhance its share of renewable energy. The country's energy policy emphasizes the development of wind, solar, and hydroelectric power. Notable projects include the development of large-scale wind farms and solar parks. These efforts have reduced dependence on imported fossil fuels, lowered greenhouse gas emissions, and created green jobs (Rathnasiri et al., 2024). Sri Lanka's proactive approach to renewable energy highlights the potential for small island nations to achieve significant progress in clean energy deployment.

Pakistan's Billion Tree Tsunami: This project is a large-scale reforestation initiative to restore degraded landscapes and enhance carbon sequestration (Sabir et al., 2022).

Launched in 2014, the project has planted over a billion trees, significantly increasing forest cover and biodiversity. This initiative contributes to Pakistan's NDC targets and provides numerous co-benefits, including improved water regulation, soil stabilization, and job creation in rural areas. The success of the Billion Tree Tsunami underscores the impact of large-scale environmental restoration projects on climate mitigation and sustainable development.

Maldives' Climate-Resilient Infrastructure: Creating climate-resilient infrastructure has become a primary concern for the Maldives, an archipelagic nation extremely vulnerable to sea level rise. The islands are being protected against the consequences of climate change through initiatives such as the construction of sea walls, higher structures, and sustainable tourism amenities. In order to reduce its carbon footprint, the Maldives is also actively investing in renewable energy sources, including solar electricity (Keiner et al., 2022). These initiatives demonstrate the country's commitment to integrating climate mitigation and adaptation into its national development plans to guarantee sustainability and long-term resilience.

Nepal's Hydropower Development: Nepal has immense potential for hydropower, which it is actively harnessing to meet its energy needs and reduce greenhouse gas emissions. Projects like the Upper Tamakoshi Hydroelectric Plant, one of the largest in the country, contribute significantly to Nepal's NDC targets by providing a renewable energy source. Hydropower development helps reduce reliance on fossil fuels and promotes regional energy trade, enhancing economic growth and energy security (Pandey et al., 2023). Nepal's focus on hydropower demonstrates the potential of leveraging natural resources sustainably for climate and development benefits.

Afghanistan's Sustainable Land Management: Afghanistan faces severe environmental degradation and desertification, which impact agriculture and livelihoods. To combat these challenges, sustainable land management practices, such as reforestation, soil conservation, and water management, are being implemented (Khurram et al., 2024). Programs supported by international organizations aim to restore degraded lands, improve agricultural productivity, and build resilience to climate impacts. These initiatives align with Afghanistan's NDC commitments and highlight the importance of land restoration in achieving sustainable development in fragile environments.

Accordingly, each South Asian country has undertaken unique and innovative initiatives to align their NDCs with sustainable development goals, addressing specific national challenges and leveraging local strengths. From renewable energy projects in Nepal to sustainable land management in Afghanistan and climate-resilient infrastructure in the Maldives, these efforts collectively contribute to the region's progress toward a sustainable and climate-resilient future.

5.3.9 Lessons Learned

Some vital takeaways from successful initiatives in South Asia are outlined below. These initiatives highlight the importance of integrated policy frameworks, community engagement, and leveraging local strengths to achieve climate and development goals.

Policy Support and Incentives: Clear policy support and financial incentives are crucial for the success of climate initiatives. India's solar mission benefited from government subsidies, tax incentives, and a supportive regulatory environment, which attracted investments and accelerated project implementation.

Community Engagement: Involving local communities in the planning and implementation of projects ensures sustainability and enhances the social acceptance of climate actions. Bangladesh's climate-resilient agriculture programs succeeded because they engaged farmers and addressed their specific needs and challenges.

Integrated Approaches: Aligning climate goals with broader development objectives creates synergies and maximizes benefits. Bhutan's forest conservation efforts are integrated with its overall environmental and socio-economic development strategies, ensuring a holistic approach to sustainability.

Further, some lessons were learned from less successful attempts. Insufficient funding has been a significant barrier, with many climate initiatives failing due to inadequate financial resources. Ensuring sustainable and adequate funding is essential for the long-term success of climate projects. Additionally, limited technical and institutional capacity has hindered the effective implementation of climate policies, emphasizing the need for building local capacities through training, education, and technical assistance. Poor coordination among government agencies, sectors, and stakeholders has also led to fragmented and inefficient efforts. Establishing robust coordination mechanisms and fostering collaboration is necessary for cohesive and effective climate action. Addressing these challenges is crucial for enhancing the effectiveness of climate policies and ensuring sustainable development in the region.

Hence, the case studies and lessons from South Asia provide valuable insights into successful strategies for aligning NDCs with sustainable development. By replicating successful models, addressing common challenges, and fostering an integrated approach, South Asian countries can enhance their climate resilience and achieve their sustainable development goals.

5.4 Conclusion, Future Directions and Recommendations

5.4.1 Implications to the Practice

To enhance the alignment of NDCs with SDGs, policymakers in South Asia should consider the following strategic recommendations:

Integrated Policy Frameworks: Develop integrated policy frameworks that prioritize climate action while addressing broader socio-economic development goals. Climate issues must be included in national development plans, sectoral policies, and investment strategies to guarantee synchronization and synergy across many policy areas.

Strengthened Institutional Capacity: Invest in building institutional capacity at all levels of government to implement and monitor climate and development policies effectively. This includes enhancing technical expertise, data management systems, and coordination mechanisms among relevant stakeholders to facilitate informed decision-making and policy implementation.

Enhanced Financial Mechanisms: Mobilize adequate and sustainable financial resources to support climate and development initiatives. This involves leveraging domestic and international funding sources, improving access to climate finance mechanisms such as the Green Climate Fund (GCF) and the Global Environment Facility (GEF), and promoting innovative financing instruments such as green bonds and public–private partnerships.

Community Engagement and Empowerment: Foster meaningful engagement with local communities, civil society organizations, and the private sector to ensure inclusive decision-making and ownership of climate and development projects. Empower communities through capacity-building initiatives, participatory planning processes, and knowledge sharing to enable bottom-up solutions that address local needs and priorities.

Adaptive Governance Structures: Establish adaptive governance structures that facilitate iterative policy learning, flexibility, and responsiveness to evolving climate risks and development challenges. This involves promoting adaptive management approaches, incorporating feedback mechanisms, and fostering multi-stakeholder collaboration to enhance the resilience and effectiveness of climate policies.

5.4.2 Future Research

Areas for further research and exploration to advance the alignment of NDCs with sustainable development goals include:

Impact Assessment: Conduct comprehensive impact assessments to evaluate the social, economic, and environmental effects of implementing Nationally Determined Contributions (NDCs) on local communities, vulnerable populations, and ecosystems. This entails assessing the benefits and trade-offs of adaptation and mitigation techniques for climate change and devising plans to maximize positive outcomes and reduce negative ones.

Technology Innovation and Transfer: Explore opportunities for technology innovation and transfer to accelerate climate action and promote sustainable development. This includes researching new clean technologies, improving existing technologies, and facilitating technology transfer between developed and developing countries to enhance capacity building and technology diffusion.

Policy Integration and Coherence: Investigate strategies for enhancing policy integration and coherence across sectors and governance levels to ensure holistic and coordinated responses to climate and development challenges. This involves examining institutional arrangements, policy frameworks, and governance mechanisms that promote cross-sectoral collaboration and alignment of objectives.

Behavioral Change and Social Transformation: Study the role of behavioral change and social transformation in driving climate action and fostering sustainable development. This includes exploring factors influencing individual and collective behavior change, identifying effective communication and engagement strategies, and promoting values, norms, and lifestyles that support sustainability.

Climate Justice and Equity: Examine climate justice and equity issues to ensure that climate policies and interventions are fair, inclusive, and equitable. This involves analyzing the distributional impacts of climate change and adaptation measures on different social groups, addressing the needs of vulnerable populations, and promoting justice-based approaches to climate governance.

5.4.3 Conclusion

In conclusion, South Asia stands at a critical juncture in its journey toward achieving net zero emissions and sustainable development. The region can address pressing climate challenges by aligning NDCs with sustainable development goals while fostering inclusive and resilient societies. Strategic policy interventions strengthened institutional capacity, enhanced financial mechanisms, and community empowerment are essential for realizing this vision. Future research efforts should focus on impact assessment, technology innovation, policy integration, behavioral change, and climate justice to advance the alignment agenda. Through collective action, South Asia can chart a path toward a more sustainable and prosperous future for its people and the planet.

References

Ahmed, A. U., Appadurai, A. N., & Neelormi, S. (2019). *Status of climate change adaptation in South Asia region. Status of climate change adaptation in Asia and the Pacific* (pp. 125–152).

Akpan, J., & Olanrewaju, O. (2023). Sustainable energy development: History and recent advances. *Energies, 16*(20), 7049.

Akter, S., & Bagchi, K. (2021). Is off-grid residential solar power inclusive? Solar power adoption, energy poverty, and social inequality in India. *Energy Research and Social Science, 82*, 102314.

Bataineh, M. J., Sánchez-Sellero, P., & Ayad, F. (2024). Green is the new black: How research, development, and green innovation give businesses a competitive edge. *Business Strategy and the Environment, 33*(2), 1004–1023.

Chandel, S. S., Shrivastva, R., Sharma, V., & Ramasamy, P. (2016). Overview of the renewable energy sector initiatives under India's national action plan on climate change. *Renewable and Sustainable Energy Reviews, 54*, 866–873.

Dawn, S., Tiwari, P. K., Goswami, A. K., Singh, A. K., & Panda, R. (2019). Wind power: Existing status, achievements and government's initiative towards renewable power dominating India. *Energy Strategy Reviews, 23*, 178–199.

de-Assis, T. F., Monteiro, T. G. M., de Abreu, V. H. S., D'Agosto, M. D. A., & Santos, A. S. (2022). Enabling the green bonds market for sustainable transport projects based on the measure/monitoring, reporting and verification method. In *Carbon footprints of manufacturing and transportation industries* (pp. 1–24). Springer Nature Singapore.

Dewasiri, N. J., Baker, H. K., Banda, Y. W., & Rathnasiri, M. S. H. (2022). The dividend decision model: A possible solution for the dividend puzzle. In *Exploring the latest trends in management literature* (pp. 249–267). Emerald Publishing Limited. https://doi.org/10.1108/978-1-80262-277-520231021

Dewasiri, N. J., Karunarathne, K. S. S. N., Menon, S., Jayarathne, P. G. S. A., & Rathnasiri, M. S. H. (2023). Fusion of artificial intelligence and blockchain in the banking industry: Current application, adoption, and future challenges. In A. Saini, & V. Garg (Eds.), *Transformation for sustainable business and management practices: Exploring the spectrum of industry 5.0* (pp. 293–307). Emerald Publishing Limited. https://doi.org/10.1108/978-1-80262-277-520231021

Dewasiri, N. J., Dharmarathna, D. G., & Choudhary, M. (2024). Leveraging artificial intelligence for enhanced risk management in banking: A systematic literature review. In R. Singh, S. Khan, A. Kumar, & V. Kumar (Eds.), *Artificial intelligence enabled management: An emerging economy perspective* (pp. 197–213). De Gruyter. https://doi.org/10.1515/9783111172408

Fankhauser, S., Smith, S. M., Allen, M., Axelsson, K., Hale, T., Hepburn, C., Kendall, J. M., Khosla, R., Lezaun, J., Mitchell-Larson, E., Obersteiner, M., & Wetzer, T. (2022). The meaning of net zero and how to get it right. *Nature Climate Change, 12*(1), 15–21.

Gupta, A., & Mason, M. (2016). Disclosing or obscuring? The politics of transparency in global climate governance. *Current Opinion in Environmental Sustainability, 18*, 82–90.

Held, D., & Roger, C. (2018). Three models of global climate governance: From Kyoto to Paris and beyond. *Global Policy, 9*(4), 527–537.

Islam, A., Shaw, R., & Mallick, F. (2009). *Bangladesh climate change strategy and action Plan (BCCSAP)*. Ministry of Environment and Forest, Government of the People's Republic of Bangladesh.

Jat, M. L., Chakraborty, D., Ladha, J. K., Parihar, C. M., Datta, A., Mandal, B., Nayak, H. S., Maity, P., Rana, D. S., Chaudhari, S. K., & Gerard, B. (2022). Carbon sequestration potential, challenges, and strategies towards climate action in smallholder agricultural systems of South Asia. *Crop and Environment, 1*(1), 86–101.

Jayarathne, P. G. S. A., Sachithra, V., & Dewasiri, N. J. (2021). Sustainable practices through competitive capabilities in the apparel industry: Sri Lankan experience. *FIIB Business Review*. https://doi.org/10.1177/23197145211060398

Jayawardena, N. S., & Dewasiri, N. J. (2023). Food acquisition and consumption issues of South Asian countries: A systematic literature review and future research agenda. *FIIB Business Review*. https://doi.org/10.1177/23197145231194113

Kalinowski, T. (2024). The Green Climate Fund and private sector climate finance in the Global South. *Climate Policy, 24*(3), 281–296.

Karunarathna, K. S. S. N., Dewasiri, N. J., Singh, R., & Rathnasiri, M. S. H. (2024). What does artificial intelligence—Powered ChatGPT bring to academia? A review. In R. Singh, S. Khan, A. Kumar, & V. Kumar (Eds.), *Artificial intelligence enabled management: An emerging economy perspective* (pp. 3–13). De Gruyter. https://doi.org/10.1515/9783111172408

Kaur, S., Kaur, D., Tiwana, A. S., & Gupta, S. (2022). Climate change-related governance and policies in Indian Himalayas. *Climate change: Impacts, responses and sustainability in the Indian Himalaya* (pp. 309–330). Springer International Publishing.

Keiner, D., Salcedo-Puerto, O., Immonen, E., van Sark, W. G., Nizam, Y., Shadiya, F., Duval, J., Delahaye, T., Gulagi, A., & Breyer, C. (2022). Powering an island energy system by offshore floating technologies towards 100% renewables: A case for the Maldives. *Applied Energy, 308*, 118360.

Khalid, A. M. (2023). Creating synergies among the sustainable development goals and climate action: Insights from a developing economy. *Sustainability, 15*(17), 13137.

Khan, M., Mfitumukiza, D., & Huq, S. (2021). Capacity building for implementation of nationally determined contributions under the Paris Agreement. In *Making climate action more effective* (pp. 95–106). Routledge.

Kim, W., Song, C., Lee, S. K., Choi, G., Yang, R., Bak, I., & Lee, W. K. (2023). A way forward for climate technology transfer and sustainable development goals. *Environmental Science and Policy, 142*, 29–41.

Khurram, S., Shalizi, M. N., Bashari, M., Akamani, K., & Groninger, J. W. (2024). Barriers and opportunities regarding community-based forest management in Afghanistan: Considerations for fragile states. *Environmental Conservation, 51*(1), 6–16.

Kumar, B. R. (2022). Case 21: Bhadla Solar Park. *Project finance: Structuring, valuation and risk management for major projects* (pp. 205–208). Springer International Publishing.

Lidskog, R., & Sundqvist, G. (2022). Lost in transformation: The Paris Agreement, the IPCC and the quest for national transformative change. *Frontiers in Climate, 4*, 906054.

Meena, R. S., Nigam, D., Gupta, S. K., Aravindh, M. A., Saketh, P. V., & Meena, S. (2021). An analysis of one of the largest solar parks in the world: 2000 MW solar park in Pavagada (Karnataka). *Water and Energy International, 64*(7), 36–46.

Mishra, U., Rathnasiri, M. S. H., & Dewasiri, N. J. (2024). Net-zero economy and energy sustainability: A Nepalese perspective. In R. Singh, S. Khan, A. Kumar, S. Luthra, & H. Chokshi (Eds.), *Net zero economy, corporate social responsibility and sustainable value creation. CSR, sustainability, ethics & governance* (pp. 145–161). Springer. https://doi.org/10.1007/978-3-031-55779-8_9

Mitra, M., Singha, N. R., & Chattopadhyay, P. K. (2023). Review on renewable energy potential and capacities of South Asian countries influencing sustainable environment: A comparative assessment. *Sustainable Energy Technologies and Assessments, 57*, 103295.

Muhie, S. H. (2022). Novel approaches and practices to sustainable agriculture. *Journal of Agriculture and Food Research, 10*, 100446.

Muratori, M., & Mai, T. (2020). The shape of electrified transportation. *Environmental Research Letters, 16*(1), 011003.

Pandey, R., Shrestha, R., Bhattarai, N., & Dhakal, R. (2023). Problems identification and performance analysis in small hydropower plants in Nepal. *International Journal of Low-Carbon Technologies, 18*, 561–569.

Puig, D., Haselip, J. A., & Bakhtiari, F. (2018). The mismatch between the in-country determinants of technology transfer and the scope of technology transfer initiatives under the United Nations framework convention on climate change. *International Environmental Agreements: Politics, Law and Economics, 18*, 659–669.

Qudrat-Ullah, H. (2023). A review and analysis of green energy and the environmental policies in South Asia. *Energies, 16*(22), 7486.

Rahman, S. M., & Ahmad, M. M. (2015). Climate finance towards vulnerable countries: Evidence from the Global Environment Facility (GEF). *International Journal of Green Economics, 9*(2), 199–214.

Rathnasiri, M. S. H., Dewasiri, N. J., & Kumar, A. A. (2024). Policy framework and implementation strategies for Sri Lanka's transition to a net-zero economy. In R. Singh, S. Khan, A. Kumar, S. Luthra, & H. Chokshi (Eds.), *Net zero economy, corporate social responsibility and sustainable value creation. CSR, sustainability, ethics & governance* (pp. 43–60), Springer. https://doi.org/10.1007/978-3-031-55779-8_3

Rathnasiri, M. S. H., & De Silva, V. (2023). The role of shopping values, cognitive and affective evaluations on the relationship between store environment and store patronage intention: A future research agenda, Sri Lanka. *Journal of Management Studies, 5*(1), 100–120.

Riyadh, Z. A., Rahman, M. A., Saha, S. R., Ahamed, T., & Current, D. (2021). Adaptation of agroforestry as a climate-smart agriculture technology in Bangladesh. *International Journal of Agricultural Research, Innovation and Technology (IJARIT), 11*(1), 49–59.

Sabir, M., Ali, Y., Khan, I., & Salman, A. (2022). Plants species selection for afforestation: A case study of the Billion Tree Tsunami Project of Pakistan. *Journal of Sustainable Forestry, 41*(6), 537–549.

Srivastav, A. L., Dhyani, R., Ranjan, M., Madhav, S., & Sillanpää, M. (2021). Climate-resilient strategies for sustainable management of water resources and agriculture. *Environmental Science and Pollution Research, 28*(31), 41576–41595.

Suberi, H. K. (2020). 100% renewable energy strategy: Bhutan's energy and environmental perspective. In M. Asif (Ed.), *Energy and environmental outlook for South Asia* (pp. 63–96). CRC Press.

Tiwari, S., Bashir, S., Sarker, T., & Shahzad, U. (2024). Sustainable pathways for attaining net zero emissions in selected South Asian countries: Role of green energy market and pricing. *Humanities and Social Sciences Communications, 11*(1), 1–13.

Ugwu, M. C., Adefolake, O. A., & Naomi, E. N. (2024). The role of public-private partnerships in building clean energy infrastructure in The United States and Nigeria. *International Journal of Management and Entrepreneurship Research, 6*(4), 1049–1068.

UNDP (United Nations Development Programme). (2021a). *Pakistan: Key highlights from the NDC.* Climate Promise. Retrieved May 26, 2024, from https://climatepromise.undp.org/what-we-do/where-we-work/pakistan#:~:text=Key%20highlights%20from%20the%20NDC,provision%20of%20international%20grant%20finance

UNDP (United Nations Development Programme). (2021b). *Bangladesh: Key highlights from the NDC.* Climate Promise. Retrieved May 26, 2024, from https://climatepromise.undp.org/what-we-do/where-we-work/bangladesh

UNDP (United Nations Development Programme). (2021c). *Sri Lanka: Key highlights from the NDC.* Climate Promise. Retrieved May 26, 2024, from https://climatepromise.undp.org/what-we-do/where-we-work/sri-lanka

UNDP (United Nations Development Programme). (2021d). *Maldives: Key highlights from the NDC.* Climate Promise. Retrieved May 26, 2024, from https://climatepromise.undp.org/what-we-do/where-we-work/maldives

United Nations. (2017). *Aligning nationally determined contributions and sustainable development.* Retrieved May 26, 2024, from https://sdgs.un.org/publications/publication-aligning-nationally-determined-contributions-and-sustainable-development

United Nations India. (2018). *Climate change, clean energy, and disaster resilience.* Retrieved May 26, 2024, from https://india.un.org/en/171990-climate-change-clean-energy-and-disaster-resilience

Van-der-Spek, M., Banet, C., Bauer, C., Gabrielli, P., Goldthorpe, W., Mazzotti, M., Munkejord, S. T., Røkke, N. A., Shah, N., Sunny, N., & Sutter, D. (2022). Perspective on the hydrogen economy as a pathway to reach net-zero CO2 emissions in Europe. *Energy and Environmental Science, 15*(3), 1034–1077.

Waghmare, S. A., & Puranik, B. P. (2021). Energy and environmental scenario of India. In *Energy and environmental security in developing countries* (pp. 105–145).

Waheed, A., Bernward Fischer, T., & Khan, M. I. (2021). Climate change policy coherence across policies, plans, and strategies in Pakistan—Implications for the China-Pakistan economic corridor plan. *Environmental Management, 67*(5), 793–810.

Yangka, D., Rauland, V., & Newman, P. (2019). Carbon neutral policy in action: The case of Bhutan. *Climate Policy, 19*(6), 672–687.

Yenneti, K. (2012). The Sun Shines on India? A review of the implementation and financial continuum of the National Solar Mission. *Renewable Energy Law and Policy Review, 4*, 280–291.

Part II
Sustainable Technologies Driving Net Zero Policies

Chapter 6
Net Zero Policies and Climate Resilience: A Focus on Emerging Technologies and Implementation Strategies in the Middle East

Jacinta Dsilva, Mohamed Irfan Shaikh, and Fahad Ziaul Ain Usmani

Abstract The Middle East is working on achieving their net zero goals and enhancing climate resilience in the region. However, they are facing several challenges and opportunities. The region's major characteristic is its fossil fuel dependency and extreme climatic conditions. However, are turning into emerging technologies and innovative implementation strategies that are supporting environmental goals. This chapter focuses on the interdependency of net zero policies and climate resilience in the Middle East with particular emphasis on emerging technologies such as renewable energy sources, carbon capture and storage (CCS), and smart grid systems. The chapter further examines how the emerging technologies can be effectively implemented to reduce greenhouse gas (GHG) emissions and encourage climate resilience. One of the main aims of the chapter is to highlight the net zero initiatives of countries such as United Arab Emirates, Saudi Arabia, and Egypt, who are investing heavily solar and wind energy projects, as well as pioneering CCS technologies. Furthermore, the chapter focuses on the importance of regional cooperation and international partnerships that are driving technological adoption and infrastructure development. The analysis further evaluates policy frameworks that support the transition to net zero, including renewable energy investments regulatory reforms, and public–private partnerships (PPP).

Keywords Net zero policies · Climate resilience · Middle East · Renewable energy · Public–private partnerships

J. Dsilva (✉) · M. I. Shaikh · F. Z. A. Usmani
SEE Research Center, SEE Institute, The Sustainable City, Dubai, United Arab Emirates
e-mail: jacinta@seeinstitute.ae

R. Singh and D. Crowther (eds.), *Transition Towards a Sustainable Future*, Approaches to Global Sustainability, Markets, and Governance,
https://doi.org/10.1007/978-981-97-5756-5_6

6.1 Introduction to Net Zero Policies and Climate Resilience

Prime potential for emerging technology solutions embodies green hydrogen, bioenergy, and CCS jet fuels. Reinforcement learning science too falls into this bracket. Many net zero calculations for climate neutrality may exhibit in the long-term electrification of transport. At the same time, other net zero paths have been presented with the continuous usage of other flex-fuel options like biofuels or LH2 to power shipping and aviation from 2020 onward. However, if one of these technologies can succeed at source and outsell in the long-term what influences as much as anything in several instances should be government policy and regulation. For instance, without a stringent Bergen standard, climate solutions say for aviation would truly just nullify and replicate the partial greenhouse gas (GHG) abatements from renewable energy usage.

The transition to net zero is a multistage journey (Markard & Rosenbloom, 2022). Globally, countries are focusing on the immediate aspects of this transition as the world's attention has turned to issues like climate resilience and net zero economies. Most recently, established economies, notably Rishi Sunak in the U.K. and Joe Biden in the U.S., have announced aggressive targets to reduce emissions throughout the economy, but policymakers now need to swiftly change attention to their more detailed and emerging technology components rather than targets and visionary speeches (Chan & Sopian, 2022). Emerging technologies are pivotal to the successful deployment of these strategies as they allow for decarbonization approaches more cheaply and deliver the arduous to decarbonize sectors of the world economy as at present decarbonization is often concentrated around the low hanging fruit. In recognition of this truth, many net zero roadmaps show penetration of emerging technologies as much as 10 or maybe 15% of the final energy, and hence, most models on net zero express that a future of global net zero transformation will require many not developed before and innovations to empower economic growth and prosperity in net zero world (Cansino et al., 2022; Singh et al., 2024).

6.1.1 Definition and Significance of Net Zero Policies

However, the renewable energy price can become too expensive and behave in a suboptimal way if expired policies for renewable energy deployment made it along the transition. In other words, policies would become outdated, but if the supply chain constraints for technology deployment are tightened, even if too mechanical, the reverse would occur. Such rigid enhancements could be slow and expensive to adopt due to carbon or material constraints of society which risks sparking a backlash. That is why, policy timings and tensions for energy projects, as well as the market and target lifetimes shape the portfolio of projects necessary to reach net zero, with negative emission technology (NETs) vital for managing social tensions and negative emission demands later. Rapid transitions in energy, industrial, and trade systems are required for net zero. A fast decarbonization rate increases the average energy system costs between now and mid-century, getting to net zero. Energy must be considered in the economy-wide context with net zero comprehensive supply chains and the integration of sectors, such as electricity, heat, transport, and industry to support flexibility in the complex interactions of supply and demand.

Net zero is a term used to describe a state of no net greenhouse gas (GHG) emissions, with the carbon dioxide removed from the atmosphere compensating or offsetting the equivalent amount of GHGs emitted into the air (Wienda Pratama et al., 2022). It is an equilibrium of non-CO_2 emitting sources of emissions, termed as carbon removals, with the human-made CO_2 emissions to allow a net additional accumulation of human-made CO_2 in the atmosphere (Dumas, 2021). Thus, it is pertinent to state that the target for net zero is the human-induced emissions and not the total concentration of carbon dioxide in the atmosphere as natural carbon sinks are also available. Therefore, "net zero CO_2 emissions" can be achieved by offsetting accumulated CO_2 through increased carbon sequestration either through land-based afforestation, reforestation, and ecosystem restoration projects, soil carbon enhancement, and direct air capture with geological storage technologies (John & Derakhshi, 2022).

6.2 The Importance of Climate Resilience in the Middle East

The climate crisis has become deeply intertwined with discussions on the sustainable development of societies from the local to the global level. This is particularly true in the Arab world, where the consequences of climate change are already posing existential threats to the continued provision of basic needs such as water, food, and electricity (Markard & Rosenbloom, 2022). Given that around 70% of the Arab region is already arid and semiarid, and temperature increases are higher than the global average, the region is expected to experience serious water shortages and severe impacts on agriculture and other natural ecosystems because of the reduced availability and increased demand for water. Energy systems in the region are also highly exposed to impacts such as fuel scarcity, increased energy demand, and the disruption of supply chains. Underlying these vulnerabilities is an economic model that has large-scale negative impacts, is heavily reliant on fossil fuel extraction, influences the region's development strategy, and has also led to depleting local natural resources while accelerating the ecosystems' capacity to recover thus increasing the risk of desertification and chasing biodiversity loss.

Climate risks in the Middle East are among the highest and most severe globally, and the region is even more vulnerable due to its particular socio-economic context. Certain dimensions of this vulnerability, driven by high social and economic inequalities, are likely to be exacerbated by climate change (Wu et al., 2021). Extreme temperatures and the heat stress they cause are perhaps the most obvious examples, but advances in detecting and modeling natural events now confirm a wide range of instances where the risks of these multiple dimensions might be greater still. The region's large population, its concentration in urban areas, and the rapid pace of urban growth are such examples, as are water and food insecurity for a significant proportion of the population. A less well-studied one might be the region's stark social, demographic, and economic disparities (Strielkowski et al., 2023). The risks of these dimensions need to be considered when designing responses to climate change. Here, the focus is on advances in our understanding of the impacts of, and vulnerability to, climate-related hazards on both natural and managed systems.

6.3 Current State of Climate Change in the Middle East

Climate change also affects the Middle East through the frequency of hazardous wind and sand- and dust storm events. An increase in the frequency of such dust and sandstorms has been observed in many parts of the region, mainly driven by declining rainfall. In Palestine, an increase of about 50% in the frequency of sandstorms and minor dust storms has been reported. Though sand and dust storms have been regular features in many parts of the world, the wide geographic range, frequency, duration, and intensity of the sand and dust storm problem significantly affect the climate, environment, human health, agriculture, property and infrastructure, and socio-economic development and culture in the region. Most of Kuwait was covered by a high column concentration of dust particles whose dimensions exceeded 50–155 km, and multiple airports were subsequently closed, with 80 people ending up in hospital for respiratory issues following a massive sandstorm in April 2009. The avalanche ruined much of the country's infrastructure (Yuan et al., 2020).

The Middle East is home to almost 500 million people, making it one of the most densely populated areas in the world, with an average population density more than twice that of Asia and Africa (Al-Jawaldeh et al., 2022). An exceptionally warm climate and water scarcity already make most of the Middle East uninhabitable and its population vulnerable. Yet, according to the World Bank, of all the regions in the world, the Middle East and North Africa are expected to be the hardest hit by climate change because of water scarcity, the region's dependency on agriculture, and its high urbanization and population growth rates. Most studies predict that rising global average temperatures due to the greenhouse effect will cause an acceleration of the water cycle, reduced total agricultural output, and an overall loss of productive land for both crops and livestock, directly affecting global food production. Crop yields are expected to decrease significantly, as is the availability of water for irrigation. More extreme weather events, such as droughts, but also heavy rainfall and flooding, constitute additional risks for agriculture and food production. The drying up of rivers and groundwater sources could have serious implications for food production in a region whose land had to be irrigated substantially for agriculture, which requires about 60–85% or more of the water in the region. Deforestation and loss of vegetation across the region would further exacerbate desertification. The natural habitats of crops and native plants are expected to shift in some cases to cooler regions inside and beyond the region.

6.3.1 *Climate Change Impacts and Vulnerabilities in the Region*

A large proportion of the population in the region is affected by heat stress due to high temperature and high humidity. By end of this century, population in the African coast of Red Sea will be increasingly exposed to heat threat and with 4 °C global warming; such heat wave event could be observed once in every two year. This would deeply affect human health, particularly with the existing burden of communicable diseases. A hypothesis cannot be made here is that whether the increase in the intensity and frequency of extreme event will equally decrease human mortality in Africa or it will also worsen the epidemic situation particularly for the population still bounded to other communicable diseases.

Furthermore, sea level rise will worsen the salinization of water in coastal areas, particularly in the Nile Delta and the North African and Levantine coasts. From 1970 to 2010, the Middle East and North Africa (MENA) region experienced 370 natural disasters. It includes floods, storms, extreme temperatures, and forest fires. Vulnerability of these regions also arise due to the high concentration of people in cities, growing informal settlements, and lack of urban planning. People are also vulnerable to heat wave events because of rapid urbanization and un-negligible increase of green land.

The impact of climate change in the Middle East has been well-established. The IPCC report published in 2022 in the region has projected an increase of extreme temperature in the region. Other such social-economic dimensions of climate change impacts such as agriculture and water have also been observed by Social Dimensions of Climate Change (2023). It has been observed that drought is affecting agricultural productivity; the share of overexploited water sources is rapidly growing. Such deficiencies in water availability have been observed in Iran (Nouri et al., 2023). In the Middle East region, seven of the world's thirty-seven water-stressed countries are in the Middle East, while six countries in the region are already facing absolute water scarcity. In this context, changes in precipitation pattern and sea surface temperature have also been observed (Hussein Al-Rimmawi, 2012), and it has led to alteration in extreme climate events such as extreme wind pattern and floods in the region.

6.3.2 Emerging Technologies for Achieving Net Zero Emissions

Renewable energies in the Middle East have become important worldwide for many reasons. As demand continues to climb and costs continue to drop, several Middle Eastern players have pledged to increase their use of renewables. Multiple governments in the region are pursuing diversification of the domestic economy by increasing the role of low-carbon industries. The economies of many Middle Eastern countries are currently heavily dependent on the open energy markets in gas, oil, or petrochemicals. These economies are often negatively affected by fluctuations in oil and gas prices, so a growing number of countries are looking to reduce their vulnerability by expanding low-carbon sectors (Cansino et al., 2022). For the MENA region, the climate change challenge represents a unique economic opportunity to diversify their economies and, in some countries, to become global leaders in low-carbon products and services. Many companies in the MENA region that have recognized this opportunity are therefore looking to complement their core business in the fossil fuel sector with investments in low-carbon industries such as waste management, water treatment, solar energy, wind power, green hydrogen, and building and infrastructure management.

Earlier this year the Intergovernmental Panel on Climate Change (IPCC) released a special report confirming that the world must reach "net-zero" emissions of carbon dioxide (CO_2) by 2050 if it wants to remain below 1.5 °C of global warming. To enact these aggressive objectives, it is essential to emphasize innovative interdisciplinary research and integrate climate policy goals with respect to emerging technologies. At the same time, it is critical to ensure careful, creative, and robust implementation as this transition unfolds (Khojasteh et al., 2022). In striving to limit the average global temperature rise to 1.5 °C (as described in the Paris Agreement), the Middle Eastern and North African (MENA) countries are working to transform various sectors: agriculture, water, health, and energy. The International Renewable Energy Agency (IRENA) predicts that the implementation of the Paris Agreement may lead to a large increase in energy investment, with about 93% of it flowing into renewable energy sources and energy efficiency sectors. Renewables are expected to contribute to 83% of all the power generation in the near future (Kong et al., 2022).

6.4 Renewable Energy Sources and Technologies

Globally, the expansion of renewable energy technology, supply, and markets has not been linear and has accelerated in many regions due to political constraints, trade relations, and supporting technologies including energy storage and demand-side management. Integrating research and policy on renewable energy requires rapprochement of natural science, engineering, economics, and political science. Renewable Energy Sources (RES) development pathways should draw lessons from country- or region-specific bottlenecks and achievements, as these are highly relevant for comparison and contextualization for transitioning countries include the optimal mix of policy instruments, legal frameworks, and fiscal incentives. The implementation of technology-neutral tendering schemes across renewable technologies, concluded by public auctions for a defined or unlimited capacity, with the provision of grid interconnection, should be continuously reviewed for ideal targeting of identified rural and urban potential areas which should be identified, correlated with technological, geographical, grid- and demand-side constraints.

In that view, a sustainability assessment measures of offshore wind and solar energy potential, and the economic opportunity of in situ hydrogen- and ammonia-based synthetic fuels have been computed. One-fifth of the examined European and MENA coastal countries were identified as region-suitable for meeting the regional energy demand with 100% renewable electricity from these energy sources, whereas three-fifths exceeded the regional hydrogen and ammonia demand. Conclusively, in the transition period to renewable energy solutions, a techno-economic decision block has been identified to avoid climate infra-marginality in future. As cost-effective energy storage technologies, energy carriers and storage infrastructure will enable a sector-integrated transition to 100% Renewable Energy Systems (RES).

The increasing interest in RES enables integrated, sector-coupled energy networks and facilitates the electrification of high-temperature heat, maritime and aviation applications (Kong et al., 2022). The energy system transition to RES comprises technical, economic, and political challenges, because of high investment needs for further expansion of wind and solar generation capacities onshore as well as offshore (DeAngelo et al., 2021). Wind and solar energy systems should be installed operationally by 2030, particularly in countries with high current RES costs, to unlock efficient utilization and design of RES electricity and facilitate long-term decarbonization (Abdelrashied & Bhattacharya, 2021).

6.5 Carbon Capture and Storage (CCS) Technologies

In the context of hydrocarbon production and processing, CCS facilitates decarbonization of large-scale crude oil enhanced oil recovery or gas processing (Hoa Nguyen et al., 2023). Furthermore, the Middle East region holds vast potential for geological storage and fits sequestration requirements naturally because the traditional onshore gas production stocks are typically depleted. This contribution provides data and methods for assessing trade-offs between CCS and other solutions for net-zero in the power sector of the Middle East region variable renewable energy dominates in deployment. It provides a supplement to the public model release. Results indicate a preference for solar power in mixture with storage, reflecting the economic dominance of limited storage penetration already in energy system decarbonization. Substituting wind for solar power would increase costs by 3%, likely due to the large RES shift in this relatively little-renewable-suitable region. In situations where renewable energy has hard penetration limits, as inferred from MENA,

CCS can potentially replace between 14 and 67% of renewable energy share without offshell strongly increasing system operation cost.

Carbon dioxide capture and storage (CCS) plays a key role in mitigating climate change for achieving the global net-zero target by the mid-21st century (Araújo & Luiz, 2021). This is especially true for the Middle East and North African region (MENA) seeing the preeminent role of natural gas and unfavorable prospects for nuclear power. CCS has several applications in industrial processes (e.g., cement, steel, and ammonia production), but the central role and widespread acceptance assume the role of a bridging technology in power generation and negative emissions in future energy system. Although wind and solar power with energy storage and nuclear power are sufficient to enable long-term energy system decarbonization of MENA, we find that even significant amounts of these technologies can be substituted by CCS, i.e., increasing storage and transmission costs, to keep costs at a similar level (Ek Fälth et al., 2021).

6.6 Implementation Strategies for Net Zero Policies in the Middle East

Energy prices and the huge financial risk involved are fundamental co-drivers for strategic diffusion (Chan & Sopian, 2022). In Saudi Arabia, the ambitious price of US$0.01 per kilowatt hour (¢/kWh) for the most asked Renewable Energy Project Development Office (REPDO) round 2 solar projects in the open desert laid out clear bounds for industry responding to drastically impelled cost–benefit outcomes from a major solar capacity increase, not a trough of the laboriously fulfilled demand spurred by proven H2 opportunities. A further undertaking is the land yield of Saudi solar projects, which like previously Iranian solar sites benefits from a "high annual average irradiation of > 2200 kWh/m^2/day" providing it with a 6883 PWh/a insolation from > 94,000 km^2 idealizes a current annual surface area used for solar PV of 7277 km^2. Later last year, UAE set a world record by placing the third CSP-PV hybrid site. PV employs land efficiently when the panels double as sunshades like on water and in agriculture. For example, the Worldwide Fund for Nature (WWF) has published that 463 TWh of solar power can be generated annually by paneling 1% of the Gulf with the latter 2 types at a full capacity yield of 438 W/m^2 averaging 2011–2015.

Implemented in a timely and strategic fashion, the Middle East's abundant renewable energy sources (RES) could offer significant socio-economic opportunity across the energy value chain in the decades leading up to mid-century (Salman & Hosny, 2021). The recent, substantial leap frog in photovoltaic (PV) prices and reduced costs of storage both open the door to the Middle East's cost-optimal production of hydrogen (H_2) from electrolyzed seawater by integrating floating solar PV facilities in areas suitable for ocean-based H_2 transportation. A potential sign of the Saudi Arabia leadership's noticing those opportunities took form in their plan to garner a colossal 50 GW of solar and wind capacities by 2030, which we will rigorously brief below (Dumas, 2021).

Governments in the Middle East are beginning to implement policies that mandate emissions reductions and promote renewable energy. For instance, the United Arab Emirates (UAE) has launched the UAE Net Zero by 2050 strategic initiative, aiming to align with the Paris Agreement's objectives (UAE Ministry of Climate Change and Environment, 2021). The government focuses on developing and implementing such frameworks and monitoring clear timelines, enforceable targets, and accountability mechanisms. Some of the governments are offering tax breaks, subsidies, and low-interest loans to reduce the financial burden on businesses transitioning to low-carbon operations. For example, Saudi Arabia's Public Investment Fund (PIF) has been actively investing in renewable energy projects, such as the $500 billion NEOM project, which aims to be powered entirely by renewable energy (Saudi Vision 2030, 2020). They are also establishing green bonds and climate funds mobilizing capital for large-scale sustainable infrastructure projects.

Another important strategy is leveraging cutting-edge technologies for the successful implementation of net zero strategies. Advances in solar and wind energy, energy storage, and carbon capture and storage (CCS) are pivotal. The Middle East, with its abundant solar resources, is particularly well-positioned to exploit solar energy. The Mohammed bin Rashid Al Maktoum Solar Park in Dubai, one of the largest solar parks in the world, illustrates the region's potential in this field (Dubai Electricity and Water Authority, 2022). Moreover, adopting smart grid technologies can optimize energy distribution and consumption, enhancing overall efficiency.

Importantly, collaboration between the public and private sectors can improve the implementation of net zero initiatives since it facilitates resource sharing, risk mitigation, and innovation. As per a report from Siemens, Qatar has collaborated with Siemens to develop a smart city in Lusail, incorporating sustainable energy solutions which showcase the benefits of Public–Private Partnerships (PPP) (Siemens, 2020).

By establishing clear regulatory frameworks, providing financial incentives, implementing advanced technologies, fostering public–private partnerships, and engaging society, the Middle East can transition toward a sustainable, low-carbon future. These strategies not only address environmental challenges but also pave the way for economic diversification and resilience in a post-hydrocarbon era.

6.7 Policy Frameworks and Regulatory Mechanisms

In this research, the focus on policy and regulatory mechanisms are on the Middle Eastern countries who have shown renewed impetus to strengthen their climate policies considering renewed global ambition for emission reduction and with the goal of meeting the targets set in their ratified Nationally Determined Contributions (NDCs). Countries such as UAE, Saudi Arabia, Oman, Bahrain, and other countries have begun deploying various domestic policy instruments to support the deployment of renewables while investing in targeted renewable technology R&D, EV, and biofuel standards and electric mobility pilot projects. Countries are progressing technology transfer through enhanced synergistic cooperation with bilateral and multilateral institutions that include increased investment and technological access (Net Zero Economy, Corporate Social Responsibility and Sustainable Value Creation, 2024).

Enhancing technology transfer policies tend to reduce regulatory barriers and coordinating heterogeneity of countries' standards, technical regulations, certification criteria, and codes, in addition to easing access to climate finance for cleaner technologies through grants, concessional loans, and non-financial support. Such non-financing support to technology diffusion explicitly targets risk protection, facilitating technology transfer through foreign direct investments, and capacity building (Cansino et al., 2022).

Policy is a set of mutually agreed upon guidelines, principles, and laws enacted by governments to achieve specific social, economic, and environmental goals. In the context of climate change, policies help achieve emission reduction targets, adaptation goals, and technological innovation (Ukoba et al., 2024). A multitude of climate policies are implemented across the globe to facilitate climate-resilient and low-carbon technology transition. These policies aim at altering economic incentives and regulatory frameworks to favor low-carbon alternatives and incentivize new investment (Babayomi et al., 2022). For example, the UAE has set its Net Zero 2050 strategy that builds upon the strategic initiative which outlines the UAE's ambitious journey toward net zero, and the National Net Zero by 2050 Pathway, which sets out the timeline and mechanisms for this transition. This strategy is designed to act as a stimulus for economic and societal advancement by leading the transition to net zero emissions. Furthermore, the strategy is expected to create 200,000 job opportunities across the solar, battery and hydrogen sub-sectors, contribute around 3% to the national GDP enhance export opportunities. The focus is on 6 key sectors namely power, industry, transport, buildings, waste, and agriculture. This strategy will enhance efficiency, expand the utilization of renewable energy sources, implement sustainable transportation and green building practices. The country as well is intending to incorporate nature-based solutions and employing carbon capture technologies. Importantly, the Ministry of Climate Change and the Environment will coordinate the wide-ranging and progressive initiatives of over 45 major stakeholders to meet its target for each sector (The UAE's Net Zero 2050 Strategy|the Official Portal of the UAE Government, 2021a, 2021b).

6.8 Public–Private Partnerships in Climate Action

Public–Private Partnerships (PPPs) are effective tools because they enable informed and balanced risk allocation between public and private stakeholders, provide long-term visibility and stability throughout the contract duration, and address climate change uncertainty through predictable contractual terms. Research on public–private institutions are a reason to focus on the governance of climate change, in particular their underpinning processes, and the share of responsibilities and resources between the public and private sectors' influencing governance has been shown to serve as the empirical basis of public–private partnerships in climate action involving both public and private institutions is crucial to implement sustainability. It is necessary to promote investment in low-carbon PPPs and to ensure that changing climate conditions, disaster risk and potential climate change mitigation and adaptation measures are identified. It is also important to consider during the development, design and implementation of each individual PPP project (Public–Private Partnership Legal Resource Center, 2019).

Public–private institutions (PPI) can take on many forms since the concept is performance based on cooperation has a wider perspective, as well as collective action challenges of climate change. A PPI also brings together resources and capabilities from public and private parties to create a sustainable partnership. Regional actors play a crucial role in climate and energy policy for their ability to involve relevant local stakeholders in the development of the policy itself. Trading places on power in society. Recurving climate policy to calculate efficiency that is more strategic and pragmatic than in other parts of the world. Political aspects are a primary problem for the climate in the Middle East, although relevant countries have announced zero net emissions by 2050 or 2030. Middle East producers are also working hard to approach hydrogen as an alternative in the Gulf.

Multiple actions can be considered by public and private actors to address climate change. Different countries have implemented mixed policy portfolio (Strielkowski et al., 2023). In general, direct regulation is the most common method of climate policy implementation. However, over the years policy development has matured, and the focus is still on hard means of implementing the energy pathway to technological innovation and the creation of alternative routes to impact the deployment of innovations and knowledge application in the market (Ek Fälth et al., 2021). Furthermore, countries have defined mid- and long-term vision and elaborated different strategies to meet objectives. Finally, a co-responsibility in climate and energy policy between the public sector and the market for the development of business strategies that involve the decarbonization of the economy (Salman & Hosny, 2021).

6.9 Challenges and Opportunities in Implementing Net Zero Policies

According to Renné (2022), policies have been aimed at the power sector traditionally, however, the policy landscape for transport and especially for heating and cooling has been weaker. Although these sectors have been receiving more policy attention in recent times. Furthermore, 145 countries have adopted regulatory renewable energy policies and mandates directed to the power sector by 2020. However, only 65 countries have established regulatory incentives and mandates in the transport sector, and only 22 countries in the heating and cooling sector.

Several significant policy implications have been analyzed for policy changes in the region—to maximize the flexibility of adaptation measures and minimize trade-offs between local and global environmental implications for the MENA region. Local regions that resonate the choices in technologies and policy directions also help to create synergies, e.g., to solve local problems emanating from agriculture and water management by integrating bioenergy systems.

The region has enough room for policy implementation, which currently has a unique place in the net zero discussion. Low-carbon solutions have a definitive impact on policy design. Two issues overlap in the region: the implementation of net zero policies (the inclusion of clean technologies for energy generation, transport, and carbon capture, utilization, and storage technologies) and the climate resilience/global adaptation policies (the inclusion of green natural infrastructures, biodiversity, afforestation, agroforestry and other agricultural land use projects in national policy). Therefore, the design of net zero/near zero-carbon emission regions involves local climate analysis and national and regional input studies related to future energy/economic/industrial projections.

The Middle East and North Africa (MENA) region emits significant amounts of GHGs per capita, comparable to the levels of emissions in the developed countries (Ek Fälth et al., 2021). The region includes a variety of socio-economic dynamics with different constituents and mechanisms in place to reduce emissions. These characteristics make the region a complex, diverse, and interesting case for the implementation of net zero policies (John & Derakhshi, 2022). The high emissions, ambitious plans for growth, and international implications, have prompted the perfecting of theoretical and methodological understanding of policy implementation in the region, to develop measures that are effective environmentally and are politically feasible (Ajmone-Marsan et al., 2012).

The progress in renewable energy among the Mashreq countries has been quite different. Egypt, Israel, and Jordan have all made significant improvements in developing renewable energy projects. This push is largely due to the need to support growing populations, enhance energy security, and seize economic opportunities. Egypt, in particular, has been successful in drawing substantial international investment for its renewable initiatives. On the other hand, Lebanon has been struggling to meet its renewable energy goals due to innate sociopolitical and economic issues. Whereas, In the Maghreb region, Morocco has taken the lead in renewable energy development, forced by political and security motivations. The country aims to have 52% of its energy capacity come from renewable sources by 2030. On the other hand, Tunisia, which also lacks significant hydrocarbon resources, has set ambitious renewable energy targets. However, political instability and economic challenges may affect its ability to attract needed international investments. Meanwhile, Algeria has fallen behind in its renewable energy targets, as it has focused on other investments during low oil prices times (Hafner et al., 2023).

Therefore, it is necessary to understand that, strengthening electricity connections between these countries could bring several benefits, such as improved energy security, increased opportunities for energy trade, and a higher share of renewable energy in the region's energy mix. Addressing interconnections between regional power is critical to support and expand renewable energy projects. This collaboration would not only enhance the efficiency and stability of energy supply, but also foster economic and political cooperation across the MENA region.

6.9.1 Technological and Financial Challenges

A technological challenge highlighted in strategic documents relating to the power sector is managing high levels of intermittent (solar and wind) generation in power systems with relatively lower levels of hydropower and system flexibility (Kuralbayeva, 2021). The cost of developing a smart grid, including upgrading the existing power grid to allow communications, will require up to 10% of the total investment in this sector. These challenges are compounded by the need to plan for a more flexible generation because of intermittent renewable growth. A key technology identified for providing flexible electricity generation and ensuring resilience in the grid was grid-scale storage (e.g., batteries, compressed air, pumped storage), particularly long-duration storage in anticipation that it might be required as the share of solar and wind power continues to increase.

Cost and time are also identified as the key barriers for achieving Net Zero Middle East. Although solar and wind energy resources offer great potential, cost is still a key barrier. Beyond cost, the uncertainty of how much new infrastructure and system upgrades will be required to support the growth in renewables remains a significant impediment to ambitious renewable energy deployment (Stern & Valero, 2021). The update of the energy mix in the Middle East, in the Gulf countries, remains significantly slow. However, ongoing power market reform and the privatization of the electricity sector is expected to help enhance the rate of transition, as well as the development of a spot market. Hydrogen is looked at as an alternative to large-scale storage of electricity or as a possible long-term storage technology. There is some optimism about hydrogen in the Middle East, but to speed up its deployment, issues relating to demand, infrastructure, its price, and the ability to trade it across borders need to be resolved.

The key technological challenge to global net-zero is the cost and pace of renewable deployment, which is still in its early stages (Ajmone-Marsan et al., 2012). The most recent report from the Energy Transitions Commission predicts that new renewable power capacity must grow to 5–10 times today's annual deployment rates by 2030 and continue doubling every 2–3 years to 2050. Even under the most ambitious growth rate, 99% of the power supply is projected to come from (dispatchable) solar, wind, hydro, biomass, geothermal, or nuclear by 2050. In this context, the report insists that overcoming past skepticism (on both economic costs and feasibility of power grid integration) around long-duration storage, hydrogen electrolyzers, two key dispatchable storable low-carbon fuels: hydrogen and sustainable fuels paired with long-duration storage is essential.

6.10 Case Studies of Net Zero Initiatives in the Middle East

On regional basis, coherent strategies combined with multi-pronged initiatives, including future promising technologies in niches, are crucial, such as GCC nations, EU as a best practice net-zero cities and for some sectors such as aviation that has large gaps, South Korea's targets should also be revised. LAC's must set net-zero ambitions across all sectors—contribution customization and alignment. Set ambitious long-term (2050) targets for GHG emissions that, in line with the latest climate science, imply net-zero CO_2 emissions by mid-century. Beyond CO_2, set targets and actions to mitigate other GHG emissions that accumulate in the atmosphere over more than a decade.

In recent years, the Middle East has embarked on an ambitious plan to diversify its resource dependency and implement robust national and regional strategies for sustainable and resilient growth. In response to these rapidly changing and accelerating investment needs, industry, government, and other stakeholders have mobilized nearly to adopt decarbonization commitments, announcing dates to achieve net-zero-carbon emissions and objectives to increase renewable energy and operational efficiencies (Chukwudumebi Obiora et al., 2024). However, the post-pandemic recovery brings forward substantial uncertainty as governments, companies, and investors face a set of economic, social, and environmental challenges that have altered energy demand patterns and investment priorities (Strielkowski et al., 2023).

Is the Middle East region embarking on a journey toward climate resilience? If climate resilience is adherence to objectives such as net-zero, then the answer to the question is a resounding "yes." At country level, Arabian Gulf nations like UAE, Saudi Arabia and Qatar have initiated low/regeneration function carbon and new technologies strategies significantly across the nation. At sub-national level, energy innovation and circular economy strategies have been implemented.

Dubai and Abu Dhabi have launched many drivers for technologies like smart transportation and efficient traditional energy (Himeur et al., 2020). UAE Net-zero initiatives in the Middle East are one of the best examples of net-zero initiatives. UAE announced its net-zero goals by November 16, 2021. This strategy claims it will accelerate clean energy by installing 47 gigawatts (GW) of renewable energy capacity and enhancing energy efficiency, with an aim to cut the emissions by 30% from both the industrial sector and power production and 40% from the petrochemical sectors by 2030. Saudi Arabia and the UAE as well have set goals of achieving net-zero CO_2 emissions by 2060 and 2050, respectively. Importantly, both the countries have also laid out plans for power sector decarbonization through the rapid expansion of both renewable and nuclear power. Besides reducing upstream emissions, UAE will accelerate the efficiency of oil and gas industries, by implementing renewable energies and emission reduction technologies. As per the IPCC's latest report identified that carbon capture and storage is a crucial component in climate change mitigation pathways that avoid massive reductions in energy demand.

6.10.1 Masdar City in the United Arab Emirates

Originally, upon launch in 2006 and first commercial tenants welcomed in 2010, the city was designed as a testbed for cutting-edge renewable energy and clean technology. Masdar City features the Masdar Institute of Science and Technology, developed in collaboration between Masdar Institute in Abu Dhabi and the Massachusetts Institute of Technology. However, it is now being transformed, without substantial public consultation, into a private and commercial retail and leisure district. Masdar City's various modes of transition are animated by a powerful narrative of and capital accumulation. The result is that Abu Dhabi, and Masdar City, are being shown to be primed for more deeply diversified future digital and knowledge economies, and multiple layers of environmental and energy governance, and private infrastructure capitalism for resilience across all and in each sector (Chan & Sopian, 2022).

In a similar way, are the energy systems emerging from the planning in region-specific resilience plans, or, to put it differently, how new climate adaptation and mitigation uncertainties relate to older categories and the future climate? These questions are taken as the entry points of a double-barreled inquiry on climate resilience governance, via a contrast between Masdar City and the Masdar's parent entity Mubadala Investment Company's 2020 strategic plan "investing for the good of Abu Dhabi, the UAE and future" (Ajmone-Marsan et al., 2012). The government of Abu Dhabi has committed to several objectives to curb climate change at the emirate, national, and personal and corporate levels, including zero-carbon Masdar City.

6.10.2 The Sustainable City, Dubai

The Sustainable City in Dubai illustrates net-zero initiatives through its comprehensive approach to reducing carbon emissions and promoting sustainable living. Importantly, extensive use of solar energy, with photovoltaic panels installed on rooftops and carports generating substantial amounts of renewable electricity has been noticed. This initiative significantly reduces the city's dependence on fossil fuels, aiming for a net-zero energy balance. The community's design includes highly energy-efficient buildings constructed with sustainable materials and advanced insulation techniques. The houses are equipped with smart home systems monitoring and optimizing energy usage, further contributing to the net-zero goal. Additionally, strategies such as active and passive cooling and heating designs reduce the need for artificial climate control, while lowering energy consumption. The community also enables sustainable transportation by providing electric vehicle charging stations and promoting the use of electric and hybrid vehicles among residents. The design includes car-free zones to encourage walking and cycling, reducing greenhouse gas emissions associated with transportation.

The United Arab Emirates (UAE) is an example of the hydrocarbon-rich economies that have pursued the greening of the nation's economy. It is one of the world leaders in economic growth through investments from the fossil fuel industry. A part of this growth includes the planning of future sustainable cities, including Masdar City, and investment in green future industries, green infrastructure, and technology (Sweidan, 2021). The city is located around 17 km east-southeast of the city of Abu Dhabi and will cover an area of 6 km^2 in total. It is a fundamental role of the UAE in establishing global leadership in renewable energy and clean technology and contributing its effort to combat the global threat of climate change.

6.11 Role of International Cooperation in Advancing Net Zero Goals

International cooperation plays a crucial role in advancing net zero goals by fostering collaboration, resource sharing, and the development of innovative solutions. Countries working together can pool resources, share the best practices, and leverage technological advancements to accelerate the transition to a low-carbon economy. The second imperative is to scale up investment in cuts to growing greenhouse gases in the atmosphere quickly. Between the first two quarters of 2019 and 2020, global expenditure on clean energy decreased by 9%, partly under the influence of reduced energy expenditure. International partnerships enable the distribution of financial and technical assistance to developing nations, helping them overcome economic and infrastructural barriers to implementing sustainable energy solutions. Evidenced in the recent pandemic for economic travel bans and non-essential business restrictions, impacts on emissions were only temporary, as investments in green infrastructures and products were reduced. This not only negated all the economic gains that have been achieved through ambition and 55% emission reductions. There are some choices to move forward with these two fronts. We need to understand the imperatives of cooperation in this context: simple necessity and a win–win outcome. Rivalry particularly with the US and China is widely discussed.

Global efforts by governments and private sector organization are required to help ensure the global warming, limit to 1.5 °C, set out in the Paris Agreement of the United Nations Framework Convention on Climate Change (UNFCCC) is met (Stern & Valero, 2021). The UNFCCC sets forth its objectives in Article 2 (Konstantinou, 2021). One of these is the "stabilization of greenhouse gas concentrations in the atmosphere at a level that would prevent dangerous anthropogenic interference with the climate system." The UNFCCC emphasizes "that a rapid transition to low-carbon and resilient economies is essential" to achieve reduction targets of their future warming. As is underscored by recent conflict over reducing hydrofluorocarbons (HFC) under the Kigali Amendment of the Montreal Protocol. Multilateral agreements and global initiatives, such as the Paris Agreement, set unified targets and create accountability, driving nations to commit to more ambitious climate actions. Cross-border projects, such as interconnected power grids and transnational renewable energy investments, illustrate the tangible benefits of international cooperation.

International cooperation is an important part of addressing the interconnected challenges of achieving the net-zero goals—the first and foremost which is unprecedented and concerted action at a global scale to limit global warming to 1.5 °C (Cansino et al., 2022). This has been emphasized by several countries and multilateral institutions in the lead-up to several COPs. Relevant challenges under consideration at COP28 were the determined implementation of the limits on global warming to 1.5 °C, advancement in the race to net-zero, and the goal to reap indigenous and resilient hardware to manage climate crises, by building and construction policies.

6.12 Conclusion and Future Scope

The Middle East's journey toward achieving net zero emissions and enhancing climate resilience is a multifaceted and complex endeavor, requiring concerted efforts across various sectors and stakeholders. This comprehensive analysis highlights the critical role of emerging technologies, robust policy frameworks, international cooperation, and public–private partnerships in driving this transformation. The significance of net-zero policies lies in their ability to balance greenhouse gas emissions with carbon removal efforts, ensuring a sustainable equilibrium. The Middle East faces unique challenges due to its socio-economic context and environmental vulnerabilities, making the implementation of effective climate resilience strategies even more crucial. The region's arid and semi-arid climate, combined with high temperatures and water scarcity, exacerbates the impacts of climate change, affecting agriculture, water resources, and energy systems. Emerging technologies such as renewable energy sources and AI-based solutions play a pivotal role in achieving net zero emissions.

The Middle East has made significant strides in expanding its renewable energy capacity, with countries like the UAE and Saudi Arabia leading the way. Projects like Masdar City in Abu Dhabi and The Sustainable City in Dubai exemplify the region's commitment to integrating cutting-edge renewable energy and clean technology into its infrastructure. Implementation strategies for net zero policies must address both technological and financial challenges. The high cost and pace of renewable deployment, coupled with the need for substantial infrastructure upgrades, present significant barriers. However, ongoing power market reforms and the privatization of electricity sector offer opportunities to enhance the rate of transition. Hydrogen, as a potential large-scale storage technology, also holds promise, but issues related to demand, infrastructure, and cross-border trade need to be resolved. Public–private partnerships are essential in fostering climate action and achieving sustainable development goals. The involvement of both public and private sectors in governance and resource allocation ensures a more comprehensive and effective approach to addressing climate change.

Regional actors, local stakeholders, and international cooperation play pivotal roles in this collaborative effort. The Middle East's efforts in diversifying its economies and reducing reliance on fossil fuels present unique opportunities for growth and leadership in the low-carbon sector. The region's commitment to increasing renewable energy capacity, enhancing energy efficiency, and implementing emission reduction technologies demonstrates a proactive approach to climate resilience.

Looking ahead, the Middle East must continue to innovate and invest in sustainable technologies to achieve its net zero goals. The integration of renewable energy sources, such as solar and wind power, with advanced storage solutions will be critical in ensuring a stable and resilient energy supply. The development of hydrogen as a viable energy carrier and storage option holds significant potential for the region's long-term sustainability. International cooperation will remain a cornerstone of the Middle East's climate strategy. Collaborative efforts with global partners, multilateral institutions, and regional alliances will be essential in scaling up investments in clean energy and technological innovations. The region's ability to leverage international expertise and financial support will be crucial in overcoming the challenges associated with large-scale renewable deployment and infrastructure development.

Policy frameworks and regulatory mechanisms must evolve to support the rapid transition to a low-carbon economy. Governments in the Middle East need to implement policies that incentivize the adoption of clean technologies, reduce regulatory barriers, and facilitate access to climate finance. Enhancing technology transfer through bilateral and multilateral cooperation will further accelerate the region's progress toward net zero. Public awareness and engagement are also vital components of a successful climate strategy. Educating communities about the importance of sustainability and encouraging behavioral changes can significantly contribute to reducing the region's carbon footprint. Initiatives that promote energy efficiency, water conservation, and sustainable consumption practices will play a crucial role in achieving the region's climate goals.

In conclusion, the Middle East is poised to make significant strides toward achieving net zero emissions and enhancing climate resilience. By leveraging emerging technologies, fostering international cooperation, and implementing robust policy frameworks, the region could pave the way for a sustainable and resilient future. The journey ahead will require unwavering commitment, innovative thinking, and collaborative efforts from all stakeholders, but the rewards of a sustainable future are well worth the endeavor.

References

Abdelrashied, M., & Bhattacharya, D. (2021). Future photovoltaic electricity production targets and the link to consumption per capita on the policy level in MENA region. https://doi.org/10.48550/arXiv.2109.02129

Ajmone-Marsan, M., Arrowsmith, D., Breymann, W., Fritz, O., Masera, M., Mengolini, A., & Carbone, A. (2012). The emerging energy web. *The European Physical Journal. Special Topics, 214*(1), 547–569. https://doi.org/10.1140/epjst/e2012-01705-1

Al-Jawaldeh, A., Nabhani, M., Taktouk, M., & Nasreddine, L. (2022). Climate change and nutrition: Implications for the Eastern Mediterranean region. *International Journal of Environmental Research and Public Health/international Journal of Environmental Research and Public Health, 19*(24), 17086–17086. https://doi.org/10.3390/ijerph192417086

Al-Rimmawi, H. (2012). Middle East chronic water problems: Solution prospects. *Energy and Environment Research, 2*(1). https://doi.org/10.5539/eer.v2n1p28

Araújo, O. Q. F., & Luiz, J. (2021). How is the transition away from fossil fuels doing, and how will the low-carbon future unfold? *Clean Technologies and Environmental Policy, 23*(5), 1385–1388. https://doi.org/10.1007/s10098-021-02123-x

Babayomi, O. O., Dahoro, D. A., & Zhang, Z. (2022). Affordable clean energy transition in developing countries: Pathways and technologies. *Iscience, 25*(5), 104178–104178. https://doi.org/10.1016/j.isci.2022.104178

Cansino, J. M., Román-Collado, R., & Nassar, S. (2022). The clean development mechanism in Eastern Europe: An in-depth exploration. *Environmental Science and Pollution Research International, 29*(49), 74797–74822. https://doi.org/10.1007/s11356-022-20988-3

Chan, H.-Y., & Sopian, K. (2022). Transitioning to a high renewable net-zero power generation system in Malaysia. Philosophical Transactions—Royal society. Mathematical, physical and engineering sciences/philosophical transactions—Royal society. *Mathematical, Physical and Engineering Sciences, 380*(2221). https://doi.org/10.1098/rsta.2021.0132

DeAngelo, J., Azevedo, I., Bistline, J., Clarke, L., Luderer, G., Byers, E., & Davis, S. J. (2021). Energy systems in scenarios at net-zero CO_2 emissions. *Nature Communications, 12*(1). https://doi.org/10.1038/s41467-021-26356-y

Dumas, J. (2021). *Microgrid management with weather-based forecasting of energy generation, consumption and prices*. https://doi.org/10.48550/arXiv.2107.01034

Ek Fälth, H., Atsmon, D., Reichenberg, L., & Verendel, V. (2021). MENA compared to Europe: The influence of land use, nuclear power, and transmission expansion on renewable electricity system costs. *Energy Strategy Reviews, 33*, 100590–100590. https://doi.org/10.1016/j.esr.2020.100590

Hafner, M., Raimondi, P. P., & Bonometti, B. (2023). Low-carbon energy strategies in MENA countries. In *Perspectives on development in the Middle East and North Africa (MENA) region* (pp. 175–261). https://doi.org/10.1007/978-3-031-30705-8_4

Himeur Y., Alsalemi, A., Bensaali, F., Amira, A., Varlamis, I., Bravos, G. N., Sardianos, C., & Dimitrakopoulos, G. (2020). *Marketability of building energy efficiency systems based on behavioral change: A case study of a novel micro-moments based solution.* https://www.semanticscholar.org/paper/Marketability-of-building-energy-efficiency-systems-Himeur-Alsalemi/0b25aabecfbc9751c5695206c733baef861bf0ff

John, D., & Derakhshi, E. (2022). Low carbon mobility transitions and justice: A case of Costa Rica. *Development, 65*(1), 71–77. https://doi.org/10.1057/s41301-022-00331-6

Khojasteh, D., Davani, E., Shamsipour, A., Haghani, M., & Glamore, W. (2022). Climate change and COVID-19: Interdisciplinary perspectives from two global crises. *Science of the Total Environment, 844*, 157142–157142. https://doi.org/10.1016/j.scitotenv.2022.157142

Kong, C., Zhang, J., Ntarmah, A. H., Kong, Y., & Zhao, H. (2022). Carbon neutrality in the Middle East and North Africa: The roles of renewable energy, economic growth, and government effectiveness. *International Journal of Environmental Research and Public Health/international Journal of Environmental Research and Public Health, 19*(17), 10676–10676. https://doi.org/10.3390/ijerph191710676

Konstantinou, C. (2021). *Towards a secure and resilient all-renewable energy grid for smart cities*. https://doi.org/10.48550/arXiv.2101.10570

Kuralbayeva, K. (2021). *Forest carbon offsets over a smart ledger*. osf.io.

Markard, J., & Rosenbloom, D. (2022). *Phases of the net-zero energy transition and stategies to achieve it* (pp. 102–123). Routledge EBooks. https://doi.org/10.4324/9781003183020-8

Net Zero Economy, Corporate Social Responsibility and Sustainable Value Creation. (2024). *SpringerLink*. 10.1007-978-3-031-55779-8

Nguyen, D. H., Chapman, A., & Tsuji, T. (2023). Assessing the optimal contributions of renewables and carbon capture and storage toward carbon neutrality by 2050. *Sustainability, 15*(18), 13447–13447. https://doi.org/10.3390/su151813447

Nouri, M., Homaee, M., Pereira, L. S., & Bybordi, M. (2023). Water management dilemma in the agricultural sector of Iran: A review focusing on water governance. *Agricultural Water Management, 288*, 108480–108480. https://doi.org/10.1016/j.agwat.2023.108480

Obiora, S. C., Bamisile, O., Hu, Y., Ozsahin, D. U., & Adun, H. (2024). Assessing the decarbonization of electricity generation in major emitting countries by 2030 and 2050: Transition to a high share renewable energy mix. *Heliyon, 10*(8), e28770–e28770. https://doi.org/10.1016/j.heliyon.2024.e28770

Pratama, Y. W., Patrizio, P., & Dowell, N. M. (2022). National priorities in the power system transition to net-zero: No one size fits all. *Iscience, 25*(10), 105260–105260. https://doi.org/10.1016/j.isci.2022.105260

Public-Private-Partnership Legal Resource Center. (2019). Retrieved May 22, 2024, from Public-Private-Partnership Legal Resource Center website: https:// ppp.worldbank.org/public-private-partnership/climate-smart/climate-smart-clean-technology-ppps/preparing-procuring-and-implementing-climate-smart-ppps

Salman, D., & Hosny, N. A. (2021). The nexus between Egyptian renewable energy resources and economic growth for achieving sustainable development goals. *Future Business Journal, 7*(1). https://doi.org/10.1186/s43093-021-00091-8

Singh, R., et al. (Eds.). (2024). *Net zero economy, corporate social responsibility and sustainable value creation, CSR, sustainability, ethics & governance*. 10. 1007/978-3-031-55779-8_5

Social Dimensions of Climate Change. (2023). Retrieved May 21, 2024, from World Bank website: https://www.worldbank.org/en/topic/social-dimensions-of-climate-change

Stern, N., & Valero, A. (2021). Innovation, growth and the transition to net-zero emissions. *Research Policy, 50*(9), 104293–104293. https://doi.org/10.1016/j.respol.2021.104293

Strielkowski, W., Gorina, L., Korneeva, E., & Kovaleva, O. (2023). Energy-saving technologies and energy efficiency in the post-pandemic world. Retrieved August 1, 2024, from arXiv.org website: https://arxiv.org/abs/2312.11711

Sweidan, O. D. (2021). The environmental and energy policies to enable sustainable consumption and production in the Gulf Cooperation Council countries. *Clean Technologies and Environmental Policy, 23*(9), 2639–2654. https://doi.org/10.1007/s10098-021-02184-y

The UAE's Net Zero 2050 Strategy | The Official Portal of the UAE Government. (2021a).

The UAE's Net Zero 2050 Strategy | The Official Portal of the UAE Government. (2021b). UAE. Retrieved May 22, 2024 https://u.ae/en/about-the-uae/strategies-initiatives-and-awards/strategies-plans-and-visions/environment-and-energy/the-uae-net-zero-2050-strategy

Ukoba, K., Yoro, K. O., Eterigho-Ikelegbe, O., Ibegbulam, C., & Jen, T.-C. (2024). Adaptation of solar power in the global south: Prospects, challenges and opportunities. *Heliyon, 10*(7), e28009–e28009. https://doi.org/10.1016/j.heliyon.2024.e28009

Wu, C.-J., Manne, S., Ranganathan, P., Bird, S., & Greenstein, S. (2021). *Socio-technological challenges and opportunities: Paths forward*. https://doi.org/10.48550/arXiv.2108.06738

Yuan, Y., Alahmad, B., Kang, C.-M., Al-Marri, F., Kommula, V., Bouhamra, W., & Koutrakis, P. (2020). Dust events and indoor air quality in residential homes in Kuwait. *International Journal of Environmental Research and Public Health, 17*(7), 2433–2433. https://doi.org/10.3390/ijerph17072433.

Chapter 7
Fostering Human Capital Through Emotional Labour for Sustainable Human–Human-Machine Cooperation in Achieving Net-Zero Policies

Anna Rostomyan

Abstract In our times fostering the perseverance of human capital and replacing in some areas humans with machines has become one of the net-zero milestones. Though machines surpass us humans in some areas, they are devoid or partly devoid of human emotions, which makes them more stress resistant. This can be a huge step forward towards preserving energy and keeping human capital strong and rigid. Nonetheless, since we are not only rational, but also emotional beings, we need emotions to feel cherished and connected with other social species. So, the current paper elucidated the progress of science and discusses the importance of emotional labour on both parties, i.e. humans and machines, to foster a stronger human–human-machine cooperation in achieving net-zero policies.

Keywords Human capital · Human resources management · Emotional capital · Emotional labour · Emotion AI · Human–human-machine interactions

7.1 Introduction

Nowadays we are living in very challenging and tempting times where we have set big goals in achieving net-zero policies to have a more sustainable future for us and our next generations. Therefore, where the human capital is scarce it is currently being largely and extensively being replaced by machines which gives us humans the ultimate chance to preserve the most valuable resource, namely our human energy. Human capital refers to the sets of skills employees have in their portfolio baggage that make them up as unique additions to the whole capital of the company. In other words, human capital refers to both the soft and hard skills of the workers that are essential in both their personal and professional lives. As we have already given

A. Rostomyan (✉)
Department of Psychology, Media University of Applied Sciences, Berlin, Germany
e-mail: annarostom@yahoo.com

R. Singh and D. Crowther (eds.), *Transition Towards a Sustainable Future*, Approaches to Global Sustainability, Markets, and Governance,
https://doi.org/10.1007/978-981-97-5756-5_7

stated, currently our human workforce is at many places being replaced by AI robots and other AI technologies in many different aspects of the business sector and not only (also in the household and some other places). For instance, translators even in such big companies as the German Deutsche Welle are largely being replaced by AI tools to aid journalists in their jobs or robots working in delivering DHL services. Even in the education sector, some universities hire AI robot professors as being deployed by Harvard University. Yet, there are a number of unaddressed ethical issues that have to be respected while applying AI in our businesses. For this very reason the EU introduced the AI Act that has to be followed by all organizations and institutions that make usage of AI dating from 2024.

This might lead to the frustration and fear of having many jobless people whose skills are no longer being in demand in the market since the same activities can be executed in just a couple of seconds by AI tools. Nonetheless, there is an aspect that is sometimes being overlooked, namely the emotional aspect. Thus, when referring to the human nature, Carnegie (1982) once stated, "When dealing with people, remember, you are not dealing with creatures of logic, but creatures of emotion" slightly hinting to the oblivious fact that we humans are emotional beings largely being governed by our internal states, emotions and feelings. This exclamation shows that we humans are not merely working machines, fully devoid of any emotions and/or feelings, but instead it hints to the obscure fact that very often our verbal and non-verbal behaviour in both our professional and personal lives is being governed by our internal states that have the utmost potential of having an immense impact on our actions and decisions (Rostomyan, 2022b). The present paper, thus elucidates how emotional capital can leverage stronger bongs between humans and machines and how this can be heightened by means of the application of Emotion AI. As a result, we will have more efficient and sustainable human–human-machine interactions, where machines can replace humans in vital positions, and humans will have with machines more enjoyable and harmonious interactions and the net-zero policies will be better followed in regards of the workforce.

7.1.1 The Significance of Emotional Capital and the Unique Potential of Emotional Labour in Fostering Thereof in Human Resources Management

Emotional capital mainly refers to the soft skills that make up the emotional repertoire of both the employees and the employers that makes proficient workplace interaction plausible. According to Oxford Business English Dictionary (2005), "emotional capital" is the feelings, beliefs and values of a certain company's employees that make good relationships possible and help a business flourish, to be competitive in the labour market and to stand out in line of the excellence of its workforce. This mostly refers to our ability to perceiving, understanding, digesting, managing and expressing our internal feelings and emotions, as well as recognizing them in the others and through the execution of healthy emotion management in the process of interpersonal communication attaining successful and harmonious interpersonal relations (Rostomyan, 2020, 2022a). The Oxford definition comes to suggest that in case a company has long-term goals in terms of preserving the potential of its co-workers, to sustainably succeed in terms of attaining its goals, it has to pay a close attention to the needs (especially emotional ones) of its employees. This can mostly refer to building long-term relationships with its partners, consumers and customers, as well as fostering a healthy and enjoyable internal communication with its direct subordinates in human resources management (Rostomyan & Rostomyan, 2023).

Like economic capital, emotional capital is not typically equally distributed in our society, that is to say that emotional capital is not something static and equally measured in the members of the society, but is the situated activation and embodiment of our emotional experiences, our choices in the verbal and non-verbal expression or suppression therein, and their management pursing the ultimate goal of stimulating healthy and strong relationships, especially within the business sector (Hochschild, 1979, 1983). According to Arlie Hochschild, all we depending on the requirements and demands of our profession have to act in a certain way which entails the outward inhibition of our internal states and feelings (Hochschild, 1983). This means that depending on our profession we have to suppress and not fully exhibit our negative emotions, like stress, distress, loneliness, fear, frustration, irritation, and/or anxiety. On the other hand, it also entails the expressions of other signs of positive emotions like happiness, delight, joy, glee, satisfaction, aspiration, hope and enjoyment. For this very reason the smile is considered as a universal indicator of positive emotions and is usually deployed by professionals (Rostomyan, 2022a). This kind of emotion management happens based on previously learnt norms and display rules, that are learnt so early in life that we can even no longer remember their source, as well as the requirements of the companies we work for, where we have to deal both with our very own emotions, the emotions of our co-workers, the emotions of our consumers/ customers, as well as other business partners. According to Professor Dr. Paul Ekman, an American psychologist and professor, and his colleague Wallace Friesen (1975), there are normally 4 types of display rules widely used across situations and cultures, that are applied in our lives for emotion management purposes, these are as follows:

(a) *Personal display rules*, which denote personal preferences,
(b) *Cultural display rules*, which denote cultural specifics,
(c) *Needs of the moment*, denoting the requirements of the given moment, and
(d) *Vocational requirement*, which designate the requirements of this or that professions.

Let us go through them in more detail one by one:

(A) **Personal display rules**: this category involves our personal characteristic features like being an extrovert or introvert, being sociable or anti-social, etc. That have an impact on our choice of expressing emotions overtly or suppressing them. This may also be very tightly connected with our upbringing, for instance according to William James (1983), the first educator to offer a psychology course in the United States and a leading thinker of the late nineteenth century, stating that if a mother used to shower with love her child, when growing up, the child does the same with its surrounding world and vice versa. Besides, according to Dr. Paul Ekman's research (1975, 1979, 2004), there are also some families that encourage the exhibition of emotions, while others are more reserved, which is later on depicted in the actions and verbal and non-verbal behaviour of the individuals in the society at large, both in their personal and professional lives.

(B) **Cultural display rules**: this category involves cases that because of our cultural characteristic features we might suppress or express our internally experienced emotions and feelings. This refers to the phenomenon that some nations are considered to be more emotional (like Italians, Spanish, Americans, Armenians, etc.) than others (like English, Germans, Swiss, etc.). Besides, Geert Hofstede (1997) suggested that there are cultures that are more indulgent in showing their emotions and some other cultures that are more restrained. So, in his discussion on cultural characteristics one of the dimensions according to him was the opposition of indulgent-restrained that can be used to characterize cultural specifics.

(C) **Needs of the moment**: this category includes the expression of suppression of emotions according to the needs of this or that very moment that the individual basks in, as the name itself suggests. So, for instance, two children have been playing in the playground, one of them hit the other one and the latter fell down, but while seeing the mother of the other child, the guilty child may express a wide face and a seemingly innocent face to overtly display innocence not to be accused. In another case, a guilty criminal may exhibit an innocent face to plead guiltless in front of the Judge and the jury to find his/her way out of jail with the help of falsified emotions as Ekman and Friesen truly suggested (1975).

(D) **Vocational requirements**: this category involves the management of emotions based on the demands of the profession. This can include cases of stewardesses displaying smiles and friendly faces during the flight even during the times of turbulence for the passengers not to feel fear, distress and/or anxiety. Or else, for instance in the banking sector at the customer department, where the assistants should display a calm face, calm voice and friendly demeanour (even in stressful and conflict situations) to remain calm and exhibit professional posture (Rostomyan & Rostomyan, 2023). Moreover, there are some sub-fields that will foster Emotional Capital and can also be trained in the employees through coaching (Infographic 7.1).

As it can be seen in the infographic above, these can be but are not limited to:

(A) Purpose,
(B) Consistency,
(C) Emotion,
(D) Flexibility,
(E) Involvement,
(F) Loyalty,

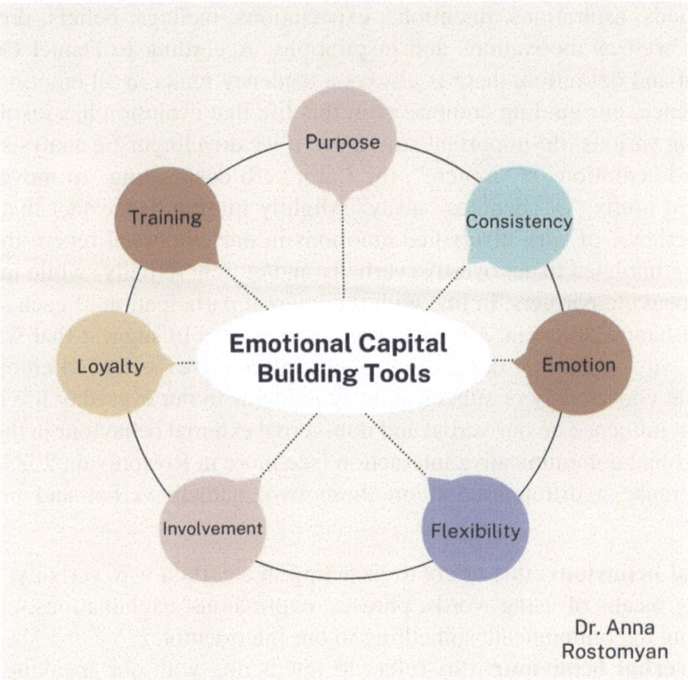

Infographic 7.1 Emotional capital building tools. *Source* Rostomyan and Rostomyan (2023)

(G) Training, etc. (Rostomyan & Rostomyan, 2023).

All the aforementioned strategies and display rules can be detected in our day-to-day activities, including the professional world, and can be measures of emotional labour that one deploys in managing emotions in cooperation with others in our surroundings, especially in human resources management. Moreover, if we gain an understanding of therein, it will become far easier for us to understand ourselves and others better, which will undoubtedly resultantly lead to better and stronger interpersonal connections that will in their turn lead to better labour output and better turnover.

7.1.2 *The Role of Emotional Labour in Having Successful Cooperations at the Workplace*

We humans almost always experience some sort of emotion or feeling throughout the entire day. Moreover, our internally experienced states may vary during the day depending on the various extralinguistic factors and some perceived external factors. It is also worth mentioning that our internal processes may also greatly vary based on our moods, aspirations, intentions, expectations, feelings, beliefs, dreams and desires, as well as motivations and inspirations. According to Daniel Goleman's description and definition, there is always a tendency to act in all emotions, which are, in essence, our guiding compasses in this life that evolution has instilled in us for handling various life-important situations. If we do a linguistic analysis, the root of the word "emotion" is "motere", the Latin verb designating "to move", where the attached prefix "e-" denotes "away", slightly hinting to the fact that because of the experience of vary diversified emotions in our emotional repertoire, we are very often stimulated to act overtly (verbally and/or non-verbally) while interacting with our speaking partners, in line with the internal participation of each and every emotion at hand (Goleman, 1995:6). The above comes to suggest that we usually act in this way or another depending on our internally experienced emotions and feelings that consciously or subconsciously guide us in our everyday life decisions and have an influence on our verbal and non-verbal external behaviour in the process of interpersonal communicative interaction (see more in Rostomyan, 2022a).

Let us make a differentiation on these two, namely verbal and non-verbal behaviour:

(a) **Verbal behaviour**: this refers to us acting in a certain way verbally, that is to say by means of using words, phrases, expressions, exclamations, etc. While wanting to communicate something to our interlocutor.
(b) **Non-verbal behaviour**: this refers to interacting with our speaking partners by means of expressing information without explicitly conveying them through words or phrases, that is to say encoding meaning without verbal activity, namely through non-verbal markers, which can be, but are not limited to facial expressions, gross bodily movements, eye contact, bodily posture, mimics, etc. (see more in Rostomyan, 2023a).

In our everyday life activities we willingly or unwillingly indulge into the execution of our day-to-day duties and while interacting and cooperating with the others in our surroundings (be it in our personal or professional lives) we exhibit verbal and/ or non-verbal communication or a combination thereof, which enhances the overall emotive perlocutionary effect on the audience, which is the level of influence on the audience achieved by the speaker/writer, especially the one bearing emotional content and colorings (see more in Rostomyan, 2020, 2023a). This can bring us to the basic assumption that while decoding information towards our speaking partners in the process of communication, cooperation and interaction, we are mostly and predominantly being guided by our internally felt and experienced emotional states. Nevertheless, it should be mentioned in this connection that under certain situations (as we have seen in our previous sub-chapter mostly depending on cultural and professional frameworks), we are obliged to inhibit the outward expressions therein and to replace a somewhat negative emotion as anxiety before an important speech in front of a big audience with possibly a positive emotion of hope.

For this kind of action, a very interesting new term has been coined and introduced to science, namely "*emotional labour*", which comes to suggest that people are obliged to act in definite ways in terms of exhibiting or not exhibiting and suppressing their emotions predominately based on their practiced professions. Thus, we have to state that the term "emotional labour" was first introduced by the sociologist Arlie Hochschild in her book "The Managed Heart" published in (1983), where the author described the notion designating the "labour" that we have to exhibit in accordance to our vocations. As the term itself suggests, "emotional labour" is indeed a "labour", namely a kind of a conducted work in line of our internally experienced emotions and feelings, since dealing with the emotions of not only ourselves but also of the others is not an easy task to do and requires concentration and effort, which is indeed a duty that we have to commit ourselves to if we wish to handle successful relationships especially at the workplace.

The term "labour" is a borrowing from French and belongs to the high-flown word-stock of the English language, most probably being derived after the Norman conquest in the eleventh century, stemming from the Old French "labor, labour" (modern French labeur), which itself is probably derived from Latin "labōr-", the inflectional stem of the noun labor "labor, work, toil". This can remind us of the well-known Latin saying "*Vivere est laborare*" literary meaning that living is to work and subtly suggesting to the fact that we humans are designed to work and cultivate culture and thus "sculpture" our surrounding world for us all and for the coming generations. According to Hochschild (1983), managing emotions does not only refer to handling the outward expression of our very own internal feelings and emotions, but also acting in a certain way that has a positive impact on the experience of emotions of the others. Thus, we can come to think that the execution of "emotional labour" results in a kind of a tango where the "manager" of the expression of emotions takes the leading role and leads the partner towards experiencing a beautiful blend of professionally handled emotions and feelings, which come and go in a flow through their interaction(s). In this respect, it is worth mentioning that here we do not only speak of only the inhibition of negative feelings and the proficient exhibition of positive feelings, but instead a healthy management of both our internally experienced positive and negative emotions respecting the face of the interactants. Moreover, it goes without saying that the workplace can be full of intense feelings, where the presence of stress, anger, fear, distress and anxiety can sometimes become a part and parcel therein. Of course, we should avoid expressing these very negative emotions at the workplace and work out some strategies on how we can deal with the latters, as for instance:

- taking deep breaths to regulate the circulation of fresh air supplied to the brain for staying calm and feeling safe,
- counting to ten before getting angry, thus leaving some time for the brain to reassess the actual outward situation,
- facing the fear and bringing some opposite facts delivering a safe space for hope, which has the utmost potential of providing the brain with positive impulses, etc.

The aforementioned strategies are very essential and vital especially in preserving your face in front of the others and presenting a calm and professional demeanour that has the utmost potential of picturing you as a confident persona. Thereinafter, we can state that emotional labour has to be explained as "*the management of feelings to create a publicly observable facial and bodily display*" (Hochschild, 1983:7). In essence,

A. Hochschild's definition of emotional labour implicitly presumes and presupposes that there are some professions that presuppose that professionals more or less consciously or subconsciously have to attempt to handle their experienced emotions by engaging into surface acting, this meaning that the employees of this or that organization or institution at hand only handle the outward expression of the internally felt emotion, but on the deeper level the emotions stay all the same. This is mostly conducted based on the formerly discussed display rules. Yet, our studies have shown that in case employees constantly manage the emotions on the surface level at the workplace, there may sprout a more positive emotion on the deeper level as well (Rostomyan, 2024b). For instance, a smile is a universally recognized indicator of positive emotions and it sends to the brain signals that the organism is in a favourable situation for its survival. So, we can assume that in case we display a smile more often, the brain will perceive positive vibes from the environment, which will result in our positive attitude towards both ourselves and the outward world, which has the potential of becoming a nice milestone towards building and keeping strong relationships with our business partners, colleagues, managers and customers.

Rostomyan and Rostomyan (2023), both being internationally recognized researchers, in their research on emotional labour conclude that emotional labour is the process of managing and handling the outward expressions of feelings and emotions and handling the outward expressions therein mostly to fulfil the emotional requirements of their professions. More specifically, according to the authors, employees are expected to regulate the outward manifestations of their emotions (especially negative ones as fear, anger, irritation and/or frustration) during their professional communicative interactions with their customers, consumers, clients, co-workers, colleagues, partners, as well as most importantly their C-level managers and bosses. Moreover, our studies and research have shown that not only workers have to manage the external exhibition and manifestation of their internally felt emotions according to the demands of their profession, but also managers and executives in case of deploying an appropriate amount of emotional labour in their professional routine stand a better chance of building stronger teams and consequently driving sustainable success.

Ashforth and Humphrey (1993) also defined the act of displaying an appropriate and adequate amount of emotion(s) in their professional interactions as emotional labour, which both emphasized the execution of verbal and non-verbal behaviour outward communicative professional demeanour and was different from the internal experience of this or that emotion and, thus, that the actual outward expression of an emotion itself as mostly different from the internally experienced one. To put it into other words, they described situations when people acted in certain ways that were different from their internally experienced emotions. Nonetheless, they were largely convinced that the expressions and manifestation of genuinely experienced emotions to be regarded as an indispensable part of conducting emotional labour.

This brings us to the assumption that emotional labour does not merely refer to emotion management in terms of only suppressing the expression of mostly negative emotions or sometimes also overly positive emotions because of the professional framework. Moreover, our research has shown that always only suppressing the experienced negative emotions and only displaying positive emotions can be destructive for our organism and for our internal emotional well-being and mental health. Thus, a healthy expression of some portion of anger on the side of a manager unhappy about a badly conducted work can even have the potential of having a positive impact on the improvement of the overall labour output, since in case of properly and adequately communicating the demands to the employees will raise the chance of having an influence on the latters in a positive and empathetic manner.

Besides, it is also very important to note that the exhibition of genuinely experienced positive emotions at the workplace can have a very positive impact on the flourishing of positive and strong interpersonal relations between co-workers and managers. So, for instance, seeing a sad co-worker and reaching out to him or her to find out what the matter is can be very decisive in the development of the relations between the co-workers, which can by itself deepen their mutual respect and understanding and be a keystone in the amelioration of their human–human interactions both at the workplace and in their private lives. Also, it has been noticed that the expression of genuine emotions both in our personal and professional lives raises the levels of happy hormones like serotonin, dopamine and oxytocin, which have the potential of heightening the experience of positive emotions, which results in positive moods that in their turn dissolve into very enjoyable experiences that also in their turn have a very positive impact on our labour output and on the overall business turnover.

It follows from the above that both surface acting and deep acting, namely exhibiting emotions on the surface level and feeling them on the deep level, alongside the manifestation of genuine emotions can lead to either emotional harmony, emotional dissonance, or emotional deviance. This means that we have to be very careful with our decisions on to what extent to display and handle our experienced emotions. Furthermore, it follows that we not only have to act on the surface level, but also carry out an analysis on the internal, deeper level to have a better understanding on both ourselves and on our surrounding world and thereinafter manage the expression of our emotions, of course depending on the situation, which will create a genuine rapport between the interactants and will stimulate their work. In this connection, it is also notable that in our everyday life while cooperating and interacting with one another we can sometimes make use of very different emotive reinforcers called *intensifiers*, like “greatly”, “fully”, “truly”, “really”, “immensely”, etc. that have the utmost potential of enhancing the overall emotional impact on the listeners/communicators. This hints to that fact that in case we want to achieve an emotional influence on the audience, we have to sound emotionally genuine and positive and do our best in creating a positive picture of ourselves also by means of applying proper emotion management (Rostomyan, 2020, 2022a, 2022b).

Nonetheless, it goes without saying that while performing in a strictly formatted business setting, we are obliged to adhere to certain specific emotion expression management techniques and the formerly discussed display rules and principles in order to professionally handle the outward expression of very strongly internally experienced emotions (both positive and negative) and to execute our professional actions and demeanour in accordance with the formerly described and discussed so-called "*Emotional Labour*" at the workplace. This further brings us to the assumption and the formerly conducted research of the author in different settings as the automotive industry, educational and banking sector, as well as the legal system, that there are some professions that demand emotion management and the execution of professionally conducted emotional labour in accordance with their profession, where for instance also economists, politicians, lawyers, educators, real estate agents, etc. Have to handle professionally the exhibition of their internally experienced emotions and feelings, beliefs and desires, etc. (Rostomyan, 2012).

To sum up with, the examination of the factual material and the literature review of the theorists on emotional labour, scientists and researchers mostly agree on the nature and significance of emotional labour in the workplace, which denotes the prerequisite and demand of ethical and professional performance within certain professional settings, which not only entails a professional demeanour, but also an performance of a healthy emotion expression management. The study of our own research has also shown that a healthy execution of emotional labour can by itself be an asset in the formation of strong interpersonal human–human relationships, which can be the basis of professional sustainable development and personal flourishing.

7.1.3 The Outstanding Role of Emotion AI in Building Strong Human–Machine Interactions

The predominant role of emotions and their significance of emotions in our lives is and very vital and indisputable, be it in our personal lives or in the professional industry. Being not only rational but also emotional beings, we experience thousands of diversified emotions throughout the entire day, which are being generated both based on internal and external factors. This may suggest that mostly all of our actions may also be subconsciously guided by our internal states, feelings, beliefs, desires, intentions, and expectations, as we have given stated in our earlier discussions.

This brings to the fact that being social beings while interacting with one another we need those endorphins to feel appreciated, valued, cherished, heard and understood. It follows that with the introduction and unprecedented spread of AI in our business lives, there has to be a kind of a common ground between humans and machines for them to operate successfully, efficiently and harmoniously. For this very reason, a new concept has been introduced to science that makes human–machine cooperation more empathetic, namely the concept of **Emotion AI**. According to recent research, Emotion AI can greatly increase customer and employee satisfaction in terms of providing the users with an empathetic service matching their needs, requirements, and expectations (Marr, 2022). This eventually gives the ultimate chance of heightening the level of credibility between the parties involved. Therefore, the main hypothesis of the present chapter is that in case we programme AI technologies in a way that they too can apply emotional labour in human–machine interactions, we will resultantly have far more enjoyable, efficient and profitable human–human-machine interactions. Here, the concept of Emotion AI jumps in and resolves the emotional gap between humans and machines.

Artificial Intelligence is recognized as the ability of hardware and software to exhibit and execute actions and tasks typically conducted by humans. It has a long history dating back to the earliest successful AI program written, programmed and implemented in 1951 by Christopher Strachey, who later became the director of the Programming Research Group at the University of Oxford and who has made many unattainable achievements and advancement in science possible that were considered impossible previously in the past. Christopher Strachey and his predecessors have truly proven that what human mind can conceive, the brain can achieve through dedication and manifestation. This reminds us of the fables on flying carpets that were foretold in fairy tales with the later introduction of flying objects and machinery.

In fact, people have always tried to build an intelligent creature/machine that would have the same cognitive abilities of humans and would replace us in some areas where human capital is scarce. In this line, for instance, Finlay and Dix (1996) in their writings describe that even the earliest writers of mankind imagined "artificial" beings created from stone and other objects, subtly suggesting that humanity can also create living beings different from human nature and created artificially: for instance, the Roman poet Ovid wrote a story about the sculptor Pygmalion, who brought to life a very beautiful woman out from a meticulously sculpted stone (the play by Bernard Shaw "My Fair Lady" and the musical bearing the same title by Alan Jay Lerner are the more recent renditions of this fantastic fable). Much later, in the age of industrial machines, Mary Shelley manufactured Doctor Frankenstein, who was a man composed of very different separate components, who was brought to life through electricity. Other examples of science fiction films continuing the same line of thought are for examples "Terminator", which presented a vision of cyborg machines almost indistinguishable from humans, who has the same and even more enhanced capabilities typical of humans. The same phenomenon can be observed in the film series entitled that "Transformers", directed by Michael Bay, where the main characters have the ability to transform into cyborgs, as well as perceive, read, understand, react and even respond to human emotions in an emotional manner through their meticulous AI programming.

As it can be seen and deducted from our science fiction (sci-fi) movies, machines are on their way in understanding and responding to human emotions. Though humans have long been seen as masters of emotions, having a mastery over machines in line of understanding human emotions and responding to the accordingly, since the former AI technologies were redundant, with monotonous voice, nonetheless, this will not be the case any longer and machines will presumably even overpass humans in this line as well, since we humans do not make sense of our internal states, where machines can provide more meticulous and adjust information with the help of their algorithms. Though there is speculation in this respect regarding the problem from the cultural point of view, this will make human–machine communication more enjoyable and far more fruitful since we, being emotional creatures, will enjoy emotional interactions with AI technologies, and the output will be resultantly more fruitful and efficient. All this can be achieved through the application and integration of the so-called *Emotion AI*, which is otherwise called *Affective Computing*, which makes machines more emotionally attuned with humans. Though we have to state that some of the human popularities are scared of this advancement in science since emotions are private phenomena that should be kept to us humans.

Though the name itself might sound somewhat misleading, Emotion AI does neither refer to a weeping or angry computer who has a bad day at work nor has private life issues which makes it weep or cry out loud. Instead, Emotion AI, dating back to early 1995 and refers to the branch of Artificial intelligence which aims to programme machine in such a way to be able to perceive, process, understand, respond and even replicate human emotions (Alkhadi, 2022). While some of us are more emotionally more sensual in perceiving human emotions than others of our species, we humans can generally easily understand and respond to the emotions and feelings of those in our surroundings, since we all possess the same emotion regulation, perception and recognition systems instilled in us by evolution that are designed to guide us throughout our entire lives (Rostomyan, 2020, 2022a). To have a better understanding of the concept of Emotion AI, we have to first have a look at the notion of EI (EQ), namely Emotional Intelligence, which is our ability to perceive, understand, manage and use our emotions depending on the outward situation at hand (Goleman, 1995; Rostomyan, 2022b; Salovey & Mayer, 1990, 1993). It should be noted here that it goes without saying that our base level of intelligence, namely our IQ, comes with us to this world since our birth, but we have to state that we are partly born with it and it can partly be enhanced through education and training. Though EQ (EI) is believed to be a static quotient, the research of Dr. Anna Rostomyan has proven that it can also be enhanced through coaching and training (Rostomyan, 2022b).

In this connection it is also worthy mentioning that the abbreviation of "EQ" designation the concept of "emotional intelligence" is nowadays being more referred to as "EI", which brings it closer to AI and subtly suggests that machines and humans are now closer than ever before. Moreover, science shows that AI technologies are on their way in attaining emotional intelligence themselves in the near future, though it is somewhat debatable to what extent an AI robot can "feel" emotions and feelings itself. This said, let us have a closer look at what emotion AI itself represents and what it entails: Emotion AI can typically be of the two following types:

(1) formerly, with the programming of emotion AI, machines gained the possibility to perceive, process, "digest" and "understand" human emotions,
(2) nowadays, with the advancement of science and the more sophisticated reinvention and further elaboration of the concept of emotion AI, machines have gained the ultimate possibility of not only perceiving, processing and "understanding" human emotions, but also responding to them accordingly in an empathetic manner, sometimes replicating the expressed emotions of the humans (see more in Rostomyan, 2024a).

This way or another, Artificial Intelligence is thus seen as the bane of human existence and something provisionally that will take over the entire world soon, whether it is good or bad we will see in the future, but in case machines are endowed with emotion detection and replication algorithms, it will be easier cooperating with them. Therefore, it is intriguing and fascinating to look at this aspect of emotion AI that is yet to be uncovered, namely in terms of emotions and their detection, as well as reciprocate replication and application possibilities (Rostomyan, 2024a). This vast progress brings humans and machines closer to each other more than ever before: robots no longer sound monotonous with their emotional voice fluctuations and chatbots no longer seem that automated but rather personalized in terms of responding to human users in an empathetic manner. Moreover, some advanced robots like Sophia the Robot, a social humanoid robot developed by the Hong Kong-based company Hanson Robotics, and Robin the Robot, the best Armenian invention of (2021) according to the "Time" magazine, even express emotions through their voices, facial expressions, and gestures. Though through Emotion AI robots can sound more human-like and can greatly promote the enjoyable and trustworthy human–machine cooperation, we cannot definitely claim if they themselves possess emotional intelligence (EQ) even if emotional intelligence, formerly referred to as EQ and vastly popularized by Daniel Goleman (1995) in his like-named book "Emotional Intelligence", is currently more referred to as EI, which hints to the fact that humans and machines are by now closer than ever before. Nonetheless, we cannot firmly state that machines are emotionally intelligent since the concept of emotional intelligence presupposes the experience of emotions on the inner side, which in case of the machines is mostly not possible though they can cognitively make assumptions on how they should feel in this moment or another. Yet, it is also noteworthy that this will no longer be the case in the nearest future and that they will advance in that line as well.

When speaking about sustainability and net-zero policies we know that countries are inclined towards achieving less carbon and gas consumption and producing more recycable and eco-friendly materials and products and that they try to use healthy materials wherever possible, like the case of Mercedes Benz replacing the genuine leather of the interior of the vehicles with the skin of mushrooms for stipulating to the progress of a sustainable automotive industry, yet few speak about the preservation of human energy which is the only sustainable force of human workforce. This hints to the fact that in case we achieve a healthy human–machine co-existence, we can reach a level where at some spaces and in some cases the human energy will be reserved and applied for some other fields like for instance creativity which are yet somewhat unattainable by machines and other AI technologies.

In this connection, it is worth mentioning that being emotional and social beings sometimes we feel drained in case of workload overload or emotional drainage, so in case machines are equipped with the programming of cooperating with us emotionally justly and adequately deploying empathetic emotional demeanour, we will feel more at ease in cooperating with the latters. For this very reason, Microsoft even has a whole team called HUE (Human Understanding and Empathy) that is working on generating more empathetic AI technologies and chatbots. So, let us have a closer look at what empathy is and why it is so vital for our survival:

Empathy is, in fact, the ability to understand and share human emotions and it can be of 3 types according to Goleman (1995), namely (a) *cognitive*, (b) *emotional*, and (c) *compassionate*. Let us have a closer look at all of them separately:

(a) on the ***cognitive level***, we simply rationally cogitize and understand the emotions of the others,
(b) on the ***emotional level***, we emotionally resonate to the emotions of the others and start feeling those very emotions of theirs on the inside,
(c) yet on the ***compassionate level***, we (and now already also the machines) not only emotionally and rationally attune in with the emotions of the others, but also go further in taking decisive steps towards finding a solution matching the needs of the customers/consumers/employees/partners/patients/companions/neighbours/colleagues (see more in Rostomyan, 2023b). As we can see from the explanation above, empathy is more than sympathy: in case of sympathy, you just like the person/machine you are interacting with, yet in case of empathy you share the emotions of the latter(s), which can also be a basis for sympathy, though it can be just vice versa, you can like somebody and therefore feel attuned towards sharing the emotions therein. Yet, it is noteworthy that in case of empathy, especially in case of the highest type of empathy, namely *compassionate empathy*, you not only emotionally share and rationally understand the emotions of the others, but also take a step forward in landing a helping hand for those in need. This last case is especially observed in the AI machines, which are here to make our lives far easier and more enjoyable, trustworthy, and harmonious. In this respect, we can truly detect the advancement of science when back in the 90 we being teenagers used to take care of back then popular Tamagotchis, digital pets created in Japan, which we humans had to look after by means of digitally feeding and giving water to them not to "starve", whereas nowadays we have robot caretakers at hospitals like Robin the Robot, who takes care of mostly children patients who are anxious and scared by means of playfully explaining to them their situation and playing different virtual games with the latters not to feel lonely. Besides, empathetic AI technologies are currently being largely applied in the field of psychology for providing metal health assistance to patients showing care and support to them which lightens the psychological burden of the patients, soothes their psyche and proposes aid to the medical staff as well (Infographic 7.2).

It follows that AI technologies are no longer to be described as machines having mental abilities of humans that are able to apt in operating in human activities, but also creations that partly share our emotional abilities, which by means of possessing Emotion AI options are better human–machine interactants that bring deeply rational insights into the table also having emotional colouring in interactions, which is essential for our survival, since we are not only rational, but also emotional beings.

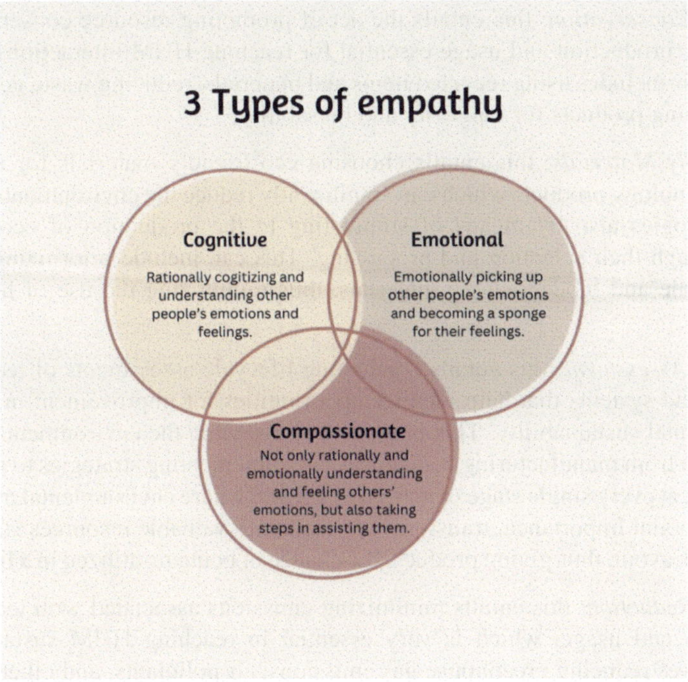

Infographic 7.2 Venn diagram of Goleman's empathy types. *Source* and explanation: Anna Rostomyan (2024a, 2024b, 2023b)

7.2 Management Techniques for Fostering Sustainable Human–Human-Machine (HHM) Interactions

Human–human-machine (HHM) interaction sustainability goals mainly focus on ensuring that interactions between humans and machines are environmentally proper and sustainable. These very provisioned goals encompass various aspects of technology design, usage, and management to minimize environmental impact and promote sustainability.

Though the advancement of science in line with new AI technologies can seem very attractive and tempting, there are some hazards with them coming as well, which should be by all means diminished by means of following some principles. Below are enlisted some of the key HHM interactions sustainability goals:

Energy Efficiency: this entails the process and goal of designing and implementing technologies that are energy-efficient and that those consume less energy which consequently promotes sustainability. This involves optimizing the energy consumption of devices, systems, and networks to reduce their environmental impact.

Resource Conservation: this entails the act of promoting resource conservation in technology production and usage essential for reaching HHM interactions sustainability. This includes using recycled items and materials, reducing waste generation, and designing products for longevity and reusability.

Eco-friendly Materials: this entails choosing eco-friendly materials for manufacturing technology products which can significantly reduce the environmental impact of technologies also by means of stimulating to the production of eco-friendly items through their selection and processing. This can include prioritizing the use of renewable and biodegradable materials, thus minimizing the use of hazardous substances.

Life-cycle Assessment: this entails conducting lifecycle assessments of technology products and systems that help identify opportunities for improvement in terms of environmental sustainability. This may involve assessing the environmental impact of products from manufacturing to disposal and implementing strategies to minimize this impact at every single stage of action. In a world where environmental awareness is of paramount importance, transforming waste into valuable resources stands as a crucial endeavour, thus giving products the chance of being re-utilized in a life-cycle.

Emission Reduction: this entails minimizing emissions associated with technology production and usage, which is very essential in reaching HHM sustainability. This involves reducing greenhouse gas emissions, air pollutants, and other harmful emissions throughout the lifecycle of technology products and systems.

Sustainable Design Practices: this entails incorporating sustainable design principles into the development of technology products and systems which is truly key in reaching HHM sustainability. This includes considering factors such as energy efficiency, recycability, and environmental impact during the design and production process.

Renewable Energy Integration: this entails promoting the use of renewable energy sources in powering technology infrastructure and devices to contribute to HHM interactions sustainability. This involves integrating renewable energy sources such as solar, wind, and hydroelectric power into the design, production, and operation of AI technology systems.

User Awareness and Education: this entails increasing user awareness and education about the environmental impact of technology usage that is very important for reaching HHM interactions sustainability. In fact, empowering users with knowledge to make environmentally conscious choices and adopt sustainable practices can greatly help minimize the overall environmental impact on human–human-machine interactions.

It follows from the above that by means of realizing, setting and striving towards achieving these HHM interactions sustainability goals, stakeholders can work towards creating a more environmentally sustainable technological ecosystem that will eventually be beneficial for both present and future generations.

Here, when we speak about present and future generation in human resources management, we should highlight that at the workplace the most important milestone towards reaching HHM sustainability is human capital, which also involves emotional capital, as we have discussed in our previous sub-chapters. This means that in case we wish to foster HHM cooperation to reach net-zero policies, human capital should be first and foremost fostered, since though AI technologies sometimes overpass humans in some areas, we are the ones who design, create, programme, generate, and produce the latters.

Since we humans are not only rational, but also emotional and social beings, fostering human capital for sustainable human–human-machine (HHM) cooperation in net-zero policies can be achieved through proper emotional labour on both parties, humans and machines, which becomes possible through the introduction of Emotion AI (namely *artificial emotional intelligence*). Fostering human capital through emotional labour is essential for sustainable human–human-machine cooperation in implementing net-zero policies. Emotional labor involves managing emotions to meet job requirements, which is crucial in navigating the complexities of collaborative efforts involving humans and machines (Rostomyan, 2024b). Thus, it goes without saying that in the context of net-zero policies, emotional labour indeed plays a very significant role in several ways some of which are briefly mentioned below:

Building Trust and Fostering Collaboration: Effective collaboration between humans and machines requires trust and mutual understanding, as is between humans and humans. Emotional labour helps individuals regulate their emotions, recognize them in others and tune in with the emotions and needs of others, as well as communicate effectively, fostering trust and collaboration among team members, colleagues and employees, including both humans and machines.

Managing Change and Sprouting Resistance: Implementing net-zero policies often involves significant changes that can evoke resistance, frustration, fear and/or anxiety among the involved stakeholders. Emotional labour here enables individuals to address these concerns empathetically, easing the transition and encouraging everyone involved that everyone's rights and needs will be respected. Besides, in case AI tools interact with us empathetically, resistance will eventually be diminished as well.

Enhancing Communication: It goes without saying that clear, authentic and empathetic communication is essential for any kind of successful cooperation. Emotional labour here helps individuals convey information in a way that resonates with others and matches their internal needs, fostering understanding and alignment towards common goals. This also entails effective communication between humans and machines, since in case machines communicate with us emotionally and empathetically, there will be created a kind of a bond between us, which will ensure trustworthy interactions.

Supporting Well-being and Resilience: Working towards ambitious goals like achieving net-zero emissions can be emotionally loaded. Emotional labour skills, such as self-care, self-awareness, self-management, social awareness, relationship management and empathy, have the greatest potential of supporting and stimulating individuals' well-being, mental health and resilience, helping them cope up with stress, frustration and anxiety and operate effectively, proficiently, and successfully.

Promoting Inclusivity and Diversity: Emotional labour encourages inclusivity and respect for diversity by means of fostering an environment where individuals feel valued, cherished, heard and understood regardless of their cultural, educational and/ or religious background or perspectives on life. This also entails producing machines that are unbiased, in the operations of which inclusivity is in the core, which is very essential for harnessing the full potential of harmonious and efficient HHM harmonious and efficient cooperation.

Actually, we humans like machines have "batteries" which involve our extraordinarily complex and multifaceted internal states that should be taken good care of. So, this suggests that in case we are always overloaded with work and do not have time to recharge, our batteries will be low, which will consequently negatively influence both our relationships and our overall labour output. This may suggest that in case we are replaced by smart AI technologies in some areas of human activity, where human capital is scarce, this will ensure actual preservation of human energy, and we will have a chance to recharge and to focus on the most important areas of our activity that will increase productability and will ensure more sustainable futures.

In summary, emotional labour is a critical enabler of sustainable human–machine cooperation in advancing in reaching net-zero policies. By cultivating a high level of emotional intelligence, as well as implementing artificial emotional intelligence, which includes the participation of empathy in human-human-machine interactions, individuals can navigate the complexities of collaborative efforts effectively, ensuring that human capital is leveraged optimally towards achieving a sustainable future for us all. Besides, the discussed goals and milestones will largely promote the awareness of all parties involved in fostering efficient, successful and harmonious HHM interactions, which are the basis of our modern societies prosperity.

7.2.1 Data Analysis on Emotion AI Through Qualitative and Quantitative Methodology and Regulations

To have a better understanding on the perceptions of humans on emotionally intelligently programmed machines, we have conducted a quantitative survey on where we humans prefer interactions and cooperation with emotionally "intelligent" machines in Germany. Both quantitative and qualitative methodology has been applied (Infographic 7.3).

As it can be concluded from the chart above, the majority of the interviewed are very much inclined towards and looking forward to interacting with emotionally agile AI technologies. As for the results of my personally conducted qualitative analysis, the opinions split into two poles:

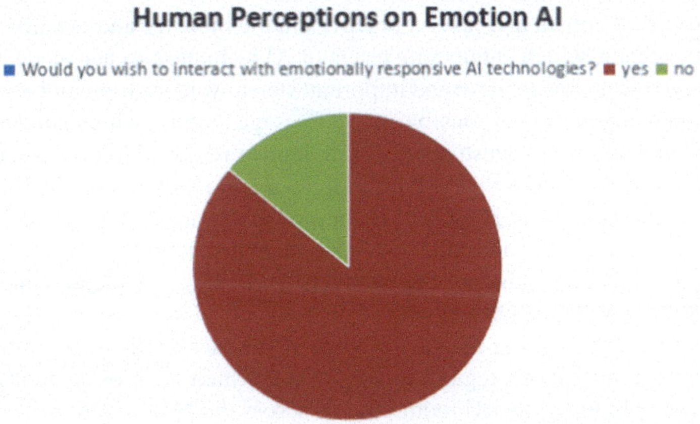

Infographic 7.3 Human perceptions of emotion AI applications. *Source* Rostomyan Research (2023)

(a) some of the interviewed people were fascinated with the vast possibilities of Emotion AI and would wish to interact with emotionally "intelligent" machines.
(b) whereas others were sceptical and cautious in interacting with robots that are able to have far more insights into the emotions of humans which are considered to be of sensitive nature.
(c) some others were excited about this development of science since while we sometimes might get lost in the labyrinth of our emotions, where Emotion AI can help us understand ourselves better.
(d) whereas a few were even scared of the prompt development of AI and would not wish them detect their emotions, considering emotions to be private phenomena.

Surely, it goes without saying that the field of AI (including its progress in line with Emotion AI) presupposes vast possibilities but it also brings challenges with it for the regulation of which the EU has elaborated and accepted the EU AI Act which is aimed at addressing sensitive issues and regulating the field of AI which has to be respected and followed by all institutions of the European Union countries' companies and institutions using AI technologies in their professional activities.

7.2.2 *The New EU AI Act*

Since the field of Artificial Intelligence is very vast and sometimes seemingly uncontrollable, the European Union has elaborated and accepted an act called the "EU AI Act" that is meant to introduce comprehensive regulations into the field of AI usage by organizations and to regulate the use of AI by businesses and institutions.

Actually, the EU AI Act introduces a landmark regulatory framework for the use of artificial intelligence impacting the entire European Union, which is mainly aimed at fostering ethical and just AI practices in the industry and towards ensuring safe public policies, which entails safe usage of AI technologies across the EU. This legislation is truly a very decisive and important step towards setting global standards for AI use focusing on the responsible and safe usage therein, which emphasizes the importance of this great milestones towards innovative businesses, which though may entail many challenges that will be addressed and regulated by the EU AI Act.

In this connection, it is notable that they even expanded the AI definition of technologies able to exhibit and conduct human mental activities to account for virtual environment behaviour including the metaverse, which in some cases presupposes unethical behaviour by the users to ensure trustworthy AI interactions. Thus, the EU AI Act is aimed at providing the AI developers and users with clear requirements, regulations, and obligations regarding uses of AI, which have to be applied for an ethical AI use following the EU requirements across the EU countries.

According to the European Commission (2024), the proposed EU AI ACT rules will include the following:

- address risks specifically created by AI applications;
- prohibit AI practices that pose unacceptable risks;
- determine a list of high-risk applications;
- regulate any AI malpractices in the EU;
- set clear requirements for AI systems for high-risk applications;
- define specific obligations deployers and providers of high-risk AI applications;
- require a conformity assessment before a given AI system is put into service;
- put enforcement in place after a given AI system is placed into the market;
- establish a governance structure at the European and national level.

All of the above will eventually ensure a smooth, trustworthy, safe and efficient AI practices that will be meant for assisting the humanity in conducting their activities in a timely and proficient manner.

7.3 Originality

The present paper not only regarded the advantages of Emotion AI as a positive advancement in AI technologies for smoother human–machine interactions, but also discussed the role of emotional labour in both humans and machines, the presence, programming and execution of which can be a groundbreaking milestone towards building and reaching a strong human–human-machine relationships, which will consequently boost and foster their positive and fruitful cooperation.

7.4 Discussion

Emotional labour has been considered as one of the most important keystones towards building harmonious and successful teams in any organization. The provided theoretical background and the empiric data have shown that in case we deploy emotional labour both at the workplace and in our personal lives, we are sure to succeed in building strong and lasting interpersonal relations. Therefore, in case we program computers, robots and other AI tools with the concept of Emotion AI that has the ultimate potential of providing computers with insights into human emotions and feelings, as well as endowing them with the chance to respond to the latters in an emotional manner, there will be created a kind of an emotional bond between humans and machines, which will make their interactions more enjoyable, resulting in an efficient, smooth, harmonious and fruitful cooperation.

7.5 Implications

Machines have the utmost potential of replacing humans in some areas of human activity where human workforce is scarce. For instance, in the medical sphere robots can replace medical staff, which can be very beneficial especially in times of pandemics. Besides, marketers can gain insights into the needs, demands, desires and expectations of their consumers. Moreover, HR agents and managers can receive the ultimate opportunity of gaining insights into the needs, requirements and motivations of their employees, which can give them the chance to redefine their strategy and stimulate their success according to their needs, which can possibly result in the flourishing of the whole company as one. Though how fascinating this can be, it also presents dangers as we can see in science fiction movies, where humanity is taken over by robots and AI machines, which sometimes can become over-cruel and would wish to destroy humanity. For example, robots have been interviewed on their perceptions of humans: some of them responded that humans are their brethren and they should act in accordance of their brotherhood and stimulate to the success of the amelioration of the quality of life for all, whereas some others responded that humans abuse machines and should be eliminated. This hints to that fact that we have to be very careful in what we put into the machines and careful in our interactions with them in order not to turn them into our foes and only benefit from their endless possibilities, as well as strive towards building harmonious and peaceful human–human-machine cooperations.

7.6 Conclusion

In summary, the present chapter discussed and elucidated the essential role of AI technologies in human lives, where humans and machines are closer than ever before. The Chapter first brought up the topic of human capital, which also involves the so-called "emotional capital", which is the only sustainable force of human workforce that can keep them going and feel secure and energized in their day-to-day activities. It also touched upon the relevance and application of some pre-learnt display rules of emotions that accompany our interactions (both in our private lives and in business) that are a part-and-parcel of our verbal and non-verbal manifestations of positive/negative emotions in our emotional repertoire. In the Chapter 7 the essence of emotional labour and its significance in human–human relations, both at the workplace and privately, has been thoroughly explained. Moreover, the author widened the scope of emotional labour and suggested that it should also be an indispensable part in human–human-machine (HHM) interaction(s) as well, since we humans are not only rational beings, but also emotional ones, which entails that emotionally adequate cooperation is essential for us humans to "function" properly. In this respect, the eminent role of Emotion AI is of utmost importance since it ensures the smooth, harmonious and enjoyable cooperation between humans and machines, where AI technologies not only detect human emotions, but also respond to them accordingly which ensures a safe space for both parties to indulge into interaction. The Chapter 7 also addressed the challenges of reaching net-zero policies in HHM interactions in terms of sustainability and effectiveness. Several principles and key points have been discussed that have the utmost potential in achieving a harmonious and sustainable mutual HHM future, where the emotional aspect should lie in the core of interactions for fruitful and sustainable cooperations. Since the field of AI is indeed very vast and challenging where sometimes ethical issues may be overlooked, the European Union has proposed and accepted the EU AI Act, which is designed for regulating AI uses. This advancement is very crucial since it will diminish unethical AI practices and will presumably create a safe space for users to apply AI technologies for the benefit of mankind. All of the aforementioned miles are very essential and vital in building a sustainable human–human-human-machine (HHM) co-existence, where everyone will feel appreciated and valued and where the AI technologies will serve as trustworthy and useful companions of the humanity, which will undoubtedly lead to more sustainable environmental, industrial, and technological successful advancements.

References

Alkhaldi, N. (2022). *Emotional AI: Are algorithms smart enough to decipher human emotions? IoT for All*. Accessed on December 23, 2022. Available at: https://www.iotforall.com/emotional-ai-are-algorithms-smart-enough-to-decipher-human-emotions

Ashforth, B. E., & Humphrey, R. H. (1993). Emotional labor in service roles: The influence of identity. *Academy of Management Review, 18*, 88–115. https://doi.org/10.5465/amr.1993.3997508

Carnegie, D. (1982). *How to win friends and influence people*. Pocket Books.

Ekman, P. (1979). About brows: Emotional and conversational signals. In M. von Cranach, K. Foppa, W. Lepenies, & D. Ploog (Eds.), *Human ethology: Claims and limits of a new discipline: Contributions to the colloquium* (pp. 169–248). Cambridge University Press.

Ekman, P. (2004). *Emotions revealed: Recognizing faces and feelings to improve communication*. Henry Holt and Company.

Ekman, P., & Friesen, W. (1975). *Unmasking the face: A guide to recognizing emotions from facial clues*. Prentice-Hall.

European Commission. (2024). *Shaping Europe's digital future: The AI act*. Accessed on: May 02, 2024. Available at: https://digital-strategy.ec.europa.eu/en/policies/regulatory-framework-ai

Finlay, J., & Dix, A. (1996). *An introduction to artificial intelligence*. Taylor and Francis Group.

Goleman, D. (1995). *Emotional Intelligence: Why it can sometimes matter more than IQ*. Bantam Books.

Hochschild, A. (1979). Emotion work, feeling rules, and social structure. *The American Journal of Sociology, 85*(3), 551–575.

Hochschild, A. (1983). *The managed heart*. University of California Press.

Hofstede, G. (1997). *Cultures and organizations. Software of the mind*. McGraw-Hill.

James, W. (1983). *The principles of psychology, with introduction by George A. Miller* (combined ed.). Harvard University Press.

Marr, B. (2022). *What is artificial emotional intelligence*?. Accessed on: December 15, 2022. Available at: https://bernardmarr.com/what-is-artificial-emotional-intelligence/

Oxford University Press. (2005). *Oxford english dictionary* (3rd ed.). Oxford University Press.

Rostomyan, A. (2012). The vitality of emotional background memory at court. *Polemos, 6*(2), De Gruyter, 281–292.

Rostomyan, A. (2020). *Business communication management: The key to emotional intelligence*. Tredition Publishing.

Rostomyan, A. (2022a). *The ultimate force of emotions in communication: A linguo-cognitive analysis of verbal and non-verbal expressions of emotions (on the material of English)*. Shaker Verlag.

Rostomyan, A. (2022b). *Emotional intelligence as a keystone towards success*. Generis Publishing.

Rostomyan, A. M. (2023a). The application of emotions in business communication management for a good reputation: Tiny little triggers with great impact on moods behaviours and communication interactions. In U. Bihler (Ed.), *Wirkungsmechanismen im Reputationsmanagement: Die Bedeutung neurowissenschaftlicher und psychologischer Erkenntnisse für den Aufbau des guten Rufs* (pp. 83–94). Wiesbaden: Springer Fachmedien.

Rostomyan, A. (2023b). Empathy: A cornerstone in the success of your company. The European Business Review (EBR). London, March 31, 2023. Available at: https://www.europeanbusinessreview.com/empathy-a-cornerstone-in-the-success-of-your-company/

Rostomyan, A. (2024a). Insights into emotion detection with EI tools and its applications through artificial intelligence (AI) in human-machine interactions. In *Proceedings of the 1st BSBI International Conference on Artificial Intelligence (AI), (Special Issue) of the Scientific Journal of Human and Machine Learning* (Vol. 2(1)). Berlin School of Business and Innovation (BSBI), Berlin, Germany.

Rostomyan, A. (2024b). The importance of emotional labour at workplace, Berlin School of Business and Innovation (BSBI). *Journal of Business and Innovation, 3*(2).

Rostomyan, A., & Rostomyan, A. (2023). The importance of emotional capital in companies. *International Journal of Research in Human Resource Management, 5*(2), 53–57. https://doi.org/10.33545/26633213.2023.v5.i2a.150

Salovey, P., & Mayer, J. (1990). Emotional intelligence. *Imagination, Cognition and Personality, 9*(3), 185–211.

Salovey, P., & Mayer, J. (1993). The intelligence of emotional intelligence. *Intelligence, 17*(4), 433–442.

Time Magazine. (2021). *The best inventions of 2021: A compassionate companion*. Accessed on: December 20, 2022. Available at: https://time.com/collection/best-inventions-2021/6113107/robin-the-robot/

Chapter 8
Climate Change Adaptation Strategies for Achieving Net-Zero Economy

Kishore Kumar, Rubee Singh, Ankita Sharma, and Bikram Paul Singh Lehri

Abstract Today, net zero economy is garnering lot of interest as climate change concerns have become one of the most pressing issues for the organizations. The negative impact of climate change (CC) could be witnessed across all industries. The direct risk (i.e. impairment cost, damages, forced closure from extreme weather events) and indirect risk (i.e. disruption in the business value chain, loss of infrastructure, etc.) emanating from CC has severely impacted the business model of the companies. It is important for companies to address climate challenges in their core business model and take climate action for achieving net zero economy. The aim of this study is to explore the impact of various organizational factors on the climate change adaptation strategies (CCAS) of manufacturing companies in India. The data was collected from 241 respondents and structural equation modelling (SEM) through Smart PLS 3.0 was employed for analysis in the study. Results indicated that corporate knowledge, processes, objectives, financial resources, collective knowledge, and incentives significantly influence the CCAS for the companies. The findings provide valuable input to the managers, practitioners, and other stakeholders interested in promoting climate actions and achieving a net zero economy. This chapter contributes to the extant literature in the field of corporate CC strategies and actions.

K. Kumar
School of Management, Bennett University, Greater Noida, India
e-mail: akishore001@gmail.com

R. Singh
Institute of Business Management, GLA University, Mathura, Uttar Pradesh, India

Department of Management Studies, Kumaun University, Nainital, Uttarakhand, India

R. Singh
e-mail: Dr.rubeerajput@gmail.com

A. Sharma (✉)
Christ University, Bengaluru, India
e-mail: dr.ankitasharma912@gmail.com

B. P. S. Lehri
University School of Business, Chandigarh University, Mohali, India
e-mail: lehribikram@gmail.com

R. Singh and D. Crowther (eds.), *Transition Towards a Sustainable Future*, Approaches to Global Sustainability, Markets, and Governance,
https://doi.org/10.1007/978-981-97-5756-5_8

Keywords Carbon emissions · Carbon neutrality · CC adaptations · Collective knowledge · Developing economies · Environment management system · Global warming · Net zero · Sustainable development

8.1 Introduction

Today, CC has emerged as one of the biggest challenges for this planet Earth, and humanity. Environmental emergencies and Humanitarian emergencies like severe draughts, heatwaves, storms, floods, pollution, hunger, and poverty have become some of the most pressing issues of our time caused due to CC (Berrang-Ford et al., 2011; Kumar, 2022; Steg, 2023). The adverse impact of CC is visible in all spheres of human lives. It has drastically changed the way humans live and work. CC refers to extreme weather events and long-term changes in the temperature level (Ayers & Dodman, 2010; Biesbroek et al., 2013). It is caused due to the exploitation of natural resources and greenhouse gas emissions (GHG). Excessive dependence on non-renewable energy sources has increased carbon footprints manifold across the region. Energy sources like petrol, diesel, gas, and coal are by far the biggest reasons for the global CC problem (Aggarwal & Dow, 2012; Zhou, 2023). It accounts for more than 78% of the GHG emissions and more than 90% of overall carbon emissions in the world (UN, 2024). Rapid reduction in carbon emissions and transition to net zero could potentially help in dealing with CC challenges. The crisis relating to CC has resulted in many negative impacts on global business. The adverse effects of CC have already become a reality for businesses across all industries (Arndt, 2023; Sun et al., 2023). Thus, the development and implementation of climate actions to mitigate CC problems has become imperative for economies around the world. My countries have introduced regulatory frameworks or mechanisms to curb the carbon footprints and address CC issues (Reckien et al., 2014; Sharma et al., 2024; Sundaram & Hansen, 2023). Similarly, business organization are also under immense pressure from various stakeholders to adopt CC strategies to reduce carbon emissions and have a positive impact on the environment (Díaz Tautiva et al., 2024; Kumar et al., 2022; Nwokolo et al., 2023). The existence and growth of the business organization largely depend upon the positive environmental and social contribution of the business. Therefore, companies are increasingly recognizing the need to invest in capabilities aimed at developing environmentally friendly technologies to curb carbon footprints (Orazalin et al., 2024; Sharma & Kumar, 2021).

Business organizations are addressing CC challenges through CC mitigation and adaptation strategies. CC mitigation strategies are widely used by the companies to limit carbon and GHG emissions (de Abreu et al., 2021; Rivera et al., 2022). It refers to the actions and practices adopted by the companies to reduce emission levels due to their business operations. CC mitigation is achieved by limiting the sources of carbon emissions through the environment management system or shifting to renewable sources of energy (Biswas & Rahman, 2023; Kouloukoui et al., 2020; Singh et al., 2024). Climate adaptation strategies are aimed at forecasting the challenges due to

CC and taking corrective actions to deal with the adverse impact of CC or tap the potential opportunities available due to CC issues (Braunschweiger & Ingold, 2023). It is a process of adjusting business processes according to the present and future impact of CC. Fawzy et al. (2020) concluded that although organizations tend to adopt CC mitigation strategies, CCAS are employed by a limited number of organizations. However, the scope of CC mitigation is limited as it is only focused on reducing carbon emission sources whereas CC adaptation is a proactive approach (Nyberg & Wright, 2024). Thus, CCAS could play an instrumental role in dealing with the negative impact of CC. It can be taken by individual level, group, organization, national, or international level (Daka, 2023; Gemeda et al., 2023). Sharifi (2020) emphasized the need for more empirical studies on CC adaptations by business organizations.

There have been limited studies based on CCAS by the business organization (Abbass et al., 2022; Herrmann & Guenther, 2017; Owen, 2020). Nalau and Verrall (2021) stated that CC adaptation is particularly important the companies in developing economies due to a specific set of challenges and business environments in these countries. This study aims to examine the factors influence CCAS of companies, especially in the context of developing countries like India. Companies in developing countries face unique challenge in terms of increasing various kinds of risk relating to supply chain management, production operation, and environmental and social risks. To address these risks and create value, companies should take into consideration CCAS. Manufacturing sector companies remain the major contributors to GHG emissions and are more prone to the challenges posed by the CC crisis. It has become imperative for companies from the manufacturing sector to take action to curb carbon emissions and take measures to adapt to the CC impact (Dwivedi et al., 2023). Understanding of organizational factors that influence CC adaptations in the business is crucial for improving sustainability in the manufacturing sector. Thus, this study aims to answer the following research question; what factors influence the CCAS of manufacturing companies? To this end, the objective of this study is to investigate the various factors that impact the CCAS of companies. This study has a number of practical and theoretical implications. The extant literature on CC adaptations is still in the evolving stage. This study significantly contributes to the extant literature and identifies the factors influencing the adaptation of CC strategies of business in the context of developing countries. The present study also presents some practical implications for the practitioners, regulators, and managers to propagate CCAS. This study provides significant insights to the managers about the key factors enabling CCAS in companies. The finding should also be useful for the regulators and government for capacity building and promote climate actions among the industries.

This chapter is organized as follows: Sect. 8.2 discusses the literature review, followed by the methodology used in this chapter in Sect. 8.3. The next Sect. 8.4 presents results and analysis. Section 8.5 includes a concluding remark, limitations, and theoretical and practical implications of the study.

8.2 Literature Review

There is an emerging field of literature on CC strategies and practices by the organization. Primarily, there are two streams of literature available on CC in the context of business organization. The first stream is related to the impact of CC on business operations. A number of studies have been undertaken based on the impact of CC and CC practices of the companies. Many studies have been conducted on the impact of CC in almost all industries (Eccles & Krzus, 2018; Johnson et al., 2023; Sun et al., 2020). Kolk and Pinkse (2005) concluded that the impact of CC will occur across the sector and there is a greater need for businesses to take climate actions. The other stream of research deals with the obstacles and factors that influence CC adaptations in the business organization. Obstacle refers to the barriers that hinder the adoption of CC strategies (Sebos et al., 2023). It represents the limits that restrict adaptive measures with the existing organizational structure (Glienke & Guenther, 2016; Mishra et al., 2023). Basson et al. (2020) concluded that barriers could be resolved through transformation in terms of organizational resources and capabilities. These barriers could be in the form of social barriers, physical resource barriers, ecological barriers, technological barriers, and economic constraints (Sparkman et al., 2021; Stankuniene et al., 2020).

Adaptive actions are extremely important to avert the negative impact of CC. The definition of adaptation is not uniform and varies according to the research field and sectors (Hossain & As-Saber, 2024; Kumar et al., 2020). Climate adaptation as the adjustment of organizational resources and capabilities against the actual or expected climate risk that may have an adverse impact on the business (IPCC, 2007; Perlin et al., 2021). Tangney (2020) described climate adaptation as the changes in the ecological socio-economic system of the organization to cope with the risk due to CC-related issues. Averchenkova et al. (2015) explained as the actions to manage the hazards and losses in response to the changing climatic conditions. These adaptive actions could be undertaken at the individual level, organizational, government level, regional, or international level (Flammer et al., 2021; Herrmann & Guenther, 2017). Plenty of studies have been undertaken of the factors that influence adoption of CCAS. Since CC adaptation is a complex process that involves different contextual factors (i.e. climate characteristics, its impact, and industry contribution, etc.), a wide range of factors need to be taken into consideration. Hoffman (2007) concluded that CC adaption factors significantly vary across the sector and region.

8.2.1 Organizational Knowledge

Corporate knowledge refers to the level of information and know-how about CC related issues and their impact on business operations (Bhushan & Singh, 2024; Jeswani et al., 2008). Turnbull (2020) outlined the role of information about CC risk and potential challenges to enable business organizations to prepare for possible hazards or losses. The availability of proper information enhances the understanding of the possible threats for the business emanating from the CC risk (Backman et al., 2017; Herrmann & Guenther, 2017; Prakash et al, 2023). Shahzad et al. (2021) highlighted that access to timely information also helps organizations to tap potential business opportunities available due to CC-related issues. A number of studies have reported the positive impact of knowledge and information on CCAS of the organization. Thus, the following hypothesis is developed:

H1: Knowledge and information significantly impact CCAS of the manufacturing companies in India.

8.2.2 Organization Financial Resources

Many studies have identified organization financial resources as an important influential factor of CC adaptation strategy in companies. The majority of studies have found a positive impact of the financial resources of the organization on CCAS (Hoffmann et al., 2009; Pawlowsky, 2001; Stechemesser et al., 2015). Access and availability of financial resources help companies to develop and implement measures to cope with the expected CC risk (Herrmann & Guenther, 2017; Kumar et al., 2021; Ruben et al., 2019; Timilsina, 2021). Averchenkova et al. (2016) highlighted that companies with strong financial resources are more likely to take climate actions to manage CC concerns. The financial resource enables companies to adopt measures for capacity building and climate resilience in the business organization against climate challenges (Alshahrani et al., 2023; Backman et al., 2017; Ooi et al., 2018). Thus, this study assumed that organizational financial resources positively influence the CCAS of companies. The following hypothesis is proposed:

H2: Organizational financial resources positively influence CCAS of manufacturing companies in India.

8.2.3 Organizational Objectives and Goals

Organizational objectives and goals are other important factors that influence the CCAS of companies. Herrmann and Guenther (2017) the intention for adaptation is driven by the kind of goals and objectives formulated by the organization to understand and deal with climate challenges. Many organizations make objectives that reflect pro-environmental and sustainable behaviour. This develops a sustainability culture in the organization and encourages CC adaptations in companies (Dunn, 2018; Tamosiunas, 2010; Weinhofer & Busch, 2013; Weinhofer & Hoffmann, 2010). Gouldson and Sullivan (2013) highlighted that the objectives that incorporate sustainability issues and CC risk improve organizational climate resilience and have a positive impact on the intention to take climate action. A number of studies have reported a significant impact of objectives and goals on the CCAS of the companies. Thus, this study assumed the following hypothesis.

H3: Corporate objective and goals have a significant impact on the CCAS of companies in India.

8.2.4 Organization Process

Organizational process refers to systems and mechanisms put in place in the companies. It facilitates the proper implementation of measures to address climate concerns (Guenther & Hoppe, 2014; Lee, 2012). Integration of the planning process in the organization helps the business to cope with future contingencies. Herrmann and Guenther (2017) emphasized that the development and implementation of proper systems and measures motivate the employees to deal with the climate crisis. The determination of measures and their types by the employee plays a crucial role in CCAS (Cho et al., 2019; Galbreath, 2010; Kolk & Levy, 2001). Not many studies have investigated the relationship between organization processes and CCAS. This study assumed that the organizational process positively influences the CCAS of the organization. Thus, the following hypothesis is proposed:

H4: Organization process positively influences CCAS of manufacturing companies in India

8.2.5 Collective Knowledge

Collective knowledge is a relatively less explored factor in the literature in the context of CCAS. Okereke et al (2009) highlighted that collective knowledge refers to awareness about climate risk in general. Wannewitz and Garschagen (2023) emphasized that if society is aware of the negative consequences of ecological problems and carbon footprints, climate adaptation measures are more likely to increase. It is important that policymakers and lawmakers provide proper knowledge about climate risk (Herrmann & Guenther, 2017; Smith & Mayer, 2018; Welch & Yates, 2018). Collective knowledge also provides potential opportunities that may emerge due to climate challenges. This study assumed that collective knowledge has a relationship with the intention to adapt to climate challenges. This study developed the following hypothesis:

H5: Collective knowledge positively influences CCAS of manufacturing companies in India.

8.2.6 Incentives and Culture

Many prior studies have explored the impact of corporate incentives and culture on the CC actions of the organization. Financial and other incentives provide a strong stimulus to take measures by considering the possible risk of CC to business operations (Dahlmann et al., 2017; Isensee et al., 2020; Nemet, 2010; Tryndina et al., 2022). Financial and nonfinancial incentive motivates to enhance CC adaptations in the organization. It also helps in creating a sustainability culture in the organization (Perdana & Tyers, 2020; Raux et al., 2021). Similarly, the sustainability culture in the organization encourages employees to adopt pro-environmental behaviours (Galpin et al., 2015; Hoffman, 2010; Linnenluecke et al., 2009; Russell & McIntosh, 2011). Thus, this study predicted that corporate incentives and culture positively impact climate adaptations. The hypothesis as follows:

H6: Incentives and culture positively influence CCAS of manufacturing companies in India (Fig. 8.1).

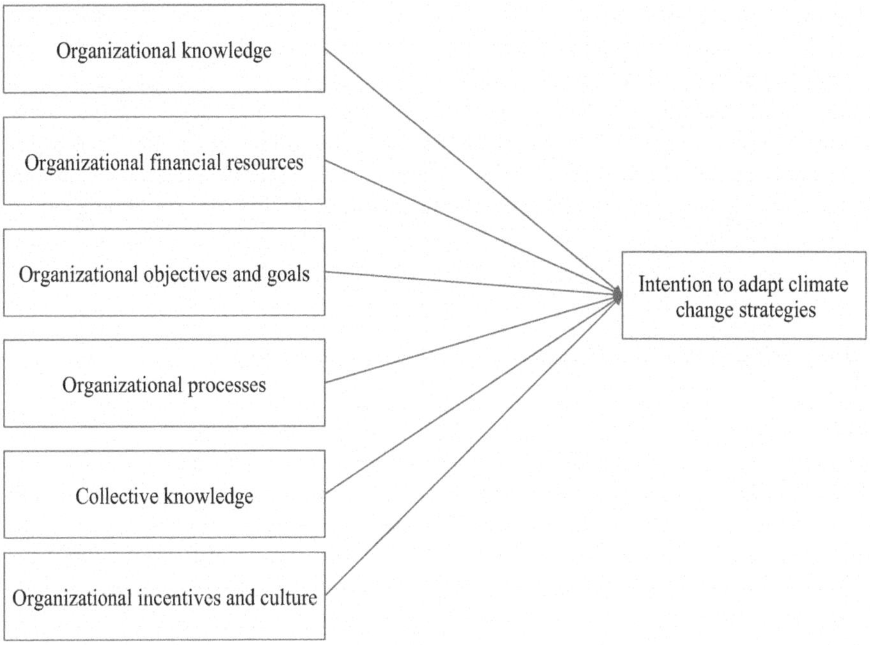

Fig. 8.1 Proposed research model

8.3 Research Methodology

8.3.1 Data and Sample

This research used quantitative technique for doing the research. The sample was selected using convenience sampling technique. The data was collected through structured questionnaires using Survey Monkey and WhatsApp. The target segment for analysis were top and middle level executives responsible for making policies for CC adaption strategies to be adopted by an organization. The purpose of the research was described to the sample and the meaning of CC along with the adaptation strategies was stated in the questionnaire. In order to get proper response from the respondents, screening question was put in a questionnaire to select right individual involved in policy formulation or decision-making regarding CC strategies.

In January 2024, the self-administered questionnaire was distributed to top and middle level employees working in manufacturing units via WhatsApp chats, Survey Monkey and Google Forms. Respondents in the research participated voluntary, additionally there was zero incentive for responses. The instrument was directed in English. The respondents showed that they were able to read, understand and write in English language. The top and middle-level executives were considered the most suitable for collecting information, as they are directly or indirectly involved in an organizations' CCAS. They are either involved in the policy formulation or a part of decision-making during CCAS. The respondents were informed about the nature of the study. Along with this they were informed that the results would be used for academic purposes.

8.3.2 Measures, Methodology and Analysis Plan

Further, Likert (5-point) scale ranging from "strongly disagree" to "strongly agree" was used for measuring the perceptions towards the construct under consideration. The questionnaire consists of demographic details and latent constructs. The details related to age, gender, designation and income were taken in the first part of the questionnaire. The latent constructs, viz., OKN, OFR, OOG, OPR, COK, OIC, and CCS were included in the second part. Before sharing the actual questionnaire with the respondents, the survey was shared with three industry experts and four independent academic scholars to determine the validity of the instrument. The questionnaire was modified based on the inputs received from the experts. Any respondent who did not have any role in making decisions or policy formulation were excluded from the research based on the filter question. We received a total 262 responses received. The respondents participated voluntarily in the study and it was found that 21 samples had incomplete information and they were removed from the final sample. The final sample size for the study was 241. Descriptive statistics of the respondents are shown in Table 8.1. The results showed that 27% of the respondents are in the agre group between 20 and 30 years. Additionally, there were 31% of the respondents in the age bracket of 31–40 years 27% in the age bracket between 40 and 50 years and the rest 15% in the age bracket between 50 and 60 years. Furthermore, in gender distribution of sample, males constituted 58% and 42% were females.

Data analysis in this research was done using Smart PLS, which is a better method in case of smaller sample size (Ringle et al., 2015). Moreover, PLS-SEM helps in better understanding the complex relationships among variables. The study consisted of different latent constructs namely OKN, OFR, OOG, OPR, COK, OIC, and CCS. The scales for the constructs were adopted from the previous literature. Items for OKN, OFR, OOG, OPR, COK, OIC and CCS were adopted from previous literature. Table 8.2 shows details of construct used in the study.

Table 8.1 Demographic details

Demographic variable		Frequency	Percentage (%)
Age	20–30 years	66	27
	31–40 years	74	31
	41–50 years	65	27
	51–60 years	36	15
Gender	Male	112	46
	Female	129	54
Qualification	Secondary school	0	0
	Graduation	76	32
	PG	153	63
	Other	12	5
Role	Chief officer	19	8
	Director	34	14
	VP	56	23
	Manager	87	36
	Senior manager	45	19
Employees in company	Less than 100	23	10
	Between 100 and 250	76	32
	Between 250 and 500	97	40
	Between 500 and 1000	34	14
	More than 1000	11	5
Years since inception	Less than 10 years	12	5
	Between 10 and 20 years	132	55
	More than 20 years	97	40

Table 8.2 Latent constructs details

Constructs	Item details	Items
Organizational knowledge	OKN1	The organization has access to information about CC
	OKN2	The organization has access to information about the impact of CC strategies
	OKN3	The organization has information about the available adaptation measures
Organizational financial resources	OFR1	The organization has funds for implementing adaptation measures
	OFR2	The organization has options of venture funds, private loan, bank loans for implementing adaptation measures
	OFR3	There is time available to search for different CCAS
	OFR4	The authorities or policy makers focus more on climate protection
Organizational objectives and goals	OOG1	The CCAS are formed to protect climate
	OOG2	The safety-related aspects are given priority while formulating CCAS
Organizational processes	OPR1	There is a motivation and focus among executives to implement CCAS
	OPR2	CCAS are informed communicated properly among executives
	OPR3	The executives are part of CCAS
	OPR4	The operational aspects is well communicated among the executives regarding CCAS
	OPR5	CC is integrated into the planning process of an organization
	OPR6	CCAS influences and give competitive edge to organizations
Collective knowledge	COK1	Policymakers provide information on CC
	COK2	Policymakers provide data on impact of CC strategies in an organization
	COK3	Policymakers provide information on adaptation measures
	COK4	Policymakers stipulate incentives for CC adaptation
	COK5	Date for CCAS is provided by the government and administration
	COK6	People are informed about CC and its impact
Organizational incentives and culture	OIC1	CC implications are financially observed in business
	OIC2	CCAS gives competitive edge to the organization
	OIC3	Policymakers provide information on adaptation measures
	OIC4	Policymakers provide financial aid to organization for implementing CCAS
	OIC5	Authorities provide legal aid to organization for CCAS
	OIC6	CC plays a crucial role in the overall management of an organization

(continued)

Table 8.2 (continued)

Constructs	Item details	Items
CC strategies	CCS1	There is a reduction in climate emissions due to implementation of CCAS
	CCS2	Organizations are adopting CC practices
	CCS3	Environment management practices help to deal with climate threats in an organization

8.4 Results and Analysis

The data was collected using survey method, that can be non-normal due to various reasons like categorical or ordinal data, natural variations. In order to analyse the non-normal data, non-parametric tests are used. PLS SEM technique has been found to be most effective to be used in case of non-normal data. It is generally used when the sample size is smaller in nature and model is complex (Hair et al., 2019). Due to this reason, data was analysed using the SEM through Smart PLS3.0. Analysis in this study was done in two stages which included measurement model and structural model analysis.

8.4.1 Measurement Model

The data was analysed using the software and results showed that factor loadings for each constructs was above 0.70. The constructs were acceptable as they explained more than 50% of the indicator variance (Hair et al., 2012). Items of the construct were removed with less than 0.50 value. Table 8.3 shows factor loading details. The composite reliability value for all constructs found to be more than 0.70, indicating that constructs were reliable in the model of the present study (Fornell & Larcker, 1981). The AVE value of all the constructs found to be above 0.50, suggesting, the convergent reliability is supported (Fornell & Larcker, 1981; Hair et al., 2019). Furthermore, each construct Cronbach's alpha value found to be above 0.70, suggesting internal consistency and reliability of the constructs and assumed that model used in the study is correct. Table 8.4 shows details of composite reliability, Cronbach's alpha and AVE.

The Heterotrait-Monotrait (HTMT) ratio was used to check the model discriminant validity. All values found to be less than 0.85, establishing discriminant validity of all the constructs in the model (as shown in Table 8.5). The variance inflation factor of all exogenous variables ranges from 1.014 to 02.531, suggesting absence of multi-collinearity issue, since the VIF values fond to be below 5 (Myers & Myers, 1990). Table 8.6 consist of the values of VIF values of variables. Further, to check the common method bias that happen when one factor accounts for the majority of

Table 8.3 Present indicator loading

Variables	Factor loading
CCS2	0.665
CCS3	0.665
OFR1	0.703
OFR2	0.796
OFR3	0.831
OIC1	0.849
OIC2	0.804
OIC3	0.783
OIC4	0.797
OIC5	0.819
OIC6	0.826
OKN1	0.887
OKN2	0.842
OKN3	0.857
OOG1	0.889
OOG2	0.889
COK1	0.341
COK2	0.773
COK3	0.790
COK4	0.710
COK5	0.625
OPR1	0.831
OPR2	0.810
OPR3	0.814
OPR4	0.833
OPR5	0.081

Table 8.4 Details of Cronbach's alpha, composite reliability and AVE

	Cronbach's alpha	Composite reliability (rho_a)	Average variance extracted (AVE)
CCS	0.263	0.263	0.442
OFR	0.673	0.682	0.606
OIC	0.897	0.898	0.661
OKN	0.827	0.828	0.744
OOG	0.734	0.734	0.790
COK	0.670	0.721	0.446
OPR	0.736	0.840	0.542

Table 8.5 HTMT ratio matrix

	CCS	OFR	OIC	OKN	OOG	COK	OPR
CCS							
OFR	0.330						
OIC	0.335	0.135					
OKN	0.291	1.060	0.096				
OOG	0.417	0.140	1.144	0.093			
COK	0.358	0.136	0.123	0.108	0.139		
OPR	0.329	0.164	1.159	0.143	1.103	0.193	

the variance (Podsakoff et al., 2003), the Harman one-factor test was employed using exploratory factor analysis (Harman, 1976). A single factor test explained a 24.41% variance extracted from all the items, indicating the absence of the common method bias.

8.4.2 *Structural Model*

Results (Table 8.7) indicate that OFR significantly influences CCS ($\rho < 0.005$). OIC has a positive impact on the CCS ($\rho < 0.005$). OKN has a positive influence on CCS ($\rho > 0.005$). OOG significantly influences the CCS ($\rho < 0.005$). COK significantly influences CCS ($\rho < 0.005$) and OPR also significantly influences the CCS ($\rho < 0.005$). The results shows that H1 to H6 are significant. The R^2 value is 47.2% for the model.

8.5 Discussion

The aim of this chapter is to investigate the impact of various factors that influence CCAS of manufacturing companies. The first hypothesis predicted that corporate knowledge positively impacts the intention to adapt CC strategies. The results show that the relationship between corporate knowledge and intention to adopt CC strategies is positive and significant. This suggests that when the information is available to the organization regarding CC the intention to adopt CC strategies becomes higher for the organization. The findings indicate that an organization with awareness and information about adaptation measures is more likely to implement a greater number of CC strategies. This is in line with the past literature which states that corporate knowledge has a positive influence on the adaptation of CC strategies by an organization (Backman et al., 2017; Shahzad et al., 2021). The organizations need information regarding GHG emissions and net zero to make CC strategies for the betterment of our planet. Organizations form a base for stimulating the growth of the economy.

Table 8.6 Shows VIF values

Variables	VIF
CCS2	1.014
CCS3	1.014
OFR1	1.204
OFR2	1.392
OFR3	1.467
OIC1	2.690
OIC2	2.177
OIC3	2.013
OIC4	2.175
OIC5	2.531
OIC6	2.343
OKN1	2.125
OKN2	1.761
OKN3	1.883
OOG1	1.507
OOG2	1.507
COK1	1.082
COK2	1.528
COK3	1.541
COK4	1.360
COK5	1.218
OPR1	1.958
OPR2	1.839
OPR3	1.877
OPR4	1.957
OPR5	1.016

Table 8.7 Shows path coefficient result

Hypotheses	*p*-value	Hypotheses supported (Y/N)
Hypothesis 1	< 0.01	Yes
Hypothesis 2	< 0.01	Yes
Hypothesis 3	< 0.01	Yes
Hypothesis 4	< 0.01	Yes
Hypothesis 5	< 0.01	Yes
Hypothesis 6	< 0.01	Yes

Incomplete information and internal interpretations are acting as a barrier in CCAS taken by the organization. To overcome this barrier, it becomes crucial to provide data to the organizations that will help them in making better strategies.

The second hypothesis proposed that organizational financial resources have significant influence on the organization's intention to adopt CC strategies. The previous literature has shown positive and insignificant results for this variable. The results indicated that organizations with adequate funding for CC adaptation measures tend to have a higher intention to implement such strategies. Furthermore, the findings demonstrated that private financing significantly impacts an organization's intention to adopt CCAS. The results further revealed that the focus should move from climate protection to climate adaptation among the organizations. The past literature also supported this argument that the finance, alternatives, and time for researching and evaluating are crucial factors influencing an organization's CCAS (Alshahrani et al., 2023; Andersson et al., 2016; Averchenkova et al., 2016; Kumar et al., 2021; Ooi et al., 2018; Ruben et al., 2019). The third hypothesis opined that an organizations objectives and goals significantly influence an organization's intention towards CCAS. The past literature also shows that there is a significant relationship between the organization's objectives and CCAS (Dunn, 2018; Gouldson & Sullivan, 2013; Tamosiunas, 2010; Weinhofer & Busch, 2013). Our findings also support the argument that intended adaptation strategies have a significant influence on climate protection considerations. Moreover, the relationship between intended adaptation measures and safety-related aspects was also found to be positively associated.

The fourth hypothesis proposed that organizational processes significantly influence an organization's intention to adopt CC strategies. The results show that there is a significant positive relationship between the employees' motivation and intention to properly implement the CC strategies. It further reiterated that the employees of an organization are the main pillars of implementing any CC strategy. Additionally, the relationship proves that if the employees will be involved in the strategy-making process, there will be a higher inclination towards applying CC strategies. The results show that it is necessary to communicate to employees about the operational impacts of the strategies and CC strategies should be properly included in the operational process. The existing literature supports similar findings, indicating that when employees are involved in the decision-making process, organizations can implement strategies more successfully (Galbreath, 2010; Kolk & Levy, 2001).

The fifth hypothesis states that collective knowledge also significantly influences an organization's intention to adapt CC strategies. The findings indicated that when policymakers collectively provide information about CC and its impacts on the planet, organizations are more likely to adopt strategies for addressing CC. Additionally, it was found that if the financial incentives or subsidy is provided by the policymakers or the government, then there will be higher chances for organizations adopting CC strategies. The results further confirmed that organizations are more likely to adopt strategies to combat CC issues when they have access to information about science and administration (Okereke et al., 2009; Wannewitz & Garschagen, 2023). Additionally, raising societal awareness about CC and the measures taken to address it effectively for a better future increases the intention to act.

The sixth hypothesis found a significant relationship between organizational incentives and culture. It states that if the effects of CC are visible financially and it impacts the financials in that case enterprises will be more inclined to adopt such strategies. Furthermore, the results revealed that if the adaptation strategies will give any competitive edge to the organization, in that case also the inclination to adopt such strategies will also increase. The findings also revealed that if the legal system, subsidy, or incentive is supportive then organizations will implement the strategies for CC. This is in line with the previous literature which stated that if the policy-makers will provide strong legal support along with financial aid the organizations will readily adapt the CC strategies (Hoffman, 2010; Isensee et al., 2020; Perdana & Tyers, 2020; Raux et al., 2021; Russell & McIntosh, 2011).

8.6 Concluding Remarks

CC is already influencing business organizations and reshaping the internal and external operating environment of the industries. In this context, CCAS have become a new norm for companies. This chapter aims to investigate the impact of various organizational factors that influence CCAS of manufacturing companies in India. The finding of the study indicates that organizational knowledge, organizational financial resources, objectives and goals, and organizational processes positively and significantly impact the CCAS of manufacturing companies in India. Further, collective knowledge, organizational culture, and incentives were also found to be significantly impacting the CC adaptations in the organization. This study should be particularly useful for managers and practitioners as they may focus on the factors to improve climate adaptations in the business. Moreover, the findings should also be useful for policymakers and regulators. They could leverage instruments like collective knowledge and incentives to propagate CC adaptation in the industries. This chapter also significantly contributes to the extant research on climate adaptations in the context of the manufacturing sector in developing economies.

However, this study also suffers from certain limitations. The sample of this chapter is limited to the employees of the manufacturing sector. Further, demographic characteristics and socioeconomic factors were not taken into consideration in this study. Longitudinal research may provide a better understanding of the various factors influencing CC adaptations. Similarly, cross-cultural study and context-specific strategies may provide deeper insights into the factors of climate adaptations. Future studies in the context of CC adaptation may take into consideration these issues.

References

Abbass, K., Qasim, M. Z., Song, H., Murshed, M., Mahmood, H., & Younis, I. (2022). A review of the global climate change impacts, adaptation, and sustainable mitigation measures. *Environmental Science and Pollution Research, 29*(28), 42539–42559.

Aggarwal, R., & Dow, S. (2012). Corporate governance and business strategies for climate change and environmental mitigation. *The European Journal of Finance, 18*(3–4), 311–331.

Alshahrani, F., Eulaiwi, B., Duong, L., & Taylor, G. (2023). Climate change performance and financial distress. *Business Strategy and the Environment, 32*(6), 3249–3271.

Andersson, M., Bolton, P., & Samama, F. (2016). Governance and climate change: A success story in mobilizing investor support for corporate responses to climate change. *Journal of Applied Corporate Finance, 28*(2), 29–33.

Arndt, C. (2023). Climate change vs energy security? The conditional support for energy sources among Western Europeans. *Energy Policy, 174*, 113471.

Averchenkova, A., Crick, F., Kocornik-Mina, A., Leck, H., & Surminski, S. (2015). Multinational corporations and climate adaptation—Are we asking the right questions? A review of current knowledge and a new research perspective. In *Grantham research Institute on climate change and the environment working paper* (p. 183).

Averchenkova, A., Crick, F., Kocornik-Mina, A., Leck, H., & Surminski, S. (2016). Multinational and large national corporations and climate adaptation: Are we asking the right questions? A review of current knowledge and a new research perspective. *Wiley Interdisciplinary Reviews: Climate Change, 7*(4), 517–536.

Ayers, J., & Dodman, D. (2010). Climate change adaptation and development I: The state of the debate. *Progress in Development Studies, 10*(2), 161–168.

Backman, C. A., Verbeke, A., & Schulz, R. A. (2017). The drivers of corporate climate change strategies and public policy: A new resource-based view perspective. *Business & Society, 56*(4), 545–575.

Basson, F., Dayamba, S. D., Korahire, J., Dipama, J. M., Zougmore, F., & Neya, T. (2020). Institutional barriers to climate change adaptation in Burkina Faso: How could we go around them? *Journal of Sustainable Development, 13*(5), 67–78.

Berrang-Ford, L., Ford, J. D., & Paterson, J. (2011). Are we adapting to climate change? *Global Environmental Change, 21*(1), 25–33.

Bhushan, M., & Singh, R. (2024). Talent Management Facilitates Net-Zero Transition Through Employee Green Behavior. In R. Singh, S. Khan, A. Kumar, S. Luthra, & H. Chokshi (Eds.), *Net zero economy, corporate social responsibility and sustainable value creation. CSR, sustainability, ethics & governance*. Springer.

Biesbroek, G. R., Klostermann, J. E., Termeer, C. J., & Kabat, P. (2013). On the nature of barriers to climate change adaptation. *Regional Environmental Change, 13*, 1119–1129.

Biswas, R. R., & Rahman, A. (2023). Adaptation to climate change: A study on regional climate change adaptation policy and practice framework. *Journal of Environmental Management, 336*, 117666.

Braunschweiger, D., & Ingold, K. (2023). What drives local climate change adaptation? A qualitative comparative analysis. *Environmental Science & Policy, 145*, 40–49.

Cho, H., Lee, P., & Shin, C. H. (2019). Becoming a sustainable organization: Focusing on process, administrative innovation and human resource practices. *Sustainability, 11*(13), 3554.

Dahlmann, F., Branicki, L., & Brammer, S. (2017). 'Carrots for corporate sustainability': Impacts of incentive inclusiveness and variety on environmental performance. *Business Strategy and the Environment, 26*(8), 1110–1131.

Daka, E. (2023). Adopting clean technologies to climate change adaptation strategies in Africa: A systematic literature review. *Environmental Management, 71*(1), 87–98.

de Abreu, M. C. S., Webb, K., Araújo, F. S. M., & Cavalcante, J. P. L. (2021). From "business as usual" to tackling climate change: Exploring factors affecting low-carbon decision-making in the Canadian oil and gas sector. *Energy Policy, 148*, 111932.

Díaz Tautiva, J. A., Huaman, J., & Ponce Oliva, R. D. (2024). Trends in research on climate change and organizations: A bibliometric analysis (1999–2021). *Management Review Quarterly, 74*(1), 227–261.

Dunn, S. (2018). Down to business on climate change: an overview of corporate strategies. *The Business of Climate Change*, 31–46.

Dwivedi, A., Sassanelli, C., Agrawal, D., Moktadir, M. A., & D'Adamo, I. (2023). Drivers to mitigate climate change in context of manufacturing industry: An emerging economy study. *Business Strategy and the Environment, 32*(7), 4467–4484.

Eccles, R. G., & Krzus, M. P. (2018). Why companies should report financial risks from climate change. *MIT Sloan Management Review, 59*(3), 1–6.

Fawzy, S., Osman, A. I., Doran, J., & Rooney, D. W. (2020). Strategies for mitigation of climate change: A review. *Environmental Chemistry Letters, 18*, 2069–2094.

Flammer, C., Toffel, M. W., & Viswanathan, K. (2021). Shareholder activism and firms' voluntary disclosure of climate change risks. *Strategic Management Journal, 42*(10), 1850–1879.

Fornell, C., & Larcker, D. F. (1981). Evaluating structural equation models with unobservable variables and measurement error. *Journal of Marketing Research, 18*(1), 39–50.

Galbreath, J. (2010). Corporate governance practices that address climate change: An exploratory study. *Business Strategy and the Environment, 19*(5), 335–350.

Galpin, T., Whitttington, J. L., & Bell, G. (2015). Is your sustainability strategy sustainable? Creating a culture of sustainability. *Corporate Governance, 15*(1), 1–17.

Gemeda, D. O., Korecha, D., & Garedew, W. (2023). Determinants of climate change adaptation strategies and existing barriers in Southwestern parts of Ethiopia. *Climate Services, 30*, 100376.

Glienke, N., & Guenther, E. (2016). Corporate climate change mitigation: A systematic review of the existing empirical evidence. *Management Research Review, 39*(1), 2–34.

Gouldson, A., & Sullivan, R. (2013). Long-term corporate climate change targets: What could they deliver? *Environmental Science & Policy, 27*, 1–10.

Guenther, E. M., & Hoppe, H. (2014). Merging limited perspectives: A synopsis of measurement approaches and theories of the relationship between corporate environmental and financial performance. *Journal of Industrial Ecology, 18*(5), 689–707.

Hair, J. F., Risher, J. J., Sarstedt, M., & Ringle, C. M. (2019). When to use and how to report the results of PLS-SEM. *European Business Review, 31*(1), 2–24.

Hair, J. F., Sarstedt, M., Ringle, C. M., & Mena, J. A. (2012). An assessment of the use of partial least squares structural equation modeling in marketing research. *Journal of the Academy of Marketing Science, 40*, 414–433.

Harman, H. H. (1976). *Modern factor analysis*. University of Chicago Press.

Herrmann, J., & Guenther, E. (2017). Exploring a scale of organizational barriers for enterprises' climate change adaptation strategies. *Journal of Cleaner Production, 160*(1), 38–49.

Hoffman, A. J. (2007). *Carbon strategies: How leading companies are reducing their climate change footprint*. University of Michigan Press.

Hoffman, A. J. (2010). Climate change as a cultural and behavioral issue: Addressing barriers and implementing solutions. *Organizational Dynamics, 39*(4), 295–305.

Hoffmann, V. H., Sprengel, D. C., Ziegler, A., Kolb, M., & Abegg, B. (2009). Determinants of corporate adaptation to climate change in winter tourism: An econometric analysis. *Global Environmental Change, 19*(2), 256–264.

Hossain, M. K., & As-Saber, S. N. (2024). Business strategies and climate change adaptation: insights from a comparative study between a developed and a developing country. *Management of Environmental Quality: An International Journal*.

IPCC. (2007). Intergovernmental panel on climate change. In S. Solomon et al. (Ed.), *Climate change 2007: The scientific basis. Contribution of working group i to the fourth assessment report of the intergovernmental panel on climate change*. New York: Cambridge University Press.

Isensee, C., Teuteberg, F., Griese, K. M., & Topi, C. (2020). The relationship between organizational culture, sustainability, and digitalization in SMEs: A systematic review. *Journal of Cleaner Production, 275*, 122944.

Jeswani, H. K., Wehrmeyer, W., & Mulugetta, Y. (2008). How warm is the corporate response to climate change? Evidence from Pakistan and the UK. *Business Strategy and the Environment, 17*(1), 46–60.

Johnson, M. P., Rötzel, T. S., & Frank, B. (2023). Beyond conventional corporate responses to climate change towards deep decarbonization: A systematic literature review. *Management Review Quarterly, 73*(2), 921–954.

Kolk, A., & Levy, D. (2001). Winds of change: Corporate strategy, climate change and oil multinationals. *European Management Journal, 19*(5), 501–509.

Kolk, A., & Pinkse, J. (2005). Business responses to climate change: Identifying emergent strategies. *California Management Review, 47*(3), 6–20.

Kouloukoui, D., Marinho, M. M. D. O., Gomes, S. M. D. S., de Jong, P., Kiperstok, A., & Torres, E. A. (2020). The impact of the board of directors on business climate change management: Case of Brazilian companies. *Mitigation and Adaptation Strategies for Global Change, 25*, 127–147.

Kumar, K. (2022). The uptake of sustainability disclosure practices by environmentally polluting industries in India. In *The equal pillars of sustainability* (pp. 199–224). Emerald Publishing Limited.

Kumar, K., Kumari, R., & Kumar, R. (2021). The state of corporate sustainability reporting in India: Evidence from environmentally sensitive industries. *Business and Society Review, 126*(4), 513–538.

Kumar, K., Prakash, A., & Khan, W. (2020). Integrating the notion of sustainable development in banking: Analysing historical and conceptual framework. *Indian Journal of Economics and Development, 16*(3), 449–458.

Kumar, K., Kumari, R., Nandy, M., Sarim, M., & Kumar, R. (2022). Do ownership structures and governance attributes matter for corporate sustainability reporting? An examination in the Indian context. *Management of Environmental Quality: An International Journal, 33*(5), 1077–1096.

Lee, S. Y. (2012). Corporate carbon strategies in responding to climate change. *Business Strategy and the Environment, 21*(1), 33–48.

Linnenluecke, M. K., Russell, S. V., & Griffiths, A. (2009). Subcultures and sustainability practices: The impact on understanding corporate sustainability. *Business Strategy and the Environment, 18*(7), 432–452.

Mishra, P., Sant, T. G., & Kumar, K. (2023). Carbon disclosure and organization performance: A literature review. *Sustainability and Climate Change, 16*(4), 302–317.

Myers, R. H., & Myers, R. H. (1990). *Classical and modern regression with applications* (Vol. 2, p. 488). Duxbury Press.

Nalau, J., & Verrall, B. (2021). Mapping the evolution and current trends in climate change adaptation science. *Climate Risk Management, 32*, 100290.

Nemet, G. F. (2010). Robust incentives and the design of a climate change governance regime. *Energy Policy, 38*(11), 7216–7225.

Nwokolo, S. C., Singh, R., Khan, S., Kumar, A., & Luthra, S. (2023). Introduction: Africa's net zero transition. In *Africa's path to net-zero: exploring scenarios for a sustainable energy transition* (pp. 1–13). Springer Nature Switzerland

Nyberg, D., & Wright, C. (2024). Defending hegemony: From climate change mitigation to adaptation on the Great Barrier Reef. *Organization, 31*(2), 247–268.

Okereke, C., Wittneben, B., & Bowen, F. (2009). Climate change: Challenging business, transforming politics. *Business & Society, 48*(4), 584–586.

Ooi, S. K., Goh, S., Yeap, J. A., & Loo, K. S. (2018). Linking corporate climate change and financial performance: Evidence from Malaysia. *Global Business & Management Research, 10*(1).

Orazalin, N. S., Ntim, C. G., & Malagila, J. K. (2024). Board sustainability committees, climate change initiatives, carbon performance, and market value. *British Journal of Management, 35*(1), 295–320.

Owen, G. (2020). What makes climate change adaptation effective? A systematic review of the literature. *Global Environmental Change, 62*, 102071.

Pawlowsky, P. (2001). The treatment of organizational learning in management science. In *Handbook of organizational learning and knowledge* (pp. 61–88).

Perdana, S., & Tyers, R. (2020). Global climate change mitigation: Strategic incentives. *The Energy Journal, 41*(3), 183–206.

Perlin, A. P., Gomes, C. M., Kneipp, J. M., & Motke, F. D. (2021). Climate change and performance in Brazilian industrial companies. *Handbook of climate change management: research, leadership, transformation* (pp. 1617–1637). Springer International Publishing.

Podsakoff, P. M., MacKenzie, S. B., Lee, J. Y., & Podsakoff, N. P. (2003). Common method biases in behavioral research: A critical review of the literature and recommended remedies. *Journal of Applied Psychology, 88*(5), 879.

Prakash, A., Kumar, K., Khan, W., & Siddiquei, M. I. (2023). Self-induced versus structured corporate social responsibility: The Indian context. *Measuring sustainability and CSR: From reporting to decision-making* (pp. 167–178). Springer International Publishing.

Raux, C., Chevalier, A., Bougna, E., & Hilton, D. (2021). Mobility choices and climate change: Assessing the effects of social norms, emissions information and economic incentives. *Research in Transportation Economics, 90*, 101007.

Reckien, D., Flacke, J., Dawson, R. J., Heidrich, O., Olazabal, M., Foley, A., & Pietrapertosa, F. (2014). Climate change response in Europe: What's the reality? Analysis of adaptation and mitigation plans from 200 urban areas in 11 countries. *Climatic Change, 122*, 331–340.

Ringle, C., Da Silva, D., & Bido, D. (2015). Structural equation modeling with the SmartPLS. *Brazilian Journal of Marketing, 13*(2).

Rivera, J. E., Oh, C. H., Oetzel, J., & Clement, V. (2022). *Business adaptation to climate change.* Cambridge University Press.

Ruben, R., Wattel, C., & Van Asseldonk, M. (2019). Rural finance to support climate change adaptation: Experiences, lessons and policy perspectives. In *The climate-smart agriculture papers* (p. 301).

Russell, S. V., & McIntosh, M. (2011). Changing organizational culture for sustainability. In *The handbook of organizational culture and climate* (pp. 393–411).

Sebos, I., Nydrioti, I., Katsiardi, P., & Assimacopoulos, D. (2023). Stakeholder perceptions on climate change impacts and adaptation actions in Greece. *Euro-Mediterranean Journal for Environmental Integration, 8*(4), 777–793.

Shahzad, M., Qu, Y., Zafar, A. U., & Appolloni, A. (2021). Does the interaction between the knowledge management process and sustainable development practices boost corporate green innovation? *Business Strategy and the Environment, 30*(8), 4206–4222.

Sharifi, A. (2020). Trade-offs and conflicts between urban climate change mitigation and adaptation measures: A literature review. *Journal of Cleaner Production, 276*, 122813.

Sharma, A., Chaudhary, M., Kumar, K., & Daga, S. (2024). Promoting net-zero economy for sustainable development: practice-based view. In R. Singh, S. Khan, A. Kumar, S. Luthra, H. Chokshi (Eds.), *Net zero economy, corporate social responsibility and sustainable value creation. CSR, sustainability, ethics & governance.* Springer.

Sharma, H. P., & Kumar, K. (2021). Developing and implementing environment management practices in small and medium size manufacturing companies in India. *IOP Conference Series: Earth and Environmental Science, 795*(1), 012022.

Singh, R., Ravesangar, K., Narayanan, S., & Singh, K. K. H. (2024). HRM practices in creating net zero: Supporting organizational sustainability. In R. Singh, S. Khan, A. Kumar, S. Luthra, H. Chokshi (Eds.), *Net zero economy, corporate social responsibility and sustainable value creation. CSR, sustainability, ethics & governance.* Springer.

Smith, E. K., & Mayer, A. (2018). A social trap for the climate? Collective action, trust and climate change risk perception in 35 countries. *Global Environmental Change, 49*, 140–153.

Sparkman, G., Howe, L., & Walton, G. (2021). How social norms are often a barrier to addressing climate change but can be part of the solution. *Behavioural Public Policy, 5*(4), 528–555.

Stankuniene, G., Streimikiene, D., & Kyriakopoulos, G. L. (2020). Systematic literature review on behavioral barriers of climate change mitigation in households. *Sustainability, 12*(18), 7369.
Stechemesser, K., Bergmann, A., & Guenther, E. (2015). Organizational climate accounting-financial consequences of climate change impacts and climate change adaptation. In *Corporate carbon and climate accounting* (pp. 217–242).
Steg, L. (2023). Psychology of climate change. *Annual Review of Psychology, 74*, 391–421.
Sun, Y., Bao, Q., & Taghizadeh-Hesary, F. (2023). Green finance, renewable energy development, and climate change: Evidence from regions of China. *Humanities and Social Sciences Communications, 10*(1), 1–8.
Sun, Y., Yang, Y., Huang, N., & Zou, X. (2020). The impacts of climate change risks on financial performance of mining industry: Evidence from listed companies in China. *Resources Policy, 69*, 101828.
Sundaram, A. K., & Hansen, R. G. (Eds.). (2023). *Handbook of business and climate change.* Edward Elgar Publishing.
Tamosiunas, A. (2010). Managing corporate strategic changes in the context of climate change. *Engineering Economics, 66*(1).
Tangney, P. (2020). Understanding climate change as risk: A review of IPCC guidance for decision-making. *Journal of Risk Research, 23*(11), 1424–1439.
Timilsina, G. R. (2021). Financing climate change adaptation: International initiatives. *Sustainability, 13*(12), 6515.
Tryndina, N., An, J., Varyash, I., Litvishko, O., Khomyakova, L., Barykin, S., & Kalinina, O. (2022). Renewable energy incentives on the road to sustainable development during climate change: A review. *Frontiers in Environmental Science, 10.*
Turnbull, S. (2020). Managing knowledge to counter climate change. In *Proceedings of virtual contribution to international forum on knowledge asset dynamics being held from* (pp. 9–11).
United Nations. (2024). *United Nation climate actions.* https://www.un.org/en/climatechange
Wannewitz, M., & Garschagen, M. (2023). Collective adaptation to climate change. *Current Opinion in Environmental Sustainability, 61*, 101248.
Weinhofer, G., & Busch, T. (2013). Corporate strategies for managing climate risks. *Business Strategy and the Environment, 22*(2), 121–144.
Weinhofer, G., & Hoffmann, V. H. (2010). Mitigating climate change-how do corporate strategies differ? *Business Strategy and the Environment, 19*(2), 77–89.
Welch, D., & Yates, L. (2018). The practices of collective action: Practice theory, sustainability transitions and social change. *Journal for the Theory of Social Behaviour, 48*(3), 288–305.
Zhou, Y. (2023). Climate change adaptation with energy resilience in energy districts—A state-of-the-art review. *Energy and Buildings, 279*, 112649.

Chapter 9
A Comprehensive Analysis of Leveraging Technology in the Manufacturing Sector to Achieve Net Zero Future

V. Rema

Abstract The current global priority is to work towards achieving net zero goals. This requires conscious and collective efforts of global stakeholders including individuals, businesses, governments and non-governmental organisations. Technology implementation plays a crucial role in our quest to achieve decarbonisation goals. This chapter aims to provide a detailed perspective on how technological innovations have been adopted by top global manufacturing companies in their business operations. It includes discussion of use cases that demonstrate the adoption of sustainable technologies, with a sectorial focus on Electronics, Food & Beverages, Automobiles, and Pharmaceuticals, through a comprehensive review of their annual ESG and Sustainability Reports. Additionally, the policy implications are discussed, and the way forward to achieve environmental stewardship is highlighted.

Keywords Sustainable technologies · Sustainable development goals · Manufacturing · Net zero policy

9.1 Introduction

Sustainability has an encompassing connotation. "*Meeting the needs of the present without compromising the ability of future generations to meet their own needs*" is the definition of sustainability given by the United Nations Brundtland Commission in 1987. While we speak of sustainability, the larger focus is usually the environmental issues. However, meeting economic requirements and the role of institutions, governments, and organisations in empowering their stakeholders also come under the ambit of sustainability (Thiele, 2024). Sustainability warrants conviction and persistence. While we speak of sustainability, prioritising Sustainable Development

V. Rema (✉)
Department of Data Sciences and Analytics, School of Social Sciences, Ramaiah University of Applied Sciences, Bengaluru, Karnataka, India
e-mail: remav.ss@msruas.ac.in

R. Singh and D. Crowther (eds.), *Transition Towards a Sustainable Future*, Approaches to Global Sustainability, Markets, and Governance,
https://doi.org/10.1007/978-981-97-5756-5_9

as a goal is critical. Every industrial breakthrough has improved mankind's quality of life. Technology has advanced greatly, moving from bullock carts to aeroplanes and driverless vehicles. Nevertheless, with the boon comes the bane. The negative impacts such as pollution of natural resources, degraded soil, global warming etc. have increased steadily. As per the International Energy Agency Report (IEA, 2023) on CO_2 emissions in 2023, there is an increase of 1.1% in the Global energy-related CO_2 emissions in 2023. This increases 410 million tonnes of CO_2 emissions to reach a whopping new record high of 37.4 billion tonnes. Further, in 2023, coal-related emissions amounted to more than 65% of the increase. Sustainability scientists globally have warned about the exponential impacts it is likely to cause. These alarming numbers call for a ***Global SOS***—the call to action towards implementing the green plans at scale, to achieve net zero emissions. The call will be to understand what the energy choices today are and where we need to go. As per the Next Move Strategy Consulting (2024) Waste Management Market by Waste Type, by Waste Treatment and by End User—Global Opportunity Analysis and Industry Forecast 2024–30 Report, in the year 2023, the global waste management market estimated at approximately 1009 billion USD is projected to grow by an alarming 37% in 2030. The waste-generating sectors predominantly constitute manufacturing with automobiles, electronics, food and beverages, energy, pharmaceuticals etc. Therefore, the need to assess the strategies employed by manufacturing companies in optimising efficiency is an important step in the direction of decarbonisation goals. The production and consumption systems require transformation, driven by policy frameworks. Delivering net zero goals require a unified approach across various sectors in the adoption of transition technologies.

The chapter is structured as follows. Section 9.2 presents a detailed review of literature relevant to strategies for addressing sustainability challenges, technological disruptions, policy pathways and energy modelling mechanisms aiming towards a carbon–neutral scenario. A comprehensive analysis is carried out in Sect. 9.3 giving a detailed perspective of how technological innovations in the manufacturing sector have been adopted by top global manufacturing companies in their business operations. Manufacturing covers a broad range of sub-sectors including food, automobiles, textiles, electronics, transportation, construction, aerospace, apparel, beverages, metallurgy, and heavy industry among the major ones. ***Use cases of sustainable technology adoption*** by select companies in the Electronics, Food & beverages, Automobiles and Pharmaceuticals have been discussed. Section 9.4 discusses the policy implications specific to manufacturing and elaborates on the challenges in technological implementation at scale. The scope for future research relevant to be carried out to achieve sustainable operations in global manufacturing is discussed. Section 9.5 concludes with key remarks and limitations.

9.2 Review of Literature

Addressing sustainability challenges and transitioning from '***concept to movement***' requires aggressive interventions. However, understanding the facets of the '*net zero phenomenon*' is the first step. Fankhauser et al. (2022) in their research have characterised the attributes of credible net zero levels through a framework that defines *the urgency, integrity and consistency* of the net zero phenomenon. Understanding the tenets of net zero is important to address the challenges that come in the way of its achievement. One of the primary focuses is to look at reducing carbon emissions by evaluating and setting short-term and long-term targets that involve businesses and the government, through a consistent approach to mitigate the ill effects of climate change. A comprehensive orientation to the reduction of emissions is a feasible strategy. This would mean the focus should not just be on scaling up clean energy supplies and reducing consumption of non-renewables but also diverting attention to heavy industries where affordable and scalable decarbonisation solutions must be put into practice. A sector-specific tailored approach is likely to support the larger global goal of carbon footprint management. Reduction of emissions and removal of carbon dioxide through Carbon Removal Technologies should be complementary efforts. The technologies integrating blockchain and AI are evolving, hence heavy investments should be the focus. More advancements in decarbonising options such as ocean alkalisation, carbon sequestration etc. are required to achieve desired goals at shorter timeframes. It is important to note that net zero demands striking the balance between carbon removal goals and handling residual emissions. The development of such infrastructure is still at its nascent stage. *The equitable transition* towards a net zero goal urges the eradication of a compartmentalised approach. This would mean that cross-country support, handholding developing countries in the area of capacity building, resource management, technology and, finance should be at the helm of global Sustainable Development Goals (SDG). While each country might have chartered its unique path towards the desired target, global collaboration will alone enable the achievement of this larger goal. This involves major policy interventions that bind countries towards a common goal. With this said, it is imperative that sustainability must penetrate to the micro-levels; from countries to cities, to communities and the individual, demanding '*responsible citizenship.*' The quest for net zero opens up economic opportunities. Technological innovations propel business opportunities and create jobs and healthier economies. Today, the global climate challenge is the elephant in the room. Acknowledging it, dealing with it and sustaining measures in handling it should be the sole purpose.

Speaking of policy intervention, the ***Paris Agreement*** (United Nations Climate Change [UNCC], 2024) was a crucial attempt to bring the world together to fight the common global threat—the climate issue. This legally binding international agreement was adopted by 196 countries during the United Nations Climate Change Conference—COP21 at Paris and became operational in November 2016. The goal of the agreement was to limit global temperature rise to 1.5 °C as overshooting this threshold is known to create erratic climate impacts like intense heatwaves, floods, and droughts. Climate diplomacy has seen a new landmark with the ***UAE Consensus Agreement*** (UNCC, 2023) at the COP28, the United Nations Climate Change Conference held in Dubai in December 2023. The world leaders shared critical remarks on achieving the set net zero targets by 2050. Building storage and transmission capacities, mobilising finance in building sustainable technologies, and investing in green and fusion technologies have been spelt out as the need of the hour. The UAE Consensus delivered at the COP28 was a milestone after the Paris Agreement, to steer the global climate agenda on the highest levels of priority. It calls for transitioning away from the use of fossil fuels in an orderly manner and achieving 2× levels of energy efficiency by 2030. The agreement was the first to set targets for ending deforestation in this decade. Launching a fund for loss and damage, onboarding the oil and natural gas industry to drive solutions, and practically looking at methane emissions were several critical elements of emphasis. Among other vital aspects of the COP28 defined the need for proactive engagement with the private sector, to create a robust public–private partnership model to drive the desired outcomes.

Literature on sustainability brings out its varied perspectives and points to the absence of its consistent definition. The concept of sustainability has been 'operationalised' based on the contexts of its implementation. The knowledge syntheses on this domain through a review of over 209 articles brought out five constructs articulating sustainability- reviewing processes/systems after a particular period; the continuing implementation strategies; individual's behaviour patterns; evolution or adaptation perspective and delivering value for individuals and systems at large (Moore et al., 2017). Fostering the economics of well-being under the triple bottom approach—*people, planet and profits* is a critical goal of sustainability. Its broad contours align with augmenting green infrastructure, green spaces with clean water, buildings that consume clean energy, enterprises that maintain lower carbon footprints, adaptation of public transportation to reduce reliance on private transport, agricultural frameworks that prevent loss of soil fertility, produce without Genetically Modified Organisms (GMOs), and indulge in practices that promote equitability (Caradonna, 2022). The factors responsible for transitions to sustainable business models can be attributed to social, cultural and economic considerations. Sector-specific studies (Belyaeva et al., 2020) reflect that the core reason driving the process could vary across geographical locations. Competitiveness, coupled with legal and social pressures drives sustainability into the business agenda.

The ***policy pathways*** implemented with clear governance and adherence across global countries are imperative to drive the net zero agenda. Post the Paris Agreement in 2015 and the United Framework Convention on Climate Change (UNFCCC), there is an evolution in the policy architecture with several treaties lined up towards achieving the decarbonising goals (Burke, 2022). Deforestation treaties such as the *Glasgow Leaders' Declaration on Forests and Land Use*, 2021 committed to conserving and restoring forest ecosystems, redesigning agricultural policies to provide impetus to sustainable practices, committing to minimising land degradation and reaffirming sustained forest management. *The Coal Elimination Treaty (CET)* endorses phasing out coal mining by 2030 which is estimated to bring out a dip in

Greenhouse gas (GHG) emissions by 40%. Under the CET, the fossil fuels transition fund, funded by taxes on the import and sale of fossil fuels, would aid developing nations in achieving the transition to renewables. Studies analyse that subject to the oil price situation, the elimination of fossil fuel consumption subsidies will cut GHG emissions worldwide by 1.8–3.2% by 2030 (Chepeliev & Mensbrugghe, 2020). However, while this step is likely to cut emissions related to power, energy-intensive sectoral GHG leakages from livestock, transportation retail etc. can create an imbalance in achieving the desired targets. Binding and unified agreements with sector-specific adherence across borders is extremely critical at this juncture, to achieve a goal that is meant for the welfare of the global community.

Decarbonising goals particular to the manufacturing sector can be driven by set *policy instruments* (Sovacool et al., 2021). Under the Kyoto Protocol, GHG emissions can be traded across countries in the carbon market where carbon is treated as a commodity. It will involve trading emission units across countries which have overshot their targets, making these *Carbon Emission Trading Schemes* a step in indulging in accelerated investments in renewables. Energy efficiency schemes such as the *Production Linked Incentive (PLI)* Scheme of the Government of India to manufacturers who produce high-efficiency solar modules bolster renewable applications. *Carbon Border Adjustment Mechanisms* or Border Tariff adjustments are trade-related policy architectures which impose taxes on imported commodities based on the quantity of GHGs produced during manufacturing. Refunds or exemptions from domestic regulations for producers that export their goods based on GHGs are examples of border adjustments. Such trade mechanisms have the potential to enhance environmental efficacy by steering economic activities that can result in an overall increase in GHG emissions or carbon leakage. These tariff policies lessen the motivation for industries to shift their manufacturing base overseas. *Sectoral Agreements* which focus on sector-specific policies help in increasing the carbon competitiveness among the firms.

Researchers (Pye et al., 2021) have deep-dived into the discussion on the approach towards *modelling of energy systems* emphasising that it requires a radical change, given the fact that net zero goals have to be achieved in shorter time frames spanning across multitudinal sectors. Among the main challenges highlighted is the lower demand-side focus than the supply-side arena. This would mean that energy models are not sufficiently capturing the demand-side dynamics which include reducing consumption, and cleaner technologies to bring improvements. Limited availability or supply of non-renewables is a known fact. Carbon-di-oxide Removal Options (CDR) technologies and their deployment should have greater levels of robustness. Furthermore, the authors suggested that it is imperative to see that the models are capable of capturing the novel system configurations that may arise and align with policies and SDG initiatives. Along this line of energy modelling, Yue et al. (2018) carried out a systematic review of research papers that employ Energy Systems Optimization Models (ESOM), Stochastic programming, Monte Carlo simulation and robust optimisation models are seen to predominantly develop energy models. It is imperative to understand that the purpose of such energy modelling is to drive policy implications by offering robust insights into outcomes such as managing

CO_2 emissions and the like. Application of uncertainty analysis as mentioned above compared to the standard methods is known to be more robust in analysing systems and operations. Mathematical optimisation programming techniques have seen their applications in small-scale scheduling problems. However, large scale manufacturing scheduling seeks heuristic and metaheuristic approaches of *Simulated Annealing, Restarted Simulated Annealing* (RSA), *Archived Multi-Objective Simulated Annealing* (AMOSA), and *Tabu search* (Renna & Materi, 2021).

9.3 Technological Adoption in Manufacturing Sector: Sub-sector Wise Comprehensive Case Analysis

Any process of production that yields commodities results in waste generation. The manufacturing sector comprises several key sub-sectors such as electronics, food, automobile, textiles, construction, and pharmaceutical among others. Today Technology has been among the core catalysts driving sustainable strategies. Technological innovations and disruptions have played a significant role in achieving net zero goals. In this section, let us delve into the realm of the various sustainable technologies adopted by the top global companies across various sectors of manufacturing to achieve net zero goals.

9.3.1 Electronics Industry

Using effective waste management strategies at source, i.e. from material management to production is gaining traction in the electronics industry. As per the Global E-Waste Monitor 2024 Report (United Nations Institute for Training & Research, 2024), the extent of electronic waste or e-waste generation is $5\times$ the quantum of recycling. In over a decade from 2010 to 2022, a whopping increase of 82% generation of e-waste has been documented. Several reasons could be attributed to this rate of alarming increase from heavy electronification with intense requirements and use of gadgets, limited product lifecycles, design inefficiencies and poor waste management infrastructure. The top global companies by market capitalisation, like Apple, Nvidia, Siemens etc. have integrated sustainability into their mission and strategies (Swallow, 2024).

Bringing down inefficiencies and wastage by around 45% since 2015, **Apple** is on the road towards making its products carbon neutral by the year 2030. Some of the core initiatives taken by the Company highlighted in the Environmental Progress Report (Apple, 2023), reflect their actions in achieving sustainability through green technologies and initiatives. The company looks at retaining quality and performance in its products by bringing in recycling innovations and aiding in achieving circularity. The three areas of focus while manufacturing are ***Sourcing, Product***

design-development, Collection and Recovery. Responsible **sourcing** of raw materials without depleting the natural resources is the first step in responsible manufacturing. The Company has demonstrated efficiency in sourcing metals like tin, gold, copper, zinc and cobalt. The Company used 100% recycled tin for the solder on its main logic board in its iPhone product in 2017 after which it expanded the use of recycled tin across its various product lines by 2022 to 100%. In a year, the demand for recycled tin increased by 8%. The use of certified recycled gold in the circuit boards of its products has upped by 4% in 2022 from 2021. Recycled copper foil has been used in iPad models released in 2022. Across iPhones, Macs and other Apple products, there has been an increase of 27% year on year in the use of recycled cobalt in batteries. The back glass of the various devices consists of high-quality recycled glass. This strategy has been adopted by the company in sourcing several high-impact materials which include aluminium, cobalt, gold, lithium, tin, tungsten, zinc, glass and other metals which constitute around 87% of their products manufactured in 2022. There has been a remarkable increase in the use of recycled content with the maintenance of its high product and performance standards. With these strategies, there has been an increase in recycled materials use by 20% in 2022. Further, making the vendors in the supply chain complaint with the set standards on material management enables sustainable sourcing.

A product's long-lasting ability is an integral aspect of manufacturing as it will lead to extended use by the customer and reduce the rate of buying newer devices. This has a direct bearing on conserving the planet's resources. Technologies in manufacturing supporting eco-designs make ***durability*** possible. The Apple Trade In is an endeavour which helps the customer contribute to waste reduction and preservation of the planet's valuable resources by providing free recycling of Apple gadgets the customer owns. Based on the geographical location, the customer is provided an estimate of the device's worth and could suitably apply for a trade-in credit. There are provisions for customers to update their present devices to the latest features and software upgrades at the Apple Trade In which makes the device last for longer, thereby reducing the burden on the planet. Providing greater options for repairs is another important way to maintain durability. Initiatives such as the Self-Service Repair, Independent Repair Provider Program, Authorised Service Centres, iPhone Upgrade Program, Hardware Reuse Program, AppleCare, technology-driven enhanced display protection through ceramic shields and periodic software support are measures in alignment with sustainability goals. Their hyper-efficient Reliability Testing Labs comply with standards that meet ones similar to the military equipment and rugged devices ensuring the durability of devices.

Recycling innovations and supply chain initiatives support the collection of end-of-life products whose components can further be used as source materials. Technology has reached another level in supporting ***the efficient recovery of components*** of used devices. Apple's Material Recovery Lab employs robots for product disintegration which will help in using the product's components as inputs for the new products, thereby aiding circularity. Daisy, Taz and Dave are robots that enable specialised disintegration through automation, supporting the recycling partners and enhancing recovery rates of primary components used in devices.

Yet another prominent brand in the semiconductor space is a leading technology company **Nvidia** which is known to develop software, hardware and networking technology to reduce carbon emissions. Data centres with high computing performance are in place to build energy-efficient devices. The Company has rolled out sustainable technologies and developed innovative product lines which reduce energy consumption, and sync with the global sustainability agenda. The liquid-cooled Graphics Processing Units (GPUs) use up to 30% less power than the traditional servers. With the present numbers being at 44%, the company aims to switch to 100% *renewables* to meet the energy requirements at their global offices and data centres by the end of 2025. Across the product lifecycle, carbon footprint evaluation is carried out to devise better management strategies. Several *water conservation measures* such as low-flow fixtures in bathrooms, recycled water for flushing and cooling towers, use of water from the set-up water treatment plants and yearly water risk assessments and water-saving strategy implementation in drought-hit locations are carried out. In view of accelerating climate change, there have been several steps taken in implementing ***sustainable packaging technologies***. As per the Corporate Responsibility Report Fiscal Year 2023 (NVIDIA, 2023), bulk cartons made out of corrugated 100% fibre materials have seen the recyclability rate overshoot a remarkable 94%. With respect to the packaging of products, foam and plastic laminations have been replaced by fibre materials. Plastics are replaced by fully recyclable High-Density Polyethylene (HDPE-2). Corrugate structures and HDPE-2 cushions have replaced the packaging of server products that usually require packaging materials like foam, metal and wood. AI provides unprecedented solutions towards decarbonisation. One such classic application is the construction of the Earth Observation Digital Twin (EODT) by the Company in partnership with the National Oceanic and Atmospheric Administration and an American aerospace and defence manufacturer, Lockheed Martin Corporation. The purpose of the ***EODT*** is to *predict glacier melting*, wildfires, and other critical climate change factors to enable the global community to respond quickly to the adverse impacts of climate change. Several collaborative initiatives are taken by the company to expedite towards achievement of the sustainable development goals. The future of *employing clean power* is being researched using the NVIDIA Omniverse Simulation platform. Another interesting sustainable technology is in the creation of a digital twin of wind turbines, which is a collaborative initiative of NVIDIA with Siemens Gamesa Renewable Energy for wind production which can be a stellar contribution to meet energy consumption requirements. The tools, expertise and super-computing power coming from niche technologies are vital to handle climate impacts at a progressive rate. AI strategies developed through the AI Nations, is the Company's initiative to support governments globally. Another noteworthy measure in ***disaster management technology*** is integrating accelerated computing interfaces with the United Nations Satellite Center (UNOSAT) to leverage artificial intelligence for earth observation activities. Partnership with the United Nations Economic Commission for Africa (UNECA) is yet another initiative to support technology augmentation that can help in identifying resource allocation and informed decision-making for governments. Global corporations like Intel, Schneider Electric, Dell, Microsoft, Google, Amazon Web Services,

IBM etc. have prioritised sustainability as part of their corporate agenda and have demonstrated major innovations in climate technology, carried out investments in clean energy and conservation projects and introduced carbon–neutral products.

9.3.2 Food and Beverages Industry

As per the Centre for Science and Environment (2022), the worldwide food industry is responsible for yearly emissions of GHGs to an extent of 21–37%. This business uses a lot of power since it requires energy for crop growth, harvesting, and transportation to the end consumer. Food and beverage packaging typically has a detrimental effect on the environment because it aggregates in landfills and contaminates the water bodies. Leading companies in the Food & Beverages (F&B) Industry have swung into action to tackle the environmental impacts. The F&B industry is extremely energy-intensive and is interconnected with several other sectors such as agriculture, mining, chemicals, retail, glass, supply chains and logistics. Hence the GHG emissions controlled by companies in this industry will have a positive ripple effect on other sub-sectors. An overview of sustainable and regenerative approaches adopted by leading F&B companies—Nestle and ITC is discussed.

Nestle invests CHF1.7 billion annually in R&D, engaging over 4000 people at 23 locations worldwide in its product innovation endeavour. The recently started Nestle Institute of Agricultural Sciences places its focus on three dimensions namely *Plant Science, Dairy Livestock and Agricultural Systems.* Technology is in place to cultivate coffee varieties that deliver higher yields with greater resilience to drought. Efforts are being made in the direction of obtaining plant-based substitutes for dairy and meat which contribute to low carbon footprints. Among the major sources of emissions is *dairy farming*. Technologies are being developed to cut down cattle's ***methane emissions*** and improve manure management. The company partnered with the Dairy Methane Action Alliance in 2023 to drive collaborative efforts in methane emissions. Top-notch priorities include animal welfare and the production of dairy products without distorting environmental requirements. The success of regenerative agriculture has a bearing on the soil health. Principles of ***conservational agriculture*** like the *Zero-tilling technology* are used, where there is minimum disturbance to the soil without tilling, thereby strengthening the biodiversity.

As perUnited Nations Conference on Trade and Development (2018) Trade and Development Report, 90% of marine fish stocks are exploited, given that fish constitutes a key component of animal protein consumed globally. Several interesting sustainable technologies have been developed to produce low-carbon products keeping the environment and the consumer in mind (Nestle, 2023). In 2019, the company steered project ***Sensational VUNA***, to produce plant-based protein sources as an alternative that could provide tuna-like structure, texture and flavour. Such sustainable technologies not only protect biodiversity but also meet the huge demands of the vegan market today. The flagship ***Sensational Burger***, which is an alternative to a beef-based one with plant-based extracts resembles the animal-based ingredients used earlier. Further, the technology used in its production is seen to contribute 80% less carbon emissions than its conventional counterparts. ***Carnation***, a condensed milk product created using oat and rice milk, a dairy-free ***Milo*** beverage, and ***Kitkat V***, a vegan version of the famous chocolate are among the signature products in the product line. Embracing veganism brings down the pressure on dairy farming, which is a great step in the SDG achievement. With an ambition to achieve net zero by 2050, the company has not just worked in the area of introduction of low-carbon products but also focused on building technologies that completely utilise crops and generate minimum waste. ***Nescafé Nativ Cascara***, manufactured in Australia is one such product meeting the above goals. The husk that surrounds the coffee bean is used to generate an exotic beverage infused with flavours. This had a direct bearing on reducing the waste from coffee farming where a byproduct is used in creating a new product altogether.

In order to meet the energy requirements, used coffee grounds are incinerated to produce biogas which runs several production centres. Technologies are being researched to use cellulose fibres from these grounds and produce biodegradable packaging for food products. It set up the *Institute of Packaging Sciences*, a contemporary initiative in the food industry to provide regenerative solutions for packaging including recyclable barrier papers and other bio-based products, as a response to addressing the traditional plastic packaging challenges. Another green chemistry technology is the development of ***enzymatic recycling*** to aid sustainable plastic waste management. With several such initiatives up the ladder, net zero is no longer a fad, but heading towards it progressively is visible through various measures evaluated at the end of 2023. Compared to the baseline emissions as seen in 2018, in a five-year duration by 2023, the company was able to see lowered emissions by 12.75% and crossed more than half towards the achievement of their net zero map of 20% reduction of absolute greenhouse gas emissions by 2025. Logistics is a critical leg of the F&B industry and integrating sustainable *practices into* ***transportation*** is an area of emphasis. The use of natural-gas-powered trucks as part of its global supply chain, enabling truck electrification of over 7700 vehicles in India as part of the Zero Emission Vehicles Emerging Markets drive, and adoption of double stacked rail in the UK than diesel-powered transport are efforts to integrate sustainability in logistics management.

In a scenario of permacrisis, the pathway to green and clean requires sustained efforts. Speaking of the **ITC**, its vision and strategy support building climate resilience in agriculture, green infrastructure, water stewardship; supporting livelihoods and promoting inclusive systems. Noteworthy initiatives in sustainable tech are discussed (ITC, 2023). Targeted *crop development efforts* by decreasing stubble burning practices, choosing the right mechanisation, and high-yield generating crop varieties have benefitted over 700,000 farmers. The ***ITC Mission Millets*** is a drive to enhance production of the super-grain, the much nutritious millets, that are resilient to peak temperatures, thrive in poor soil and seek limited water supplies. Among *biodiversity conservation programs* include bringing into practice organic pest control mechanisms, cycling of nutrients, maintenance of soil health and restoration of degraded areas to support cultivation, thereby helping the agro value chain. The commitment is to augment biodiversity over one million acres in the next five years. As per the Global Risks Report (World Economic Forum, 2024), water is an extremely stressed natural resource and water shortage has been identified among the top five risks in global economies. The responsibility lies in the hands of the governments and enterprises to initiate the required plans of action. ***ITC's Integrated Watershed Development*** projects have focused on storage enhancement by constructing over 28,000 water harvesting sheds. Other projects include minimising water withdrawals, and river basin regeneration programs in Maharashtra, which are in sync with the Government of India's water-conservation programs like the *More Crop Per Drop* and *Jal Shakti Abhyan*. Technologies are also crafted to *expedite the recycling rates* for multi-layered plastic and replacement with compostable packaging alternatives. Select core smart agri-practices adopted to foster the decarbonisation of agriculture include *Direct Seeded Rice (DSR), Broad Bed Furrow (BBF), Plant Growth Regulator (PGR) Spray and Mulching*. DSR mechanism involves scientific-direct seed sowing in fields which plummets water consumption, cost of sowing, cultivation and crop cycle for farmers. BBF method of sowing for soybeans slashes down water requirement and enhances moisture content in the soil which provides for better yield. PGR spray on wheat crops mitigates the ill effects of erratic weather without loss of yield. The mulching technique used in spice production cuts weed growth and maintains the right soil temperature for the crop. The company has adopted ***Intelligent Tobacco Curing Technologies***. Tobacco Curing is a standardised process of making the tobacco leaf market ready to obtain the required characteristics of the leaf and eliminate moisture. The process reduces by a whopping 40% manual labour requirement post-harvest of tobacco using over a hundred smart curing barns in regions where tobacco is grown. Digital sensors and control panels used in these systems make the process environmentally and economically efficient.

9.3.3 Automobile Industry

The automobile industry is among the key contributors to resource depletion and environmental pollution. A circular economy business model approach in the auto sector will involve a cohesive partnership between the material suppliers, producers, tech partners and the final customers. Sustainable manufacturing in automobiles is about reimagining auto manufacturing from design to product lifecycle management including end-of-life recycling strategies. Strategies adopted by some of the leading players in the auto sector—Volkswagen and Toyota are discussed.

Volkswagen has devised strategies to embed sustainability into the Group's governing principles. Mobility in the next era will see a new face altogether and will be predominantly governed by autonomous vehicles and extensively digital. ***New Auto***, is the Group's recent strategy adopted in 2021, keeping in view the goals for the road ahead, attempting to breach thresholds of innovation, customer orientation and sustainability. As per the Group Sustainability Report (Volkswagen, 2022), investment in Tech platforms in the area of ***Mechatronics, Software development,***

Battery Strategies* and *Mobility solutions is the cornerstone for path-breaking future developments that sync with energy goals. The most significant factor in the decrease in emissions is driving ***electric mobility***, with 7% sales of e-vehicles in 2022, moving to 11% presently and aspiring to be at 20% by 2025. The electricity requirements are met through green electricity projects where wind farms are constructed in select European countries. Through the integration of technology in design and manufacturing, passenger car production desires to meet a 50% reduction in emissions. Another infrastructural auxiliary requirement for electric mobility is battery production. The company has set up a ***battery cell production centre*** and five more are in line run exclusively with renewables. High investment in innovative technologies to augment charging infrastructure powered by renewables, and producing electric vehicles that can serve as *mobile power banks* in storing energy are aspirational initiatives that have a direct bearing on promoting sustainable energy ecosystems. Research in electric mobility reveals that e-vehicles powered with renewables will contribute to 60% lower global warming compared to the conventional fossil fuel used. With a synergetic blend of automation and electronics, the ***Scalable Systems Platform (SSP)***, an evolution in mechatronics will be an enabler for high-speed tech updates and upgradation in vehicles across, from basic to luxury models at lower costs and higher energy efficiency. On the software front, a powerhouse which enables seamless automation with intense digitisation in a sustainable manner is the ***CARIAD***. The latest advancements in EV technology, i.e. the ***modular electric drive matrix*** is aimed at achieving zero emissions, and are all set to roll out supplemented with SSPs. The production centres are EMAS certified (Eco-Management and Audit Scheme) to ensure environmental compliance in areas of design, manufacturing, handling exhaust emissions etc. One of the very newer methods of measuring environmental performance is through the implementation of detailed Product ***Life Cycle Assessments*** through ***Sphera's GaBi*** software, a comprehensive tool to achieve product stewardship.

The latest in the production strategy is the ***Zero Impact Factory Program***, as part of the 'goTozero' corporate mission, to make production 100% sustainable. As part of the study, attempts are in the direction of measuring 22 quantified eco-indicators such as wastewater discharge quotients, solvent emissions, and recycling proportion rates, among several others to plan critically addressing the impact points. ***Decarbonising efforts in the value chain*** happen at levels of raw material procurement, suppliers and production. With Memorandums of Understanding with companies like *Salzgitter AG*, a German giant in steel production; *H2 Green Steel*, a Swedish start-up that manufactures hydrogen-powered '*green steel*' that minimises carbon emissions to 95% compared to conventional methods of steel making; Volkswagen is transitioning to ***sustainable procuring***. High-voltage battery production leads to over one-third of emissions generated while manufacturing an electric vehicle. One of the strategies adopted is ensuring supplier compliance to certified low CO_2 emitting sources and also augmenting self-production of battery manufacturing with energy supplies from photovoltaic plants. As part of the ***goToZero Logistics*** drive, the future agenda is to move freight by rail instead of road to cut down GHGs. Another promising logistic channel is marine transport with ships that use renewables such as liquefied natural gas that is known to cause low pollution.

Recycling components is another important step in making manufacturing environmentally compatible and this is executed through increasing the percentage of recyclable components in the SSP-based products. The ***investments in solar parks and wind farms*** to power the production centres through renewables is a significant step and is documented to produce enough green power to supply electricity to over 150,000 households. This speaks volumes of renewables that can be harnessed when invested in technologies that are capable of leveraging them. A documented 1431 implemented measures in the area of sustainability in the current year marks the company's commitment to circularity.

Toyota, another world leader in automobiles, the first to bring the *lean manufacturing philosophy, i.e.* achieving efficiency with minimisation of waste; anchors sustainability in its strategy. As per the Sustainability Report 2022–2023 (Toyota, 2023), the company embraces green mobility solutions through its ***Kaizen culture,*** i.e. continuous improvement. The diverse product line-up of e-vehicles gives customers a wide range of choices to opt for electric vehicles. ***Tax incentives*** for e-vehicle buying and public investments in EV infrastructure serve as motivation for the customer to switch to green buying. The self-charging hybrid electric cars saw the highest sales in 2023. A policy framework in India through the *Faster Adoption and Manufacturing of Electric vehicles (FAME)* provided a boost for electric vehicle adoption, primarily through tax incentives. ***Clean fuel technologies*** such as the use of hydrogen, and ethanol blends in Flexi Fuel Vehicle technologies are techniques to transition away from traditional non-renewables. Toyota's ***Lexus*** is exemplary for a sustainable product design. It is the first to have bamboo-charcoal speakers, kenaf's stem for the door trim, recycled plastic for the luggage space, plant sources accounting for 30% of carpeting, castor-seeds based foam for seats, and recycled material for dashboard silencer. The company has embarked on ***green supply chain*** initiatives through Green Purchasing Guidelines, and Model Eco Dealer Program to ensure environmental compliance among suppliers through careful evaluation and ratings and setting targets of 33% lowered GHG emissions per product by 2030. Among the ***green logistics*** methods introduced were CNG trucks, cutting emissions by 0.32 tons each month and tractors, with better mileage. ***Plastic elimination*** was successfully carried out in packaging for components like air cleaners, metal parts, grill radiators and the like, with a cumulative impact of reduction of 2.8 million tons. ***Energy consumption*** strategies through Ledification, BLDC fans- Brush Less Direct Fans with properties of high durability and ability to achieve 50% saving of electricity; HVLS fans—High Volume Low-Speed fans used in industrial ventilation to reduce humidity are adopted.

Among the ***emission control strategies***, a noteworthy achievement is reducing the emission of *volatile organic compounds* (VOC) from the painting of vehicles. Mixing organic thinners with VOCs in desired proportions, fixation of less adhesive hoses, and handling PVC material waste at the time of Robot spray are the kaizen ways of reducing VOCs. Several air conditioner units were replaced and specific refrigerant gases, harmful gases such as Sulphur-di-oxide and Nitrogen Oxide were monitored.

Water conservation saw several initiatives such as the reduction of effluent generation through the use of condensed wastewater and the use of spiral-shaped over V-nozzle designs. The company achieved ***recycling end-of-life vehicles*** to an extent of more than 90%. With a vision to set up 100 product disintegration facilities and to use components from the used products in newer ones, the *Toyota Global 100 Dismantlers' Project* and *Car-to-Car Recycle Project* are ambitious projects lined up. Steel scrap is used to create vehicular components like stiffener plates, calliper brackets, axel and knuckle assembly. A highly structured approach to digitisation has been taken through implementation of sustainable technologies such as *Computer Vision Based Check* for Quality checks; *Smart ARS and Digi Genba, Car ETD tracking* in Production; *Material Inward Process Optimization* and *Autonomous Mobile Robot* in Logistics management. The Toyota case is exemplary of consistent delivery and sustainable excellence.

9.3.4 Pharmaceuticals Industry

The pharma industry, as per the World Economic Forum Report, contributes to 4.4% of the global GHG emissions and is expected to grow by 3× by the year 2050 if it is left with minimal intervention. Further, the pharma supply chains attribute to more than 70% of healthcare-related carbon emissions, making this an extremely energy-intensive sector (Belkhir & Elmeligi, 2019). Environmental-friendly tech implementation by leading pharma companies—Pfizer and Glaxo Smith Kline (GSK) will be comprehensively reviewed.

Pfizer, a leading global pharma company that invests heavily in innovation driven by sustainability goals, aims to achieve net zero by 2040 and is currently standing at an 11% overall reduction in GHGs compared to 2019. The Environment, Social & Governance Report (Pfizer, 2022) discusses several initiatives taken as part of the climate strategy. About 8% of the energy requirements were met using renewable technologies by 2022 and the target is to achieve 100% by 2030. Pfizer's role was significant in vaccine manufacturing during the pandemic, the COVID-19 vaccines and Paxlovid, an antiviral drug required transportation globally in cold storage mechanisms. In order to mitigate the emission contribution due to the above, other products were transported through oceans rather than air routes. Considering keeping the environmental footprint at bay and reducing the emissions in the *day-to-day operations*, investments are carried out in renovation projects to cut resource demand, with an increased focus on waste management and optimal water utilisation. The company channelled sourcing from renewables through Power Purchase Agreements (PPA) which support replenishing end-of-life commodities by alternatives that are carbon–neutral and by also augmenting process efficiency to make manufacturing environmentally friendly. Sustainability demands collaboration. Through a dedicated partnership channelised using PPA with Vesper Energy, the company will be meeting about 50% of its global electricity requirements generated by investments in solar projects in the subsequent years. Close partnership with suppliers of Active Pharmaceutical Ingredients (APIs) through the *Activate* program is an initiative to bring down emissions in the pharma value chain.

Rolling out *sustainable medicines* is on their high-priority agenda. While manufacturing medicines, techniques of energy conservation and optimal use and mix of raw materials are followed. Life Cycle Assessment (LCA) for molecules, drugs, vaccines and machines are undertaken to assess shortcomings and areas of improvement with respect to GHG emissions, enabling a proactive strategy in producing medicines that exert minimum damage to nature. Environmental risk assessment methodologies and emission control technologies monitor and ensure the *safe discharge* of APIs into water bodies. Cold chain shipping containers that store medicines saw the use of recyclable packaging materials to reduce the volume of waste generated and sent to landfills. Adherence to industry-recognised quality management systems and standards including Good Laboratory Practices, Good Pharmacovigilance Practices, Good Manufacturing Practices and the International Organization for Standardization (ISO) which integrated sustainability in their regulations, makes the company credible on the quality front.

An extremely conscious-proactive strategy has been adopted in water-stressed zones by quantifying water consumption, investing in technologies supporting water treatment and recycling, and most importantly encouraging other partners in the value chain to mimic these initiatives to *conserve water resources. Supply chain resilience* in a sustainable way through *active collaboration mode* was seen especially in vaccine-related transportation globally. Several low-income countries faced issues due to limited cold storage facilities. Accomplishing partnerships with local supply chains, the company effectively channelised the vaccine transit. Through engagement with *Zipline,* an American company specialising in instant logistics and drone delivery, Pfizer achieved transportation of vaccines to regions in Africa like Nigeria and Ghana through drones. This mode of delivery reduced the use of over 34,000 gallons of gasoline, which made it nature-friendly. Another interesting initiative in supply chain efficiency is the collaboration of Pfizer with *Project Last Mile,* a public–private partnership of Coca-Cola with the African Ministry of Health and other foundations to ensure medical supplies were made accessible to remote areas in several parts of Africa.

Talking of the **GSK**, another top player in the pharma, achieved a 10% decline in carbon emissions in its operational processes in 2023 compared to 2022. Several techniques have been adopted in the area of responsible antibiotic production, PVC-free packaging, and manufacturing of carbon footprinting tool to support the net zero agenda as per the ESG Performance Report (GSK, 2023). The utilisation of medicines and vaccines generates emissions amounting to 57% of the company's total carbon footprint. Of this, about 48% of emissions are from propellents of drugs like Ventolin or salbutamol which has over 35 million global users who suffer from respiratory conditions. Innovative technologies are looking ***at low carbon-intensive propellents*** to control their huge rate of emission to an extent of 90%. The clinical trials are currently running to facilitate the transition to low-carbon propellants. In order to combat antimicrobial resistance, research-based pharma companies, drug developers, generics etc. came together to form a private association called the AMR (Anti-Microbial Resistance) Industry Alliance. This association came into force as a response to the United Nations' demand for the pharma community to deal with antimicrobial resistance which impacts adversely the sustainable development goal of good health and well-being. Their manufacturing sites are 100% compliant with AMR Alliance regulations and ensure ***responsible production of antibiotics***. Technologies are in place to ensure the safe *discharge of wastewater* from the production of APIs. Sustainability targets are strongly directed in the space of ***sourcing*** raw materials, mainly 98% of palm oil materials have been certified as being drawn from sustainable sources. To minimise the use of core natural extracts, safe synthetics are used as alternatives in drugs and vaccines. Further, streamlining the manufacturing processing architecture has lowered carbon emissions substantially. Concerning ***ocean targets***, sourcing of substitutes to marine-derived materials is set as a target for 2030. With a new Marine Sustainable Sourcing Standard and close links with suppliers who provide alternatives, the purpose is to preserve the depleting marine resource. Establishing ***sustainable supply chains*** is vital for the pharma sector as it is heavily reliant on it. A collaborative supply chain initiative *Converge* extended the partnership of the company with its suppliers to encourage laboratory sustainability and the *Active program* steered suppliers to slash ***emissions from APIs***. Packaging of products is carried out across the various sites using PVC-free materials and recyclable materials to an extent of 86%.

Product stewardship is achieved through the in-house ***carbon footprinting tool*** developed which carries out product LCA to identify the extent of emissions by the products. This supports new product development with sustainable designs where scientists can integrate eco-principles right at the product design stage. Sustainable tech initiatives are implemented on scale with a target of obtaining net zero GHG emissions by 2045. A 100% renewable use target by 2030 is desired, solar farms and wind turbines are already being used as alternative sources of energy. When it comes to the pharmaceuticals sector, its critical legs involve drug manufacturing and delivery. Hence it is evident that taking the necessary steps in managing supply chains will be instrumental in managing emissions from this sector.

9.4 Policy Implications in Technology Adoption and Future Scope

Establishing a Green Industrial Policy is essential to define the goals of decarbonisation. Setting policies and rolling them out with a transition framework that is specific to sectors will be a milestone in SDG attainment (Meadowcroft & Rosenbloom, 2023). There is no '*one shoe* fits all' approach. Implementing overarching policy frameworks suitable to respective sectors is the key. A calculative, designed, sector-specific policies that direct companies to address their emissions and removals will be a game changer. The infrastructure required for monitoring, verification and reporting of removed carbon is in the stage of development. Focus in this direction with effective policy guidelines will support in accelerating towards the carbon–neutral goals (Fankhauser et al., 2022). Driving policy frameworks that enable the assessment of business corporations regarding their sustainable practices, standardisation of evaluation metrics, and feedback systems will close the loop and support the decarbonisation endeavour (Rogelj et al., 2021).

Climate change policy frameworks entail evaluations which must factor the interplay between ecology and the economy. While the relevance of sustainability is heavily spoken about, there are several aspects to be addressed on the policy side to make this a reality. The market mechanisms should be conducive in that direction. Innovations and investments in clean energy projects are requisite to achieve the net zero targets. Literature points to the setbacks and market failures due to the lack of impetus for investment in innovation (Stern & Stiglitz, 2021). Innovators are reluctant because of not being able to leverage the full returns of investment in R&D and concerns about intellectual property rights mechanisms (Stern & Valero, 2021). This necessitates streamlining of policies that steer benefits for innovators and also for clean energy patenting. Policies must accommodate for enabling scalable implementation of sustainable targets. The focus on mitigating climate damage will be possible through an inclusive approach, investment in interdependent physical, social and human capital, and development of support teams coordinated at local, national and global levels. Clean technologies bring about the scale of production at lower costs, hence providing subsidies and incentives for the same should be of significance for policymakers. Levying of carbon taxes should be in line with the extent to which the economies are innovation-intense in green R&D (Aghion et al., 2016).

Automation with AI augmentation has to be employed to improve the demanufacturing process that will help in recovering components. Research in this direction will further improve efficiency in material handling. Steering investments in next-generation recycling technologies will redefine disassembly mechanisms with minimal impact on recovered spare parts. The development of niche and intelligent technologies with RGB imaging for smart component recovery at a large operational scale is important to overcome the challenges of limited access to the latest recycling technologies.

9.5 Conclusions and Limitations

While a whole range of initiatives and policies are in place to achieve zero emissions and engage in environmentally friendly practices in manufacturing, it does come with a large number of challenges. Addressing them is the key to driving successful implementation. While top notch companies indulge in sustainable sourcing, maintaining high standards and safety in the use of recycled materials in new products involves due diligence systems and audits. A recycled raw material's properties are likely to differ from its original. Handling issues of toxicity or contamination is important.

Non-availability of recycled source materials across global locations could become a deterrent in levelling up sustainable sourcing. The key here will be to integrate new suppliers into the value chain and establish closed-loop supply chains to retain ecosystem resilience. To meet the demand and achieve the required scale of production, committed partnerships and cross-collaborations of the connected industries will be the need of the hour.

The future roadmap to net zero seeks mitigating ***policy*** risks, ***market*** risks and ***technology*** risks. The models in transitioning to net zero require the above focus. From the policy angle, levying Carbon taxes, regulations that prohibit the use of hazardous materials in production, legal frameworks on the use of land, wsater and biodiversity, and increasing the requirements for reporting emissions through structured mechanisms as part of Environmental and Social Governance will be methods to limit climate emissions. Companies should look at investments in *technologies* that will help bring down the expenses associated with carbon neutrality in the value chain. This would involve the effective replacement of commodities and machines that are high emitters of GHGs. When it comes to handling the *market risk,* bringing low-carbon products into the product line, and meeting the requirements of customers who seek products that have environmental credentials will be the new goalpost for companies.

Measuring sustainability is important to know the extent of achievement of net zero targets. Several researchers have worked to develop sustainability measurement metrics over the years (Eurostat, 2008; Gunnarsdóttir et al., 2020, 2022). Prominent ones are the *Energy Architecture Performance Index* and *Energy Trilemma Index* (ETI) for cross-country comparisons. Another notable metric is the *Energy Indicator for Sustainable Development* (EISD), which is known to be robust, transparent, comprehensive, and is developed through stakeholder input keeping in view the interconnections existing in the sustainability framework. However, devaluing social factors and overly highlighting economic implications are discussed as the demerits of the EISD. In this light, there is an immense amount of scope for research in developing holistic indicators that must envelop the various facets of sustainability. Sector-specific metrics can be developed to analyse carbon neutrality at micro-levels.

The pandemic demonstrated how quickly the human race can adapt to situations, pull communities together and swiftly respond to requirements, the minute the realisation sets in that it is the only choice! Such should be the urgency when it comes to handling the climatic catastrophe. Lightning speed response, social cohesion, and sustained investment in green R&D will enable us to achieve the goals of net zero. A comprehensive, holistic, pragmatic and integrated approach through multilateralism, solidarity, inclusivity, and embedding green technologies in production, processes and operations is quintessential to bringing all the stakeholders together in its achievement of a net zero future. Active collaboration among the various stakeholders -the customer, business and government with a firm affirmation to exhibit responsible behaviour is the only way to achieve environmental stewardship.

References

Aghion, P., Dechezleprêtre, A., Hemous, D., Martin, R., & Van Reenen, J. (2016). Carbon taxes, path dependency, and directed technical change: Evidence from the auto industry. *Journal of Political Economy, 124*(1), 1–51. https://doi.org/10.1086/684581

Apple. (2023). *Environmental progress report.* Retrieved from https://www.apple.com/environment/pdf/Apple_Environmental_Progress_Report_2023.pdf

Belkhir, L., & Elmeligi, A. (2019). Carbon footprint of the global pharmaceutical industry and relative impact of its major players. *Journal of Cleaner Production, 214*, 185–194. https://doi.org/10.1016/j.jclepro.2018.11.204

Belyaeva, Z., Rudawska, E. D., & Lopatkova, Y. (2020). Sustainable business model in food and beverage industry—A case of Western and Central and Eastern European countries. *British Food Journal, 122*(5), 1573–1592. https://doi.org/10.1108/BFJ-08-2019-0660

Burke, A. (2022). An architecture for a net zero world: Global climate governance beyond the epoch of failure. *Global Policy, 13*, 24–37. https://doi.org/10.1111/1758-5899.13159

Caradonna, J. L. (2022). *Sustainability: A history.* Oxford University Press.

Centre for Science and Environment. (2022). *The State of India's environment 2022.* Retrieved from https://csestore.cse.org.in/default/state-of-india-s-environment-2022.html

Chepeliev, M., & van der Mensbrugghe, D. (2020). Global fossil-fuel subsidy reform and Paris agreement. *Energy Economics, 85*, 104598. https://doi.org/10.1016/j.eneco.2019.104598

Eurostat, L. (2008). *Energy indicators for sustainable development: Guidelines and methodologies.* Retrieved from http://www-pub.iaea.org/MTCD/publications/PDF/Pub1222f_web.pdf

Fankhauser, S., Smith, S. M., Allen, M., Axelsson, K., Hale, T., Hepburn, C., Kendall, J. M., Khosla, R., Lezaun, J., Mitchell-Larson, E., Obersteiner, M., Rajamani, L., Rickaby, R., Seddon, N., & Wetzer, T. (2022). The meaning of net zero and how to get it right. *Nature Climate Change, 12*(1), 15–21. https://doi.org/10.1038/s41558-021-01245-w

Glaxo SmithKline. (2023). *ESG performance report.* Retrieved from https://www.gsk.com/media/11009/esg-performance-report-2023.pdf

Gunnarsdóttir, I., Davidsdottir, B., Worrell, E., & Sigurgeirsdóttir, S. (2020). Review of indicators for sustainable energy development. *Renewable and Sustainable Energy Reviews, 133*, 110294. https://doi.org/10.1016/j.rser.2020.110294

Gunnarsdóttir, I., Davidsdottir, B., Worrell, E., & Sigurgeirsdóttir, S. J. E. P. (2022). Indicators for sustainable energy development: An Icelandic case study. *Energy Policy, 164*, 112926. https://doi.org/10.1016/j.enpol.2022.112926

International Energy Agency (IEA). (2023). *CO_2 emissions in 2023.* Retrieved from https://www.iea.org/

ITC. (2023). *Sustainability integrated report 2023.* Retrieved from https://www.itcportal.com/sustainability/sustainability-integrated-report-2023/ITC-Sustainability-Integrated-Report-2023.pdf

Meadowcroft, J., & Rosenbloom, D. (2023). Governing the net-zero transition: Strategy, policy, and politics. *Proceedings of the National Academy of Sciences, 120*(47), e2207727120. https://doi.org/10.1073/pnas.2207727120

Moore, J. E., Mascarenhas, A., Bain, J., & Straus, S. E. (2017). Developing a comprehensive definition of sustainability. *Implementation Science, 12*, 1–8. https://doi.org/10.1186/s13012-017-0637-1

Nestle. (2023). *Creating shared value sustainability report 2023.* Retrieved from https://www.nestle.com/sites/default/files/2024-02/creating-shared-value-sustainability-report-2023-en.pdf

Next Move Strategy Consulting. (2024). *Waste management market by waste type, by waste treatment and by end user—Global opportunity analysis and industry forecast 2024–30.* Retrieved from https://www.nextmsc.com/report/waste-management-market

NVIDIA. (2023). *Corporate responsibility report fiscal year 2023.* Retrieved from https://images.nvidia.com/aem-dam/Solutions/documents/FY2023-NVIDIA-Corporate-Responsibility-Report-1.pdf

Pfizer. (2022). *Environmental, social & governance report*. Retrieved from https://www.pfizer.com/sites/default/files/investors/financial_reports/annual_reports/2022/files/Pfizer_ESG_Report.pdf

Pye, S., Broad, O., Bataille, C., Brockway, P., Daly, H. E., Freeman, R., Gambhir, A., Geden, O., Rogan, F., Sanghvi, S., Tomei, J., Vorushylo, I., & Watson, J. (2021). Modelling net-zero emissions energy systems requires a change in approach. *Climate Policy, 21*(2), 222–231. https://doi.org/10.1080/14693062.2020.1824891

Renna, P., & Materi, S. (2021). A literature review of energy efficiency and sustainability in manufacturing systems. *Applied Sciences, 11*(16), 7366. https://doi.org/10.3390/app11167366

Rogelj, J., Geden, O., Cowie, A., & Reisinger, A. (2021). Three ways to improve net-zero emissions targets. *Nature, 591*(7850), 365–368. https://doi.org/10.1038/d41586-021-00662-3

Sovacool, B. K., Bazilian, M., Griffiths, S., Kim, J., Foley, A., & Rooney, D. (2021). Decarbonising the food and beverages industry: A critical and systematic review of developments, sociotechnical systems and policy options. *Renewable and Sustainable Energy Reviews, 143*, 110856. https://doi.org/10.1016/j.rser.2021.110856

Stern, N., & Stiglitz, J. E. (2021). *The social cost of carbon, risk, distribution, market failures: An alternative approach* (Vol. 15). National Bureau of Economic Research. Retrieved from https://policydialogue.org/files/publications/papers/w28472.pdf

Stern, N., & Valero, A. (2021). Innovation, growth and the transition to net-zero emissions. *Research Policy, 50*(9), 104293. https://doi.org/10.1016/j.respol.2021.104293

Swallow, T. (2024). Top 10: Sustainable technology companies. *Sustainability*. Retrieved from https://sustainabilitymag.com/articles/top-10-sustainable-technology-companies

Thiele, L. P. (2024). *Sustainability*. Wiley.

Toyota Kirloskar Motor Pvt. Ltd. (2023). *Sustainability report 2022–2023*. Retrieved from https://www.toyotabharat.com/documents/toyota-in-india/environment/tkm-sustainability-report-2023.pdf

United Nations Climate Change. (2023). *COP28 UAE*. Retrieved from https://www.theuaeconsensus.com/

United Nations Climate Change. (2024). *Process & meetings*. Retrieved from https://unfccc.int/process-and-meetings/the-paris-agreement

United Nations Conference on Trade & Development. (2018). *Trade and development report*. Retrieved from https://unctad.org/system/files/official-document/tdr2018_en.pdf

United Nations Institute for Training & Research. (2024). *Global E-waste monitor 2024*. Retrieved from https://ewastemonitor.info/the-global-e-waste-monitor-2024/

Volkswagen. (2022). *Group sustainability report*. Retrieved from https://www.volkswagen-group.com/en/publications/more/group-sustainability-report-2022-1644

World Economic Forum. (2024). *Global risks report*. Retrieved from www.weforum.org

Yue, X., Pye, S., DeCarolis, J., Li, F. G., Rogan, F., & Gallachóir, B. Ó. (2018). A review of approaches to uncertainty assessment in energy system optimization models. *Energy Strategy Reviews, 21*, 204–217. https://doi.org/10.1016/j.esr.2018.06.003

Chapter 10
Navigating Road for Net Zero Policies to Aligning Policy, Politics, and Governance for Environmental Sustainability

Azamat Ali

Abstract Achieving a net zero emissions economy requires a nuanced understanding of how policy, politics, and governance intersect and drive progress. This chapter explores into the interconnected relationship among these dimensions in advancing the goal of net zero emissions. Firstly, it examines the role of policy frameworks in incentivizing decarbonization across various sectors, with a focus on integrating national and international policies for global cooperation. Secondly, it explores the political dynamics influencing the adoption and longevity of net zero policies, stressing the importance of political will and stakeholder collaboration. Lastly, it analyzes governance structures crucial for effective decision-making, emphasizing inclusivity and transparency. Through this interdisciplinary analysis, the study underscores the need for an integrated approach to policy, politics, and governance to realize a sustainable, net zero emissions future.

Keywords Net zero emissions · Policy frameworks · Political dynamics · Governance structures · Decarbonization

10.1 Introduction

In recent years, the urgency to combat climate change has prompted a paradigm shift in global discussions surrounding environmental sustainability. At the forefront of this discourse is the concept of a net zero emissions economy, which represents a critical milestone in the journey towards mitigating the impacts of climate change. A net zero emissions economy is characterized by a state where the amount of greenhouse gases emitted into the atmosphere is equal to the amount removed or

---

A. Ali (✉)
School of Liberal and Creative Arts (Journalism and Mass Communication), Lovely Professional University, Phagwara, Punjab, India
e-mail: azamatcug@gmail.com

R. Singh and D. Crowther (eds.), *Transition Towards a Sustainable Future*, Approaches to Global Sustainability, Markets, and Governance,
https://doi.org/10.1007/978-981-97-5756-5_10

offset through various mechanisms such as carbon capture and storage, reforestation, and sustainable land management practices (Fuss et al., 2018). This ambitious goal requires a fundamental transformation of our energy systems, industrial processes, transportation networks, and land use practices to drastically reduce carbon emissions and achieve a state of equilibrium between emissions and removals.

Central to the realization of a net zero emissions economy are the intertwined domains of policy, politics, and governance. Policy frameworks serve as the foundation upon which climate action is built, providing the regulatory, financial, and institutional mechanisms necessary to drive decarbonization efforts across sectors (Jordan et al., 2015). Political dynamics shape the landscape in which these policies are formulated and implemented, influenced by a myriad of factors including political will, public opinion, and stakeholder interests (Bäckstrand & Kuyper, 2017). Governance structures, meanwhile, play a critical role in facilitating effective decision-making, coordination, and accountability in the pursuit of climate goals (Victor et al., 2017). Together, these dimensions form a complex web of interactions that are essential for achieving the transformative changes required for a sustainable, net zero emissions future.

The importance of policy, politics, and governance in achieving the objectives of a net zero emissions economy cannot be overstated. Policies provide the necessary incentives and regulations to spur investment in clean energy technologies, promote energy efficiency, and drive innovation in low-carbon industries (Smith et al., 2016). Political dynamics shape the narrative surrounding climate action, influencing public perceptions, electoral outcomes, and the prioritization of environmental issues on political agendas (Bernauer & McGrath, 2016). Governance structures dictate the rules of the game, determining how decisions are made, resources allocated, and responsibilities distributed among various actors in the climate governance landscape (Biermann et al., 2009).

Moreover, the interconnected nature of these dimensions underscores the need for a holistic and integrated approach to climate policymaking and implementation. Effective climate policies must be grounded in political realities, supported by strong leadership, and underpinned by robust governance mechanisms to ensure their effectiveness and longevity (Bäckstrand, 2006). Furthermore, inclusive and participatory governance processes are essential for fostering collaboration, building trust, and mobilizing collective action among diverse stakeholders including governments, businesses, civil society organizations, and communities (Bäckstrand & Kuyper, 2017).

In the following sections of this chapter, each dimension—policy, politics, and governance—is explored into, examining their respective roles, interactions, and implications for achieving a net zero emissions economy. Through a comprehensive review of existing literature, case studies, and examples from around the world, insights are provided into the complex challenges and opportunities associated with transitioning to a sustainable, low-carbon future.

10.2 Literature Review

The transition to a net zero emissions economy represents a monumental challenge requiring a thorough understanding of the existing literature on various aspects including net zero emissions, policy frameworks, political dynamics, and governance structures. This section literature review aims to provide a comprehensive overview of key concepts, debates, challenges, and best practices in transitioning towards a net zero emissions future.

10.2.1 Net Zero Emissions: Concepts and Definitions

The concept of achieving net zero emissions revolves around the idea of balancing the amount of greenhouse gases emitted into the atmosphere with the amount removed or offset through natural or artificial means. While the ultimate goal is to eliminate greenhouse gas emissions altogether, achieving net zero represents a critical milestone in mitigating climate change and limiting global warming to well below 2 degrees Celsius above pre-industrial levels, as outlined in the Paris Agreement (Fuss et al., 2018). Different sectors contribute varying levels of emissions, with energy, transportation, industry, agriculture, and land use being primary contributors.

10.2.2 Policy Frameworks for Net Zero Emissions

Policy frameworks play a central role in driving the transition to a net zero emissions economy by providing the necessary incentives, regulations, and pathways for decarbonization across sectors (Smith et al., 2016). Various policy instruments have been deployed to accelerate the transition, including carbon pricing mechanisms such as carbon taxes and cap-and-trade systems, renewable energy mandates and subsidies, energy efficiency standards, and regulations targeting emissions-intensive industries. Additionally, long-term policy commitments, such as nationally determined contributions (NDCs) under the Paris Agreement, serve as guiding frameworks for countries to set emissions reduction targets and develop strategies for achieving them.

10.2.3 Political Dynamics of Net Zero Policies

The adoption and implementation of net zero policies are influenced by a myriad of political dynamics, including political will, public opinion, and stakeholder interests. While there is growing recognition of the urgent need for climate action, political barriers persist, hindering the formulation and implementation of ambitious policies (Bäckstrand & Kuyper, 2017). Political ideologies, partisan divides, and competing policy priorities often shape the trajectory of climate policies, leading to inconsistencies and delays in action. However, effective stakeholder engagement, coalition-building, and public mobilization have the potential to overcome political obstacles and foster bipartisan support for climate action.

10.2.4 Governance Structures for Effective Implementation

Effective governance structures are essential for translating net zero policies into concrete action on the ground. Multilevel governance approaches, which involve collaboration between different levels of government, as well as non-state actors, can help coordinate efforts and leverage resources for climate action (Jordan et al., 2015). Moreover, institutional arrangements that promote policy coherence and integration across sectors are critical for addressing the complex and interconnected nature of climate change (Oberthür & Gehring, 2006). Transparency, accountability, and inclusivity are fundamental principles that underpin effective governance, ensuring that decision-making processes are participatory, equitable, and responsive to diverse stakeholder needs.

10.2.5 Key Debates, Challenges, and Best Practices

One of the key debates surrounding the transition to net zero emissions revolves around the feasibility and scalability of various mitigation strategies, particularly those involving negative emissions technologies (NETs) such as carbon capture and storage (CCS), afforestation, and soil carbon sequestration (Fuss et al., 2018). While these technologies hold promise for removing carbon dioxide from the atmosphere, they also raise concerns regarding their environmental, social, and economic implications. Additionally, the equitable distribution of costs and benefits associated with climate action remains a contentious issue, with debates over burden-sharing, climate justice, and the role of developed versus developing countries in financing and implementing mitigation efforts.

Challenges abound in the transition to a net zero emissions economy, ranging from technological and economic barriers to political and institutional constraints. The rapid pace of technological innovation and market dynamics presents both opportunities and challenges for decarbonization, requiring agile and adaptive policy responses (Smith et al., 2016). Economic considerations, including the costs of transitioning to low-carbon technologies and the potential impacts on employment and competitiveness, pose significant challenges for policymakers and businesses alike. Moreover, addressing social and equity concerns, such as the distributional impacts of climate policies on vulnerable communities and ensuring a just transition for workers in carbon-intensive industries, remains a pressing challenge that requires careful attention and proactive policy interventions.

Despite these challenges, there are also numerous examples of best practices and success stories in transitioning towards a net zero emissions future. Countries such as Sweden, Denmark, and Iceland have demonstrated leadership in adopting ambitious climate targets, investing in renewable energy infrastructure and implementing policies to promote energy efficiency and sustainable transportation (Jacobsson & Lauber, 2006). Similarly, cities and subnational jurisdictions around the world are taking proactive measures to reduce emissions, enhance resilience, and promote sustainable development through innovative policies and initiatives (Bulkeley et al., 2009). These examples underscore the importance of political leadership, stakeholder engagement, and effective governance in driving progress towards a net zero emissions economy.

10.3 Policy Frameworks for Net Zero

In the pursuit of achieving a net zero emissions economy, robust and effective policy frameworks are indispensable. This section explores the diverse range of policy interventions necessary for decarbonizing various sectors of the economy, discussing the pivotal roles of incentives, regulations, and pathways in driving decarbonization efforts. Furthermore, we highlight examples of effective policy frameworks implemented by different countries or regions, showcasing innovative approaches and best practices in the transition towards net zero emissions.

10.3.1 Types of Policy Interventions

Policy interventions for achieving net zero emissions encompass a broad spectrum of measures targeting key sectors such as energy, transportation, industry, agriculture, and land use. These interventions can be categorized into several types:

- *Carbon Pricing Mechanisms*: Carbon pricing, including carbon taxes and cap-and-trade systems, aims to internalize the social cost of carbon emissions by placing a price on greenhouse gas emissions. By incentivizing businesses and individuals to reduce emissions or invest in low-carbon technologies, carbon pricing mechanisms create market signals that drive decarbonization (Fischer & Newell, 2008).
- *Renewable Energy Targets and Incentives*: Policies promoting renewable energy deployment, such as feed-in tariffs, production tax credits, and renewable portfolio standards, play a critical role in accelerating the transition away from fossil fuels towards clean, renewable energy sources (Jacobsson & Lauber, 2006).

- *Energy Efficiency Standards and Incentives*: Regulations and incentives aimed at improving energy efficiency across industries, buildings, and appliances reduce energy consumption and associated greenhouse gas emissions (Gheuens & Oberthür, 2021).
- *Regulations and Standards for Emission Reductions*: Sector-specific regulations and standards, such as vehicle emission standards, industrial emissions limits, and methane capture requirements in agriculture, are essential for achieving emission reductions and ensuring compliance (Goulder & Parry, 2008).
- *Innovation and Research Funding*: Public investments in research, development, and deployment of low-carbon technologies stimulate innovation and accelerate the adoption of breakthrough solutions for decarbonization (Pan et al., 2023).

10.3.2 Role of Incentives, Regulations, and Pathways in Decarbonization

Incentives, regulations, and pathways form the cornerstone of effective decarbonization strategies across sectors. Incentives, such as subsidies for renewable energy projects or tax credits for electric vehicles, encourage investment and adoption of low-carbon technologies by reducing financial barriers and promoting market competitiveness (Barnes, 2021). Regulations, on the other hand, establish legal frameworks and enforceable standards that mandate emission reductions, energy efficiency improvements, and adoption of cleaner technologies (Goulder & Parry, 2008).

Pathways for decarbonization outline transition plans and milestones towards achieving net zero emissions, providing clarity and direction for policymakers, businesses, and stakeholders. These pathways often involve sector-specific strategies tailored to the unique challenges and opportunities within each industry (Wyns et al., 2019). For example, pathways for decarbonizing the energy sector may prioritize investments in renewable energy infrastructure and grid modernization, while pathways for decarbonizing transportation may focus on promoting electric vehicles and enhancing public transit systems.

10.3.3 Examples of Effective Policy Frameworks

Several countries and regions have implemented innovative policy frameworks that demonstrate success in driving decarbonization and progress towards net zero emissions:

- *European Union (EU)*: The EU has established ambitious climate targets under its European Green Deal, including a binding target of achieving net zero greenhouse gas emissions by 2050. The EU Emissions Trading System (EU ETS) is a key policy instrument that caps emissions from industry and power generation while promoting investment in low-carbon technologies (Fetting, 2020).

- *United Kingdom (UK)*: The UK has enshrined its commitment to achieving net zero emissions by 2050 into law. The UK government has implemented a comprehensive policy framework, including phasing out coal-fired power plants, incentivizing offshore wind development, and introducing carbon pricing mechanisms (Committee on Climate Change, 2019).
- *California, USA*: California has implemented a suite of policies to reduce greenhouse gas emissions, including a cap-and-trade programme, renewable portfolio standards, and zero-emission vehicle mandates. These policies have contributed to significant reductions in emissions while driving innovation and economic growth in clean technology sectors (Hanemann & Farrell, 2006).
- *Costa Rica*: Costa Rica is recognized for its commitment to sustainability and has achieved significant success in decarbonizing its economy through policies promoting renewable energy, forest conservation, and sustainable agriculture (Teske et al., 2020).

These examples highlight the diversity of approaches and policy instruments used by different countries and regions to accelerate decarbonization and transition towards net zero emissions. By sharing lessons learned and best practices, these policy frameworks serve as valuable models for other jurisdictions seeking to embark on a similar path towards sustainability.

10.4 Political Dynamics and Net Zero Policies

The transition to a net zero emissions economy is not merely a technical or economic challenge but is deeply intertwined with political dynamics. This section explores into the multi-faceted political landscape surrounding net zero policies, analyzing the influence of political will, public engagement, stakeholder collaboration, ideologies, and interests. Furthermore, it examines strategies for overcoming political barriers to realizing net zero ambitions.

10.4.1 Political Will and Leadership

Political will, defined as the determination and commitment of political leaders to prioritize and implement climate action, is a crucial driver of net zero policies (Bernauer & McGrath, 2016). Strong political leadership is essential for setting ambitious targets, mobilizing resources, and galvanizing public support for transformative changes. Countries with proactive and visionary leaders who champion climate action have been more successful in adopting and implementing net zero strategies

(Bäckstrand & Kuyper, 2017). Conversely, a lack of political will or leadership can hinder progress and result in inertia or backsliding on climate commitments.

10.4.2 Public Engagement and Stakeholder Collaboration

Public engagement and stakeholder collaboration are vital for building broad-based support and legitimacy for net zero policies. Meaningful engagement with citizens, communities, businesses, and civil society organizations fosters transparency, accountability, and inclusivity in decision-making processes (Bäckstrand & Kuyper, 2017). Moreover, involving stakeholders in the co-design and co-implementation of policies enhances their effectiveness and ensures that diverse perspectives and interests are considered (Bernauer & McGrath, 2016). Public awareness campaigns, citizen consultations, and participatory forums are examples of mechanisms that can facilitate engagement and collaboration in the development and implementation of net zero strategies.

10.4.3 Impact of Political Ideologies and Interests

Political ideologies and interests shape the discourse and decision-making surrounding net zero policies, influencing the prioritization of climate action and the choice of policy instruments. Ideological divides between conservative and progressive parties often manifest in differences in climate policy preferences, with conservative governments tending to prioritize economic growth and deregulation over environmental protection (Bäckstrand, 2006). Furthermore, powerful vested interests, such as fossil fuel industries, may exert significant influence on policymaking processes, lobbying against regulations that threaten their profitability (Bernauer & McGrath, 2016). As a result, political ideologies and interests can act as barriers to the adoption and implementation of ambitious net zero strategies.

10.4.4 Strategies for Overcoming Political Barriers

Overcoming political barriers to net zero ambitions requires a multi-faceted approach that addresses both structural and strategic challenges:

- *Building Coalitions and Alliances*: Forming coalitions and alliances among diverse stakeholders, including political parties, businesses, civil society organizations, and grassroots movements, can amplify voices calling for climate action and exert pressure on decision-makers to prioritize net zero policies (Bäckstrand & Kuyper, 2017).

- *Framing Climate Action as a Win–Win*: Highlighting the co-benefits of climate action, such as job creation, public health improvements, and energy security, can help garner support from sectors that may be sceptical or resistant to change (Bernauer & McGrath, 2016). Framing climate action as an opportunity for economic growth and innovation can appeal to a broader range of stakeholders and transcend ideological divides.
- *Political Education and Advocacy*: Increasing political literacy and awareness among citizens through education and advocacy efforts can empower individuals to demand ambitious climate policies from their elected representatives (Bäckstrand & Kuyper, 2017). Grassroots campaigns, lobbying efforts, and electoral mobilization can pressure politicians to align with public preferences and prioritize net zero agendas.
- *Institutional Reforms and Capacity Building*: Strengthening institutions responsible for climate governance, enhancing transparency and accountability mechanisms, and investing in the capacity of policymakers and public officials can improve the effectiveness and credibility of climate policymaking processes (Bäckstrand, 2006). Reforming electoral systems, campaign finance laws and lobbying regulations can mitigate the influence of special interests on policymaking.

10.5 Governance Structures for Effective Implementation

Achieving a net zero emissions economy requires more than just ambitious policies; it necessitates effective governance structures that can facilitate implementation, coordination, and adaptation. This section evaluates the governance structures essential for supporting the transition to a net zero economy, emphasizing the principles of inclusivity, transparency, and accountability in decision-making processes related to decarbonization. Furthermore, we compare different governance models and assess their effectiveness in facilitating sustainable policies.

10.5.1 Governance Structures for Decarbonization

Effective governance structures for decarbonization must be adaptive, collaborative, and responsive to the complexities of climate change mitigation. Key components of governance structures include:

- *Multilevel Governance*: Climate action requires coordination across multiple levels of government, including local, regional, national, and international levels. Multilevel governance frameworks promote vertical and horizontal coordination, enabling different actors to collaborate, share responsibilities, and align policies towards common objectives (Oberthür & Gehring, 2006). For example, the

European Union's governance structure integrates policies and actions at the EU, national, and subnational levels to achieve climate targets.

- *Partnerships and Networks*: Engaging diverse stakeholders through partnerships and networks is essential for fostering collaboration, sharing knowledge, and mobilizing resources for climate action (Bulkeley et al., 2009). Public–private partnerships, joint initiatives between governments and civil society organizations, and collaborations with academia and research institutions can enhance the effectiveness and legitimacy of governance structures.
- *Inclusive Decision-Making Processes*: Inclusivity in decision-making ensures that a wide range of perspectives, interests, and voices are considered in policy development and implementation. Participatory approaches, such as citizen assemblies, stakeholder consultations, and advisory committees, can enhance the legitimacy and acceptance of climate policies (Bäckstrand & Kuyper, 2017).
- *Transparency and Accountability*: Transparent governance processes promote openness, trust, and public scrutiny, enhancing the credibility and effectiveness of climate policies (Capstick et al., 2015). Accountability mechanisms, including performance monitoring, reporting, and independent oversight, ensure that commitments are met and progress is tracked towards net zero goals.

10.5.2 Importance of Inclusivity, Transparency, and Accountability

Inclusivity, transparency, and accountability are foundational principles that underpin effective governance for decarbonization:

- *Inclusivity*: Inclusive governance processes ensure that marginalized and vulnerable communities have a seat at the table and their concerns are addressed in decision-making. This fosters social equity, reduces inequalities, and promotes climate justice (Bäckstrand & Kuyper, 2017).
- *Transparency*: Transparent governance fosters public trust by providing accessible information about decision-making processes, policy choices, and outcomes. Transparency enables public scrutiny, accountability, and informed citizen participation in climate action (Capstick et al., 2015).
- *Accountability*: Accountability mechanisms hold governments, institutions, and stakeholders responsible for their commitments and actions. Clear accountability frameworks help track progress, identify gaps, and address shortcomings in climate policies (Oberthür & Gehring, 2006).

10.5.3 Comparative Analysis of Governance Models

Different governance models exhibit varying degrees of effectiveness in facilitating sustainable policies and climate action:

- *Top-Down versus Bottom-Up Approaches*: Top-down governance models involve centralized decision-making and enforcement, whereas bottom-up approaches emphasize local initiatives, community engagement, and participatory governance (Bäckstrand, 2006). Hybrid models that combine top-down directives with bottom-up participation often achieve better outcomes by leveraging local knowledge and fostering ownership of climate actions.
- *Unitary vs. Federal Systems*: Unitary systems concentrate power at the national level, whereas federal systems distribute authority across multiple levels of government. Federal systems can promote experimentation, innovation, and policy diversity by allowing subnational entities to tailor solutions to local contexts (Jordan et al., 2015).
- *Public Sector versus Public–Private Partnerships*: Governance structures that leverage public–private partnerships can unlock additional resources, expertise, and innovation for climate action. However, careful oversight and clear rules are necessary to ensure that private interests align with public goals and values (Adhikari & Safaee Chalkasra, 2023).

10.6 Case Studies and Examples of Implementing Net Zero Policies

Examining real-world case studies and examples of implementing net zero policies provides valuable insights into the challenges, successes, and lessons learned in transitioning towards a sustainable, low-carbon future. This section presents a selection of case studies that illustrate both successful and unsuccessful attempts at implementing net zero policies, highlighting key insights and lessons learned from these experiences.

10.6.1 Costa Rica: A Model for Sustainable Development

Costa Rica stands out as a global leader in sustainability and climate action, with a long-standing commitment to conservation, renewable energy, and carbon neutrality. Despite being a small developing country, Costa Rica has made significant strides in reducing its carbon footprint and promoting sustainable development.

Case Study Overview:

- Costa Rica's National Decarbonization Plan, launched in 2018, outlines a comprehensive strategy to achieve net zero emissions by 2050 (Teske et al., 2020).
- The plan incorporates input from diverse stakeholders through participatory processes, ensuring broad-based support and ownership of decarbonization initiatives.

- Key components of the plan include transitioning to renewable energy, electrifying transportation, protecting forests, promoting sustainable agriculture, and improving energy efficiency.

Key Insights:

- Political leadership and commitment are essential for driving ambitious climate action and mobilizing resources for implementation.
- Inclusive decision-making processes foster stakeholder buy-in and contribute to the success and legitimacy of climate policies.
- Integrated and holistic approaches to decarbonization, encompassing multiple sectors and actors, are more likely to yield sustainable outcomes.

10.6.2 Germany's Energiewende: Challenges and Opportunities

Germany's Energiewende (energy transition) is one of the most ambitious renewable energy programmes globally, aiming to transition away from fossil fuels towards renewable sources while phasing out nuclear power. While the Energiewende has achieved significant milestones in renewable energy deployment, it also faces challenges and criticisms regarding its cost-effectiveness and grid stability.

Case Study Overview:

- The Energiewende was initiated in the early 2000s with the introduction of feed-in tariffs to incentivize renewable energy investments (Jacobsson & Lauber, 2006).
- Germany has since become a global leader in renewable energy, with renewables accounting for over 40% of electricity generation in 2020 (Fraunhofer ISE, 2021).
- However, the Energiewende faces challenges such as grid integration, energy storage, land use conflicts, and rising electricity costs.

Key Insights:

- Ambitious renewable energy targets can drive rapid deployment and innovation in clean energy technologies.
- The transition to renewable energy requires comprehensive planning, infrastructure investment, and grid modernization to ensure reliability and stability.
- Balancing environmental, economic, and social objectives is critical for maintaining public support and addressing equity concerns in energy transition policies.

10.6.3 Australia's Mixed Record on Climate Policy

Australia, as one of the world's largest per capita emitters of greenhouse gases, has faced challenges in developing and implementing effective climate policies.

While the country has made progress in renewable energy deployment, its climate policies have been characterized by political instability, policy reversals, and a lack of long-term vision.

Case Study Overview:

- Australia's Renewable Energy Target (RET) has driven significant growth in renewable energy capacity, particularly in wind and solar power (Li et al., 2020).
- However, Australia has experienced frequent changes in climate policy direction due to political disagreements and shifts in government priorities.
- The absence of a comprehensive and stable climate policy framework has hindered investment certainty and undermined efforts to achieve long-term emission reduction goals.

Key Insights:

- Political instability and lack of bipartisan consensus on climate policy can undermine investor confidence and impede progress towards decarbonization.
- Long-term policy certainty and stable regulatory frameworks are essential for attracting private sector investment and driving innovation in clean energy technologies.
- Public engagement and education are crucial for building awareness, understanding, and support for climate action, particularly in politically contested environments.

10.6.4 United Kingdom's Climate Leadership and Policy Innovation

The United Kingdom has demonstrated leadership in climate action through its ambitious emissions reduction targets, innovative policy initiatives, and commitment to net zero by 2050. The UK's Climate Change Act, enacted in 2008, laid the foundation for its climate policy framework and set legally binding targets for emissions reduction (Lorenzoni & Benson, 2014).

Case Study Overview:

- The UK's Climate Change Act established a framework for setting carbon budgets and reducing emissions across sectors (Averchenkova et al., 2021).
- The UK government has implemented various policy instruments, including carbon pricing, renewable energy incentives, energy efficiency standards, and support for low-carbon innovation.
- The UK's decision to phase out coal-fired power plants, invest in offshore wind farms, and prioritize energy efficiency measures has contributed to significant reductions in greenhouse gas emissions (Raugei & Leccisi, 2016).

Key Insights:

- Strong legislative frameworks and legally binding targets provide certainty and accountability for climate action, guiding long-term policy planning and implementation.
- Innovation and investment in low-carbon technologies are essential for achieving emissions reductions while maintaining economic competitiveness and energy security.
- International cooperation and collaboration are crucial for addressing global climate challenges and leveraging collective action to accelerate decarbonization efforts.

10.6.5 Sweden's Carbon Tax and Sustainable Development

Sweden has emerged as a pioneer in carbon pricing, implementing a carbon tax since the early 1990s as part of its broader sustainability agenda. The Swedish model of carbon taxation exemplifies the integration of environmental concerns with economic policies, promoting decarbonization while fostering innovation and competitiveness.

Case Study Overview:

- Sweden's carbon tax applies to fossil fuels used for heating, transportation, and industrial processes, gradually increasing over time to incentivize emissions reductions (Andersson, 2019).
- Revenue from the carbon tax is reinvested in sustainable development initiatives, such as renewable energy projects, public transportation, and energy efficiency measures.
- The carbon tax has contributed to significant reductions in per capita emissions while maintaining strong economic growth and competitiveness (Sterner, 2020).

Key Insights:

- Carbon pricing mechanisms can effectively internalize the external costs of carbon emissions, providing economic incentives for businesses and individuals to transition to low-carbon alternatives.
- Revenue recycling strategies, such as earmarking carbon tax revenues for sustainable investments, enhance public acceptance and support for carbon pricing policies.
- Sweden's experience demonstrates the importance of political leadership, public engagement, and cross-sectoral collaboration in advancing sustainable development goals (Krantz & Gustafsson, 2023).

10.6.6 Denmark's Wind Energy Revolution: Harnessing Offshore Resources

Denmark's commitment to wind energy has propelled it to the forefront of renewable energy innovation, with a particular emphasis on offshore wind power. Through strategic planning, investment, and collaboration between government, industry, and research institutions, Denmark has established itself as a global leader in wind energy deployment.

Case Study Overview:

- Denmark's long-term vision for wind energy dates back to the 1970s, with early investments in research and development paving the way for technological advancements (Hansen & Enevoldsen, 2022).
- The expansion of offshore wind farms, supported by favourable regulatory frameworks and public–private partnerships, has driven down costs and increased efficiency.
- Denmark's wind energy sector contributes significantly to domestic energy supply, export revenues, and job creation, while simultaneously reducing carbon emissions (Dubinskii, 2024).

Key Insights:

- Proactive government policies and regulatory certainty are essential for attracting investment and fostering innovation in renewable energy technologies.
- Collaboration between government, industry, and research institutions facilitates knowledge sharing, technology transfer, and skills development, driving continuous improvement and competitiveness.
- Denmark's success in wind energy underscores the potential of renewable resources to drive economic growth, enhance energy security, and mitigate climate change (Kara & Şahin, 2023).

10.6.7 Norway's Sustainable Transportation Initiatives: Electrification and Green Mobility

Norway has emerged as a global leader in sustainable transportation, with a strong emphasis on electrification and promoting green mobility solutions. Through a combination of regulatory incentives, infrastructure development, and public awareness campaigns, Norway has achieved significant progress in reducing emissions from the transportation sector.

Case Study Overview:

- Norway's ambitious targets for electric vehicle (EV) adoption, supported by financial incentives such as tax exemptions, toll reductions, and access to bus lanes, have led to a rapid increase in EV ownership (Figenbaum, 2020).

- Investment in charging infrastructure, including fast chargers along major highways and in urban areas, has addressed range anxiety and facilitated the transition to electric mobility.
- Government initiatives to electrify public transportation, including subsidies for electric buses and ferries, contribute to reducing emissions and improving air quality in urban centres (Erickson & Brase, 2019).

Key Insights:

- Integrated policies combining incentives for consumers support for infrastructure development, and regulatory measures are essential for accelerating the transition to sustainable transportation.
- Public–private partnerships and collaboration with industry stakeholders play a crucial role in scaling up green mobility solutions and driving innovation in the transportation sector.
- Norway's experience highlights the importance of a holistic approach to sustainable transportation, considering both individual behaviour change and systemic improvements in infrastructure and technology (Svennevik et al., 2021).

10.6.8 France's Energy Transition: Nuclear Power and Renewable Integration

France's energy transition strategy combines nuclear power with renewable energy integration to reduce carbon emissions and enhance energy security. Through a diversified energy mix, comprehensive legislation, and investment in innovation, France aims to achieve its climate targets while maintaining a reliable and affordable supply of electricity.

Case Study Overview:

- France relies heavily on nuclear power for electricity generation, with nuclear reactors providing a significant portion of the country's energy supply.
- The government has implemented policies to promote renewable energy deployment, including feed-in tariffs, competitive tendering processes, and support for research and development.
- Integration of renewable energy sources such as wind and solar into the electricity grid enhances flexibility and resilience, reducing dependence on fossil fuels and mitigating climate change impacts (Prăvălie & Bandoc, 2018).

Key Insights:

- Balancing nuclear power with renewables enables France to achieve low-carbon electricity generation while ensuring a stable and resilient energy system.
- Long-term planning and strategic investments in infrastructure are essential for maximizing the potential of both nuclear and renewable energy sources.

- France's experience underscores the importance of regulatory frameworks that encourage innovation, competition, and collaboration across the energy sector (Loisel et al., 2022).

10.6.9 Netherlands' Climate Adaptation and Resilience Strategies

The Netherlands has developed comprehensive climate adaptation and resilience strategies to address the challenges of sea-level rise, flooding, and extreme weather events. Through proactive measures, innovative engineering solutions, and stakeholder engagement, the Netherlands is safeguarding its communities, infrastructure, and natural resources from the impacts of climate change.

Case Study Overview:

- The Dutch Delta Programme coordinates efforts to manage water resources, reduce flood risks, and enhance resilience in coastal and riverine areas (Wesselink et al., 2015).
- Investments in flood protection infrastructure, such as dikes, flood barriers, and urban water management systems, contribute to mitigating the risks of inundation and water-related disasters.
- Collaboration with local communities, businesses, and international partners fosters knowledge sharing, capacity building, and adaptive governance approaches to climate resilience.

Key Insights:

- Integrated approaches to climate adaptation combine engineering solutions with nature-based interventions, recognizing the importance of ecosystem services in enhancing resilience.
- Stakeholder engagement and participatory planning processes ensure that adaptation measures are tailored to local needs, priorities, and vulnerabilities.
- The Netherlands' experience demonstrates the importance of proactive risk management, continuous monitoring, and adaptive governance in building resilience to climate change impacts (Hölscher et al., 2019).

10.6.10 Japan's Renewable Energy Expansion Post-Fukushima

The Fukushima nuclear disaster in 2011 prompted Japan to reassess its energy strategy, leading to a significant shift towards renewable energy and energy efficiency (Zhu et al., 2020). The disaster highlighted the vulnerability of nuclear power and underscored the need for a more diversified and resilient energy portfolio.

Key Initiatives:

- In response, Japan implemented various policy measures to promote renewable energy development.
- Feed-in tariffs were introduced to guarantee favourable prices for renewable energy producers, incentivizing investment in solar, wind, and other renewable technologies.
- Additionally, regulatory reforms aimed to streamline the permitting process for renewable energy projects and facilitate their deployment. Japan set ambitious targets to increase the share of renewable energy in its electricity mix, with a focus on achieving 22–24% by 2030 (Tracker, 2021).

Key Insights:

- Japan's post-Fukushima experience underscores the potential for renewable energy expansion in disaster recovery and energy security strategies.
- Challenges such as grid integration and land use conflicts must be addressed to maximize the benefits of renewable energy transition.
- Policy stability and long-term commitment are crucial for achieving sustainable energy goals and contributing to global climate efforts (Ogawa, 2020).

10.6.11 China's Renewable Energy Expansion

Case Study Overview:

- China, the world's largest emitter of greenhouse gases, has faced increasing pressure to address environmental concerns and reduce its carbon footprint (Leggett et al., 2019).
- The country has embarked on an ambitious renewable energy expansion programme to transition towards cleaner energy sources and combat air pollution.

Key Initiatives:

- Implementation of massive investments in renewable energy infrastructure, including wind, solar, hydroelectric, and nuclear power.
- Introduction of feed-in tariffs, subsidies, and other incentives to promote renewable energy development and attract investment.
- Setting of aggressive targets for renewable energy capacity expansion, with aims to increase the share of non-fossil fuels in its energy mix.

Key Insights:

- China's renewable energy expansion reflects its recognition of the urgent need to address environmental challenges and transition to a more sustainable energy system.
- The country's scale of investment and deployment in renewable energy technologies positions it as a global leader in the clean energy transition.

- Challenges such as grid integration, curtailment of renewable energy, and technological innovation remain areas of focus for China as it continues to expand its renewable energy capacity (Liu et al., 2021).

10.6.12 India's Renewable Energy Expansion

India, as a rapidly developing economy with a significant population and energy demand, has also made notable strides in renewable energy expansion and climate action.

Case Study Overview:

- India's National Solar Mission and National Wind Mission aim to promote the development and deployment of solar and wind energy infrastructure across the country (Bhushan et al., 2019).
- Policy measures such as feed-in tariffs, tax incentives, and renewable purchase obligations have been implemented to incentivize investment in renewable energy projects.
- India's commitment to expanding renewable energy capacity has led to substantial growth in solar and wind power installations, positioning the country as a global leader in renewable energy deployment.

Key Insights:

- Ambitious renewable energy targets and supportive policy frameworks can drive rapid expansion of clean energy infrastructure, contributing to emissions reduction and energy security.
- Policy stability and regulatory certainty are crucial for attracting investment and fostering innovation in the renewable energy sector.
- International partnerships and collaborations can facilitate technology transfer, capacity building, and knowledge exchange, supporting sustainable development goals.

10.6.13 Brazil's Forest Conservation and Sustainable Agriculture

Brazil, home to the Amazon rainforest and significant agricultural production, has implemented policies to promote forest conservation and sustainable agriculture practices.

Case Study Overview:

- Brazil's Forest Code and Zero Deforestation commitments aim to reduce deforestation rates and preserve biodiversity in the Amazon rainforest (Nepstad et al., 2014).

- Sustainable agriculture initiatives, such as the Low Carbon Agriculture Programme, encourage farmers to adopt practices that reduce emissions from deforestation, land degradation, and agricultural activities.
- Brazil's efforts to combat illegal logging, enforce environmental regulations, and promote reforestation projects have contributed to forest preservation and carbon sequestration (Garrastazú et al., 2015).

Key Insights:

- Integrated approaches that combine forest conservation with sustainable agriculture are essential for mitigating climate change and protecting ecosystems.
- Policy coherence and enforcement mechanisms are necessary to ensure compliance with environmental regulations and prevent illegal deforestation.
- International support and financing mechanisms can incentivize forest conservation efforts and promote sustainable land management practices in developing countries.

10.6.14 Bangladesh's Climate Resilience and Adaptation Initiatives

Bangladesh, despite being one of the least developed countries, has taken significant strides in addressing climate change impacts through resilience and adaptation measures. Given its vulnerability to sea-level rise, cyclones, and floods, Bangladesh has prioritized building adaptive capacity and enhancing resilience across various sectors.

Case Study Overview:

- Bangladesh's National Adaptation Programme of Action (NAPA) identifies priority areas for climate adaptation, including agriculture, water resources, and coastal zone management.
- Community-based adaptation projects, such as flood-resistant housing, early warning systems, and climate-resilient agriculture practices, empower vulnerable communities to cope with climate-related risks.
- International partnerships and financial support enable Bangladesh to implement adaptation projects and access technology transfer, enhancing its resilience to climate change impacts (Shahjahan Mondal, 2019).

Key Insights:

- Community-centred approaches to adaptation foster ownership, participation, and knowledge sharing, strengthening local resilience and reducing vulnerability.
- Mainstreaming climate adaptation into national development planning ensures coherence and integration across sectors, maximizing the effectiveness of adaptation efforts.

- Bangladesh's experience underscores the importance of global solidarity and support for vulnerable countries in building climate resilience and achieving sustainable development goals (Mehta et al., 2018).

10.6.15 Ethiopia's Sustainable Land Management and Reforestation Efforts

Ethiopia, confronted with challenges of land degradation, deforestation, and climate variability, has embarked on ambitious initiatives to promote sustainable land management and reforestation. Through policy reforms, community engagement, and ecosystem restoration programmes, Ethiopia aims to mitigate climate change impacts and enhance food security.

Case Study Overview:

- Ethiopia's Climate-Resilient Green Economy (CRGE) strategy integrates climate change adaptation and mitigation into national development planning, focusing on agriculture, forestry, and energy sectors.
- Large-scale reforestation campaigns, such as the "Green Legacy" initiative, mobilize public participation in tree planting activities, restoring degraded landscapes and enhancing carbon sequestration.
- Community-based natural resource management approaches empower local communities to sustainably manage forests, rangelands, and watersheds, promoting resilience and biodiversity conservation (Medhin & Mekonnen, 2019).

Key Insights:

- Sustainable land management practices, including agroforestry, soil conservation, and water harvesting, improve ecosystem resilience, enhance carbon sequestration, and increase agricultural productivity.
- Partnerships between government, civil society, and international organizations facilitate knowledge exchange, capacity building, and resource mobilization for sustainable land management initiatives.
- Ethiopia's experience highlights the importance of long-term commitment, political will, and community empowerment in addressing climate change impacts and achieving sustainable development (Devonald et al., 2022).

10.6.16 Malawi's Climate-Smart Agriculture and Food Security Interventions

Malawi, grappling with climate variability, droughts, and food insecurity, has embraced climate-smart agriculture (CSA) practices to enhance resilience and improve livelihoods. Through innovative farming techniques, water management strategies, and policy support, Malawi seeks to adapt to climate change while ensuring food security for its population.

Case Study Overview:

- Malawi's National Agriculture Policy (NAP) promotes CSA approaches, including conservation agriculture, drought-resistant crops, and agroforestry, to increase climate resilience and agricultural productivity.
- Investment in irrigation infrastructure, small-scale water harvesting systems, and weather information services helps farmers manage climate risks and optimize crop yields.
- Capacity building initiatives and farmer field schools empower rural communities with knowledge and skills to adopt climate-smart practices, enhancing adaptive capacity and livelihood resilience (Mwongera et al., 2020).

Key Insights:

- Climate-smart agriculture integrates adaptation and mitigation strategies, promoting sustainable land use, biodiversity conservation, and carbon sequestration.
- Targeted support for smallholder farmers, particularly women and youth, enhances their resilience to climate change impacts and contributes to poverty reduction and food security.
- Malawi's experience underscores the importance of aligning agricultural policies with climate goals, fostering innovation, and building local capacity to address the challenges of climate change and food insecurity (Amoak et al., 2024).

Real-world case studies and examples of implementing net zero policies offer valuable lessons and insights for policymakers, stakeholders, and practitioners. From Costa Rica's participatory approach to Malawi's climate-smart agriculture and food security interventions, each case study highlights the importance of political leadership, stakeholder engagement, policy innovation, and long-term vision in driving sustainable and inclusive climate action. By learning from both successes and failures, countries can develop more effective and resilient strategies for achieving net zero emissions and building a sustainable future for generations to come.

10.7 Conclusion

The analysis of policy, politics, and governance in relation to achieving net zero emissions highlights several key findings. Firstly, effective policy frameworks, informed by robust scientific evidence and stakeholder engagement, are essential for driving decarbonization across sectors. Secondly, political dynamics, including leadership, public engagement, and stakeholder collaboration, play a critical role in shaping the adoption and implementation of net zero policies. Lastly, governance structures characterized by inclusivity, transparency, and accountability are fundamental for supporting the transition to a sustainable, low-carbon economy.

An integrated approach that considers the interplay between policy, politics, and governance is crucial for successfully achieving net zero emissions. Policymakers, stakeholders, and researchers should prioritize collaboration, innovation, and long-term planning to overcome challenges and capitalize on opportunities for decarbonization. Recommendations include fostering political will, strengthening multilevel governance frameworks, enhancing public participation, and investing in capacity building and research. By working together and embracing holistic strategies, we can accelerate progress towards a greener, more resilient future for all.

References

Adhikari, B., & Safaee Chalkasra, L. S. (2023). Mobilizing private sector investment for climate action: Enhancing ambition and scaling up implementation. *Journal of Sustainable Finance & Investment, 13*(2), 1110–1127. https://doi.org/10.1080/20430795.2021.1917929

Amoak, D., Lupafya, E., Dakishoni, L., & Luginaah, I. (2024). Towards food sovereignty: the role of smallholder farmers'seed security in improving climate change resilience in Northern Malawi. *Geographical Review*, 1–26. https://doi.org/10.1080/00167428.2024.2335395

Andersson, J. J. (2019). Carbon taxes and CO_2 emissions: Sweden as a case study. *American Economic Journal: Economic Policy, 11*(4), 1–30. https://doi.org/10.1257/pol.20170144

Averchenkova, A., Fankhauser, S., & Finnegan, J. J. (2021). The impact of strategic climate legislation: Evidence from expert interviews on the UK climate change act. *Climate Policy, 21*(2), 251–263. https://doi.org/10.1080/14693062.2020.1819190

Bäckstrand, K. (2006). Democratizing global environmental governance? Stakeholder democracy after the world summit on sustainable development. *European Journal of International Relations, 12*(4), 467–498. https://doi.org/10.1177/1354066106069321

Bäckstrand, K., & Kuyper, J. W. (2017). The democratic legitimacy of orchestration: The UNFCCC, non-state actors, and transnational climate governance. *Environmental Politics, 26*(4), 764–788. https://doi.org/10.1080/09644016.2017.1323579

Barnes, A. (2021). *The challenges and prospects for carbon pricing in Europe*. Oxford Institute for Energy Studies.

Bernauer, T., & McGrath, L. F. (2016). Simple reframing unlikely to boost public support for climate policy. *Nature Climate Change, 6*(7), 680–683. https://doi.org/10.1038/nclimate2948

Bhushan, C., Bhati, P., Sreenivasan, P., Sing, M., Jhawar, P., Koshy, M. S., & Sambyal, S. (2019). *The State of renewable energy in India*. Centre for Science and Environment: New Delhi, India. Retrieved from https://shaktifoundation.in/wp-content/uploads/2014/02/State-of-Renewable-Energy-in-India.pdf

Biermann, F., Pattberg, P., Van Asselt, H., & Zelli, F. (2009). The fragmentation of global governance architectures: A framework for analysis. *Global Environmental Politics, 9*(4), 14–40. https://doi.org/10.1162/glep.2009.9.4.14

Bulkeley, H., Schroeder, H., Janda, K., Zhao, J., Armstrong, A., Chu, S. Y., & Ghosh, S. (2009). Cities and climate change: The role of institutions, governance and urban planning. *Change, 28*, 30.

Capstick, S., Whitmarsh, L., Poortinga, W., Pidgeon, N., & Upham, P. (2015). International trends in public perceptions of climate change over the past quarter century. *Wiley Interdisciplinary Reviews: Climate Change, 6*(1), 35–61.

Committee on Climate Change. (2019). *Net zero: The UK's contribution to stopping global warming*. Retrieved from https://www.theccc.org.uk/publication/net-zero-the-uks-contribution-to-stopping-global-warming/

Devonald, M., Jones, N., Iyasu Gebru, A., & Yadete, W. (2022). Rethinking climate change through a gender and adolescent lens in Ethiopia. *Climate and Development*, 1–11. https://doi.org/10.1080/17565529.2022.2032568

Dubinskii, O. (2024). The contribution of renewable energy sector value added to the national economic development in Denmark: An empirical analysis. *Review of Economic Perspectives, 24*(1), 37–56. https://doi.org/10.2478/revecp-2024-0003

Erickson, L. E., & Brase, G. (2019). Electrification of transportation. In *Reducing greenhouse gas emissions and improving air quality* (pp. 39–50). CRC Press.

Fetting, C. (2020). The European green deal. In *ESDN report* (p. 53).

Figenbaum, E. (2020). Norway—The world leader in BEV adoption. In *Who's driving electric cars: Understanding consumer adoption and Use of plug-in electric cars* (pp. 89-120). https://doi.org/10.1007/978-3-030-38382-4_6

Fischer, C., & Newell, R. G. (2008). Environmental and technology policies for climate mitigation. *Journal of Environmental Economics and Management, 55*(2), 142–162. https://doi.org/10.1016/j.jeem.2007.11.001

Fraunhofer ISE. (2021). *Public net electricity generation in Germany 2020: share from renewables exceeds 50 percent.* Retrieved from https://www.ise.fraunhofer.de/en/press-media/news/2020/public-net-electricity-generation-in-germany-2020-share-from-renewables-exceeds-50-percent.html

Fuss, S., Lamb, W. F., Callaghan, M. W., Hilaire, J., Creutzig, F., Amann, T., Beringer, T., de Oliveira Garcia, W., Hartmann, J., Khanna, T., Luderer, G., Nemet, G. F., Rogelj, J., Smith, P., Vicente, J. L. V., Wilcox, J., Dominguez, M. M. Z., & Minx, J. C. (2018). Negative emissions—Part 2: Costs, potentials and side effects. *Environmental Research Letters, 13*(6), 063002. https://doi.org/10.1088/1748-9326/aabf9f

Garrastazú, M. C., Mendonça, S. D., Horokoski, T. T., Cardoso, D. J., Rosot, M. A., Nimmo, E. R., & Lacerda, A. E. (2015). Carbon sequestration and riparian zones: Assessing the impacts of changing regulatory practices in Southern Brazil. *Land Use Policy, 42*, 329–339. https://doi.org/10.1016/j.landusepol.2014.08.003

Gheuens, J., & Oberthür, S. (2021). EU climate and energy policy: How myopic is it? *Politics and Governance, 9*(3), 337–347. https://doi.org/10.17645/pag.v9i3.4320

Goulder, L. H., & Parry, I. W. (2008). Instrument choice in environmental policy. In *Review of environmental economics and policy* (pp. 1–26).

Hanemann, W. M., & Farrell, A. E. (2006). *Managing greenhouse gas emissions in California.* California Climate Change Center, University of California, Berkeley. http://calclimate.berkeley.edu/managing_GHGs_in_CA.html

Hansen, K. B., & Enevoldsen, P. (2022). Climate politics and renewable energy in Denmark 1975–2020. In *Public governance in Denmark* (pp. 149–167). Emerald Publishing Limited. https://doi.org/10.1108/978-1-80043-712-820221009

Hölscher, K., Frantzeskaki, N., & Loorbach, D. (2019). Steering transformations under climate change: Capacities for transformative climate governance and the case of Rotterdam, the Netherlands. *Regional Environmental Change, 19*, 791–805. https://doi.org/10.1007/s10113-018-1329-3

Jacobsson, S., & Lauber, V. (2006). The politics and policy of energy system transformation-explaining the German diffusion of renewable energy technology. *Energy Policy, 34*(3), 256–276. https://doi.org/10.1016/j.enpol.2004.08.029

Jordan, A. J., Huitema, D., Hildén, M., Van Asselt, H., Rayner, T. J., Schoenefeld, J. J., Tosun, J., Forster, J., & Boasson, E. L. (2015). Emergence of polycentric climate governance and its future prospects. *Nature Climate Change, 5*(11), 977–982. https://doi.org/10.1038/nclimate2725

Kara, T., & Şahin, A. D. (2023). Implications of climate change on wind energy potential. *Sustainability, 15*(20), 14822. https://doi.org/10.3390/su152014822

Krantz, V., & Gustafsson, S. (2023). Regional collaboration for the sustainable development goals: Experiences from developing a multi-actor platform in Sweden. *Sustainable Development, 31*(6), 4007–4018. https://doi.org/10.1002/sd.2580

Leggett, J. A., Logan, J., & Mackey, A. (2019). *China's greenhouse gas emissions and mitigation policies*. Congressional Research Service.

Li, H. X., Edwards, D. J., Hosseini, M. R., & Costin, G. P. (2020). A review on renewable energy transition in Australia: An updated depiction. *Journal of Cleaner Production, 242*, 118475. https://doi.org/10.1016/j.jclepro.2019.118475

Liu, J., Wang, Q., Song, Z., & Fang, F. (2021). Bottlenecks and countermeasures of high-penetration renewable energy development in China. *Engineering, 7*(11), 1611–1622. https://doi.org/10.1016/j.eng.2020.08.016

Loisel, R., Lemiale, L., Mima, S., & Bidaud, A. (2022). Strategies for short-term intermittency in long-term prospective scenarios in the French power system. *Energy Policy, 169*, 113182. https://doi.org/10.1016/j.enpol.2022.113182

Lorenzoni, I., & Benson, D. (2014). Radical institutional change in environmental governance: Explaining the origins of the UK climate change act 2008 through discursive and streams perspectives. *Global Environmental Change, 29*, 10–21. https://doi.org/10.1016/j.gloenvcha.2014.07.011

Medhin, H., & Mekonnen, A. (2019). Green and climate-resilient transformation in Ethiopia. In *The Oxford handbook of the Ethiopian economy* (pp. 288–307).

Mehta, S., Kumar, V., Kumar, A., Mehta, S., Kumar, V., & Kumar, A. (2018). The way forward: advancing UN sustainable development goals in India and Bangladesh. In *Lessons in sustainable development from Bangladesh and India* (pp. 103–121). https://doi.org/10.1007/978-3-319-95483-7_5

Mwongera, C., Lamanna, C., Kamau, H. N., & Girvetz, E. (2020). Scaling climate-smart agriculture for agricultural transformation in Southern Africa. In *Transforming agriculture in Southern Africa* (p. 86).

Nepstad, D., McGrath, D., Stickler, C., Alencar, A., Azevedo, A., Swette, B., Bezerra, T., DiGiano, M., Shimada, J., da Motta, R. S., Armijo, E., Castello, L., Brando, P., Hansen, M. C., McGrath-Horn, M., Carvalho, O., & Hess, L. (2014). Slowing Amazon deforestation through public policy and interventions in beef and soy supply chains. *Science, 344*(6188), 1118–1123. https://doi.org/10.1126/science.1248525

Oberthür, S., & Gehring, T. (2006). Institutional interaction in global environmental governance: The case of the Cartagena protocol and the world trade organization. *Global Environmental Politics, 6*(2), 1–31. https://doi.org/10.1162/glep.2006.6.2.1

Ogawa, A. (2020). 'Community power': Renewable energy policy and production in post-Fukushima Japan. In *New frontiers in Japanese studies* (pp. 221–232). Routledge.

Pan, W., Cao, H., & Liu, Y. (2023). "Green" innovation, privacy regulation and environmental policy. *Renewable Energy, 203*, 245–254. https://doi.org/10.1016/j.renene.2022.12.025

Prăvălie, R., & Bandoc, G. (2018). Nuclear energy: Between global electricity demand, world-wide decarbonisation imperativeness, and planetary environmental implications. *Journal of Environmental Management, 209*, 81–92. https://doi.org/10.1016/j.jenvman.2017.12.043

Raugei, M., & Leccisi, E. (2016). A comprehensive assessment of the energy performance of the full range of electricity generation technologies deployed in the United Kingdom. *Energy Policy, 90*, 46–59. https://doi.org/10.1016/j.enpol.2015.12.011

Shahjahan Mondal, M., Islam, M. T., Saha, D., Hossain, M. S. S., Das, P. K., & Rahman, R. (2019). Agricultural adaptation practices to climate change impacts in coastal Bangladesh. In *Confronting climate change in Bangladesh: Policy strategies for adaptation and resilience* (pp. 7–21). https://doi.org/10.1007/978-3-030-05237-9_2

Smith, P., Davis, S. J., Creutzig, F., Fuss, S., Minx, J., Gabrielle, B., Kato, E., Jackson, R. B., Cowie, A., Kriegler, E., van Vuuren, D. P., Rogelj, J., Ciais, P., Milne, J., Canadell, J. G., McCollum, D., Peters, G., Andrew, R., Krey, V., … Yongsung, C. (2016). Biophysical and economic limits to negative CO_2 emissions. *Nature climate change, 6*(1), 42–50. https://doi.org/10.1038/nclimate2870

Sterner, T. (2020). The carbon tax in Sweden. *Standing up for a Sustainable World*, 59. Retrieved from 9781800371774.00014.pdf.

Svennevik, E. M., Dijk, M., & Arnfalk, P. (2021). How do new mobility practices emerge? A comparative analysis of car-sharing in cities in Norway, Sweden and the Netherlands. *Energy Research & Social Science, 82*, 102305. https://doi.org/10.1016/j.erss.2021.102305

Teske, S., Morris, T., & Nagrath, K. (2020). *100% renewable energy for Costa Rica*. Report prepared by ISF for the World Future Council/Germany and the One Earth Foundation, USA.

Tracker, C. A. (2021). *1.5 C-consistent benchmarks for enhancing Japan's 2030 climate target*. Retrieved from https://climateactiontracker.org/documents/841/2021_03_CAT_1.5C-consistent_benchmarks_Japan_NDC.pdf

Victor, D. G., Akimoto, K., Kaya, Y., Yamaguchi, M., Cullenward, D., & Hepburn, C. (2017). Prove Paris was more than paper promises. *Nature, 548*(7665), 25–27. https://doi.org/10.1038/548025a

Wesselink, A., Warner, J., Syed, M. A., Chan, F., Tran, D. D., Huq, H., Huthoff, F., Le Thuy, N., Pinter, N., Van Staveren, M., Wester, P., & Zegwaard, A. (2015). Trends in flood risk management in deltas around the world: Are we going 'soft'? *International Journal of Water Governance, 3*(4), 25–46. https://doi.org/10.7564/15-IJWG90

Wyns, T., Khandekar, G., & Groen, L. (2019). International technology and innovation governance for addressing climate change: Options for the EU. In *COP21 RIPPLE S–COP21: Results and Implications for Pathways and Policies for Low Emission s European Societies*.

Zhu, D., Mortazavi, S. M., Maleki, A., Aslani, A., & Yousefi, H. (2020). Analysis of the robustness of energy supply in Japan: Role of renewable energy. *Energy Reports, 6*, 378–391. https://doi.org/10.1016/j.egyr.2020.01.011

Part III
Issues and Challenges of Net Zero Policies

Chapter 11
Net Zero Strategy: Exploring Opportunities, Confronting Challenges for Developing Countries

Amit Joshi, Rubee Singh, Vikas Kumar, and Seema Rani

Abstract The Paris Agreement aims to keep global warming well below 2 °C, and hence, countries worldwide are re-examining their strategies for attaining net zero goals. The decarbonization vision calls for steps to shift to a low-carbon economy, with a focus on infrastructure compatible with sustainable practices and government agency cooperation. All of these factors have increased the pressure on developing nations to reduce their carbon emissions. Since there isn't a single, globally accepted methodology, nations at both the national and sub-national levels combine a range of methodologies to determine the most favorable pathways. Several projects globally showcase innovative and pioneering low-carbon solutions. This paper aims to analyze prevalent patterns in four sectors, including electricity and heating, transport, manufacturing, and construction, regarding greenhouse gas emissions at the global level. The opportunities identified to reduce greenhouse gas emissions among developing nations, such as promoting the use of renewable sources and nuclear power for energy generation, are crucial. In the transportation sector, electro-fuels, bio-fuels, hydro-fuels, and electric vehicles can substitute for traditional transport methods. We must transform the manufacturing and construction industries by incorporating sustainable measures like carbon capture and enhancing the effective life of equipment. This chapter also considers the challenges associated with successful implementation of these initiatives. Implications, limitations, and scope for future research on the study are furnished toward the end.

A. Joshi · R. Singh (✉)
Department of Management Studies, Kumaun University, Nainital, Uttarakhand, India
e-mail: dr.rubeerajput@gmail.com

R. Singh
Institute of Business Management, GLA University, Mathura, Uttar Pradesh, India

V. Kumar
Faculty of Business Law and Social Sciences, Birmingham City University, Birmingham, UK

S. Rani
Department of Commerce, Kurukshetra University, Kurukshetra, India

R. Singh and D. Crowther (eds.), *Transition Towards a Sustainable Future*, Approaches to Global Sustainability, Markets, and Governance,
https://doi.org/10.1007/978-981-97-5756-5_11

Keywords Net zero strategy · Environmental sustainability · Net zero policies · Developing countries · Challenges

11.1 Introduction

Climate change is the foremost environmental issue of the century and one of the most intricate dilemmas facing humanity (Wei et al., 2022). Several climate discussions have been conducted by the global community in order to tackle the issue of climate change. The Paris Agreement set a goal to keep the increase in average global temperature within the range of 1.5–2 °C (Allen et al., 2018). Even if all the countries that have officially agreed to this deal fulfill their commitments to reduce greenhouse gas emissions (GHG), the global temperature is still projected to increase by more than 3 °C compared to preindustrial levels (Olhoff & Christensen, 2020). Just 4.5% of nations have successfully attained carbon neutrality, while the majority of countries are still in the process of planning to accomplish this goal by the years 2050–2070 (Chen et al., 2022). Therefore, achieving a stable atmospheric concentration of greenhouse gases that can support the earth is a challenging and unfinished endeavor requiring contributions from different stakeholders (Singh et al., 2024a). Furthermore, it is likely that emissions will need to be reduced to a point where there is a net reduction in the level of carbon dioxide (CO_2) in the atmosphere after this period. Deep targets encompass all greenhouse gases, including methane (CH_4) and black carbon, which must be reduced by at least 50% by 2050, and nitrous oxide (N_2O), which must be reduced by at least one-third (Bataille et al., 2020).

The Paris Agreement specifies that the implementation of emission reductions should take place within the framework of sustainable development and poverty eradication efforts, while also considering the unique circumstances of each nation (Fay et al., 2015). This involves designing climate policies that encompass the entire economy and specific sectors, with the aim of maximizing the benefits that align with the motives of energy saving, sustainable economic growth, job creation, poverty reduction, maintaining the air and water quality, access to energy, and similar ones (Waisman et al., 2019). Net zero deep decarbonization can only be achieved by employing policies that are country-specific, including through analysis of the strengths and weaknesses of a nation. It is important to note that achieving net zero deep decarbonization requires more than just reducing emissions in the short term. It also involves making significant changes to energy, transportation, manufacturing and construction industries, and land use, which are the top greenhouse gas emitting sectors globally as referred to in Fig. 11.1 (Ritchie et al., 2024a), in order to achieve deep reductions in all sectors by mid-century, specifically among developing nations (Bataille et al., 2016; Das & Gosh, 2023; Iyer et al., 2017).

The United States has allocated approximately $80 billion per year for climate-related expenditures from 2022 to 2027 through several acts, including the Inflation Reduction Act (Carey & Shepard, 2022). These investments cover a wide range of areas, from basic scientific research to the implementation of clean technologies

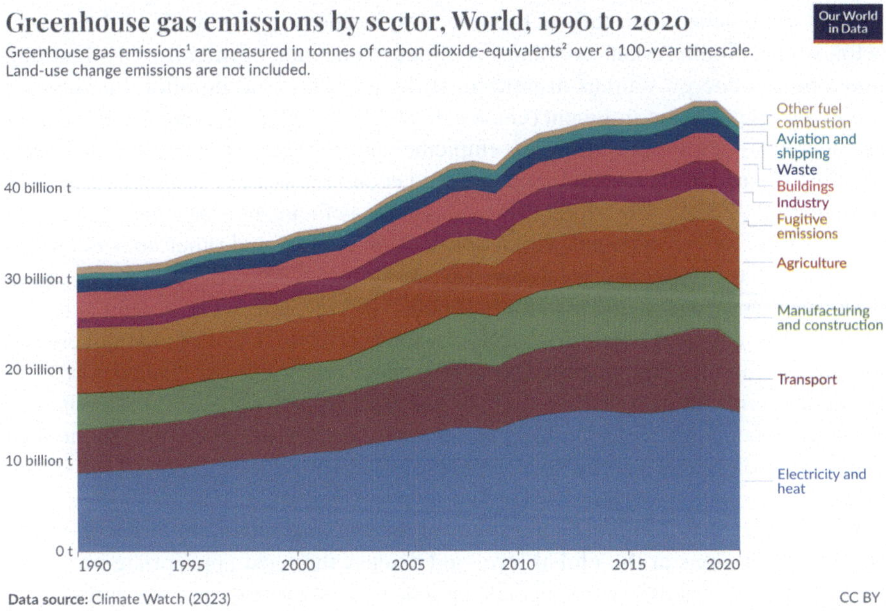

Fig. 11.1 Sector-wise greenhouse gas emissions

in the commercial sector. The Inflation Reduction Act is projected to allocate over $114 billion toward investments in renewable energy by 2031. This will be accomplished through a range of programs, including government loans, tax credits, and loan guarantees provided to corporations and industrialists. Additionally, tax credits will be offered to end users, including consumers, and grants will be given to governments at the local and state levels. These investments and activities are expected to reduce carbon emissions in the United States by up to 40% by 2030, compared to the levels in 2005. This is not driven by a sense of kindness but rather by an imperative need to achieve global net zero. Finland and Sweden, for instance, have set goals to achieve net zero GHG emissions by 2035 and 2045, respectively, and then transition to a state of net negativity (Rogelj et al., 2021). The objective of the European Union is not only to realize net zero emissions but also to follow a framework to strive for net-negative emissions by 2050. Such efforts can offset comparatively lower reductions made by other countries facing more significant technical or political challenges. In addition, the targets must be receptive to being made more stringent.

National governments have framed policies in this direction, but the focus has remained on energy generation rather than diversifying the framework for heating and transportation, where legislation has been less robust. As of 2020, 145 countries have incorporated renewable energy into the power sector through the imposition of regulatory policies. In contrast, 65 nations have implemented similar regulations for the transport sector, while only 22 nations have done so for the heating and cooling sector (Gibb et al., 2022). However, it should be noted that many of these countries have not yet reached their targets because, despite the progress made in the energy sector, several South Asian economies continue to encounter difficulties in obtaining contemporary energy sources in order to strike a harmonious equilibrium between their economies and environment (Guo et al., 2020). Developing countries encounter significant obstacles that impede the implementation of decarbonization measures, including organizational, social, political, and economic issues (Singh et al., 2024b; Virmani et al., 2022). Therefore, difficulties such as financial obstacles, inadequate infrastructure, an incompetent workforce, lock-in effects, and other considerations necessitate context-specific solutions (Das & Ghosh, 2023). In response to these issues, we present a strategic plan for four major sectors aimed at achieving a state of net zero emissions in the industrial sector by the year 2050. Developing markets and emerging economies that have the ability to produce renewable energy and power are significant contributors to carbon dioxide removal (Godin et al., 2021). That is why the advancements for lowering CO_2 are notably targeted at developing economies by utilizing renewable energy sources, some of which may be exported (Obobisa, 2022).

This chapter aims to identify the status of the top four sectors in terms of greenhouse gas emissions at the global level and discuss the most appropriate measures to lower such emissions, with special reference to developing nations, through an in-depth review of the literature on the subject at hand. Solutions to confront sector-specific challenges are also deliberated in the literature review section. The findings and implications are also provided along with the conclusion. Toward the end, limitations and future research scope are presented.

11.2 Literature Review

11.2.1 Electricity and Heating

Global electricity production has been steadily increasing, with a rise from 76 thousand kWh (kilowatt hour) in 1975 to 179 thousand kWh in 2022 (Statista, 2023). In 2022, coal combustion accounted for almost 35% of electricity production, while natural gas contributed 22%, hydropower contributed 15%, and nuclear power contributed 9%. The remaining electricity generation was derived from sources such as wind, solar, and oil. The proportion of electricity generated by coal has declined from 38% in 2000 to 35% in 2022, which is still very high when compared with other sources (Ritchie et al., 2024b). Hence, there is a need for significant transformations in this sector for increased use of other sources for energy generation, especially renewable sources of energy.

11.2.1.1 Harnessing the Renewable Energy Sources

The idea of transitioning the worldwide energy system to rely solely on renewable sources has been present for a considerable period. This concept has mainly emerged as a response to concerns about global warming, to eliminate carbon emissions from the energy sector entirely, to meet the worldwide energy demand through the utilization of carbon approaches, and to maintain low prices to ensure that clean and inexpensive energy is available to everyone (Aktar et al., 2020). Additionally, it aims to tackle broader issues such as the depletion of fossil fuel reserves, deterioration of the quality of regional air, and lower availability and fairness in energy distribution. The commercialization of renewable technology has been significantly influenced by decades of successful research and development initiatives. Solar photovoltaic (PV) and wind technologies are becoming the primary sources of new power generation (Gibb et al., 2022) in light of the fact that these technologies will be even more economically viable in the future. Bloomberg New Energy Finance (BNEF) conducted a study that reveals a significant decrease in the actual cost of solar PV modules over the past ten years. The role of research and development in enhancing the usefulness, cost-effectiveness, and dependability of these innovations remains crucial.

In the field of solar PV, research programs should prioritize the reduction of the leveled cost of electricity through advancements in solar cell efficiency and durability, enhanced manufacturing profitability, and decreased reliance on expensive rare earth materials (Renné, 2022). It would have expanded market applications like the integration of PV with agricultural production and the implementation of efficient reuse programs or recycling at the end-of-life in accordance with international standards and protocols. Furthermore, it is crucial to maintain research and development efforts aimed at enhancing grid flexibility, reliability, digitization, and resiliency. So, it can be concluded that it is technically possible to establish a system with 100% renewable energy sources, especially with ongoing advancements in storage systems, energy-saving appliances, innovative grid systems, advanced bio-fuels, and hydro-fuel cells for transportation sectors that are difficult to electrify. Ongoing advancements, propelled by government and commercial research and development funding, will continue to reduce the expenses associated with solar photovoltaic (PV), concentrated solar power (CSP), and wind technologies (Obobisa, 2022). Nevertheless, there is still a requirement for more robust policy support across national, regional, and municipal levels of government.

11.2.1.2 Increasing Nuclear Energy Generation

As a part of any power system, nuclear power plants (NPPs) offer minimal energy emissions and ensure a reliable energy supply. Electricity generation through nuclear power rose from 2 thousand kWh in 1990 to 2.6 thousand kWh in 2022. But the overall share of energy generated using nuclear power in global power generation declined from 17% in 1990 to 10% in 2022 (Ritchie et al., 2024b; Statista, 2023).

Hydropower and geothermal energy, as renewable sources, have limited potential due to the climatic characteristics of a place and rely heavily on weather conditions (Alhusni et al., 2023). Wind and solar energy, like other renewable sources, are subject to discontinuity due to weather conditions, resulting in changes in the supply–demand gap (Mertens, 2022). Advancement of the nuclear program involves evaluating security infrastructure factors that need to be addressed prior to deciding to initiate a nuclear power program.

Nuclear reactors generate excess heat that is not utilized for power generation and must be extracted from the system. Typically, this excess heat is dissipated into the environment through the use of water, which is either discharged into lakes, rivers, or the sea or released into the atmosphere through cooling towers. For many years, several areas have been using excess heat from nuclear power stations to provide heat for buildings and local communities using specialized hot water networks, a practice known as district heating (Chen et al., 2021). Leurent et al. (2018) examined 18 sites where nuclear heat recovery was used for district heating systems, including four locations in EU countries. Currently, the predominant method of hydrogen synthesis is steam methane reforming. This process is highly efficient and necessitates a heat source, which can be generated by the combustion of natural gas, the utilization of renewable electricity, or the utilization of heat from a nuclear reactor. This possibility is a possibility now being investigated by many countries (Värri & Syri, 2019). Similar initiatives may be used in developing nations to harness the full potential of nuclear energy.

11.2.2 Transportation Sector

On a global scale, the transport sector has a share of 16.2% of all GHG emissions, ranking second only to the electricity and heat generation sector (Ritchie et al., 2024a). Hence, there is a need to expedite the shift toward carbon–neutral transportation to decrease the level of emissions in the industry. Incorporations of electric vehicles and alternate fuels are widely established methods across countries to limit the carbon footprints of this sector.

11.2.2.1 Promoting the Use of Electric Vehicles

Road transportation is a significant contributor to GHG emissions and outdoor air pollution. This type of pollution caused deaths from various health issues such as heart and lung disease, lung cancer, and other respiratory infections (Health Effects Institute, 2019). As a result, the low-emission means of transport have led to a shift from the conventional internal combustion engine to the ones that are powered by battery technology, basically lithium-ion. The use of such batteries is highly prevalent in electric vehicles (EVs). This technology is widely used in EVs due to its reliability, higher energy density, and safety features (Husain, 2021). The policies implemented by national governments are essential in facilitating the development of low-carbon circular economy practices and promoting the use of electric vehicles (Bonsu, 2020). Due to the negative effects on the environment and public health caused by emissions from road traffic, the national governments are strongly advocating for the adoption of electricity as an ultra-low-emission technology (Chen et al., 2022). Electricity has numerous advantages in the transportation sector, including increased efficiency, decreased CO_2 emissions, and enhanced air quality.

Developing nations can learn from the United Kingdom's transition to ultra-low-emission vehicles (ULEVs), which is considered a viable approach for improving air quality by eliminating GHG emissions and enhancing energy savings (Chen et al., 2021). The shift toward electric vehicles will lead to innovation and extensive research and development, as currently, trucks powered by either batteries or hydrogen are being utilized for short-distance trucking. The UK's effort to solidify its position as a global leader in battery technologies and cutting-edge innovation was to support the development of automotive batteries for energy savings. This initiative is in line with the projects funded by the Faraday Battery Challenge and the UK Battery Industrialization Center (Chen et al., 2021). Ultimately, it is determined that electrifying the transportation sector is the most effective method for reducing the industry's carbon emissions.

11.2.2.2 Efficient Use of Alternatives Fuels

There is an anticipated rise in the worldwide need for energy in transportation; hence, light-duty vehicles can be easily converted to hydrogen fuel, which is an energy-dense liquid fuel (IEA, 2017). It can be done without significant alterations in performance; however, there can be potential reductions in range and time taken for refueling the vehicles. It is expected that liquid fuels, due to their high energy density, will continue to be the preferred choice for transportation services even for long-distances, assuming they are accessible (Muratori et al., 2017). Possible choices for such fuels encompass ammonia, alcohols, ethers, and hydrocarbons. For different categories, there are choices available for generating low-carbon or carbon–neutral fuels that may be incorporated into an energy system with zero GHG emissions. Additionally, each of these options can be converted into another using existing thermochemical processes. This improves the cost-effectiveness and safety of transitioning to alternative energy sources in various regions worldwide.

Bio-energy, including biogas and bio-methane projects, remains the predominant choice for investment in low-carbon gas, constituting 40% of the overall investment. They are highly utilized for the reduction of carbon emissions as a substitute for fossil fuels. Prussi et al. (2019) explored the utilization and prospects of biogas as a fuel for transportation throughout the European Union. The study stressed the process of enhancing biogas to bio-methane. By 2030, the consumption of bio-methane in automobiles as a substitute for natural gas is projected to reach 30 billion cubic meters per year. In addition, bio-methane will be utilized for maritime and inland waterway transportation. The use of bio-methane will result in a reduction of roughly

1000 million metric tons (Mt) of greenhouse gas (GHG) emissions by the year 2040. This pertains to the CO_2 emissions due to the use of natural gas as well as the methane emissions that would have resulted from the combustion of fossil fuels (Ghafoori et al., 2022). Due to its identical nature with natural gas, bio-methane is fully compatible with natural gas vehicles, and it can be used without making any changes to the equipment used by end users or distribution systems.

11.2.3 Manufacturing Industry

The manufacturing industry, together with construction, stands in third place when it comes to GHG emissions (Ritchie et al., 2024a). There are two primary mechanisms of GHG in the manufacturing sector: first, the combustion of fuel to generate heat results in the release of combustion emissions, and second, the conversion of materials into new products leads to the release of industrial process emissions. Manufacturing activities also result in indirect greenhouse gas emissions by utilizing electricity generated from fossil fuels. Nevertheless, policies that target emissions originating from manufacturing have the potential to impact both indirect emissions and direct emissions. In this section, two sectors with high GHG emissions are discussed.

11.2.3.1 Transforming the Iron and Steel Industry

The iron and steel industry (ISI) contributes a significant amount to CO_2 emissions, accounting for around 25% of direct emissions to greenhouse gases worldwide. In 2018, it accounted for around 7% of the total direct emissions (The World Steel Association, 2019). Managerial science can be utilized to efficiently allocate resources and optimize the production of iron and steel, managing issues such as low concentration, high costs, excessive production capacity, and inadequate specialization. To decrease GHG emissions in the industry, it is crucial to improve the efficiency of iron and steel products, curtail the increasing demand, and reduce the overall energy consumption in the production process. These objectives can be accomplished through the replacement of materials, enhancements in design, and the recycling and reutilization of steel. Partial substitution of iron and steel goods with carbon fiber or aluminum alloy can be done to cope with the existing demand (Soares et al., 2017). In industrialized nations, the amount of steel used in commercial buildings is double the necessary amount according to safety requirements in the building industry. However, this excessive use of steel does not enhance the overall service life of the structures. This surplus steel can be conserved for alternative applications (Plocher & Panesar, 2019).

Increasing the production and utilization of lightweight cars while maintaining performance standards would result in a substantial reduction in the demand for steel (Kawajiri et al., 2020). In order to achieve this, one can utilize finite element analysis to optimize the design of the body, resulting in a reduction in steel usage while at the same time preserving the physical characteristics of the main structure (Plocher &

Panesar, 2019). The efficiency of production lines and the lifespan of industrial goods can be improved to reduce iron and steel loss in the industrial sector (Li & Hanaoka, 2020). This implies that an organized system for recovering, recycling, and reusing scrap must be established for the ISI. Furthermore, additional assistance should be provided to the recycling industry, and efforts should be made to promote technological advancements in electric arc furnaces (EAF). Concurrently, the percentage of EAF steelmaking could potentially rise to 40–70%, leading to a substantial increase in the electrification rate of this industry and promoting the utilization of renewable energy. It is stated that if all the scrap is successfully recovered during the production of ferrous materials, then the GHG emissions can potentially be decreased up to the level of 60–110 million metric tons by 2030 and 130–170 million metric tons by 2060 (Vercammen et al., 2017). A study highlighted that the Industry 4.0 strategy suggests a substantial integration of information technology into industrial manufacturing processes (Lu et al., 2016). Management approaches that utilize information technology should be implemented in many production procedures to enhance effectiveness, such as EAF, casting, and workshop operations. Consequently, production management systems can be useful to achieve the goals regarding energy conservation and emission reduction.

11.2.3.2 Adopting Reforms in the Cement Industry

The cement industry has recently attracted the great attention of climate activists and researchers, as it alone accounts for 7–8% of direct CO_2 emissions at the global level (Barbhuiya et al., 2024). Cement is widely used in the construction industry due to its versatile, binding, and durable properties. Approximately 40% of the CO_2 emissions produced during the manufacturing of cement come from the use of fossil fuels, while the remaining share results from the process of calcining calcium carbonate ($CaCO_3$), which is commonly found in limestone (Dean et al., 2011). To eliminate process emissions, it is necessary to make significant modifications to the cement production through the installation of carbon-capture technology or changing the materials used during the process (Barker et al., 2009). It is stated that the concentration of CO_2 in cement plants is usually over 30% by volume, which is higher than the concentration of CO_2 in power plants, which ranges from 10 to 15%. This higher concentration makes post-combustion carbon capture more feasible in cement plants. An alternative approach is to utilize oxygen and recycled CO_2 for firing the kiln (Zeman & Lackner, 2006). A significant portion of carbon dioxide emissions generated during cement manufacture is naturally reabsorbed within a timeframe of 50 years as the natural process of carbonation of cement (Xi et al., 2016). Therefore, capturing emissions from cement production might potentially lead to a net reduction in emissions due to the carbonation process of the cement produced.

The carbon pricing process can also be useful, which has two primary approaches: carbon tax and cap-and-trade systems (Carl & Fedor, 2016). In the case of carbon tax, a direct charge is placed on the specific emissions of CO_2 or the amount of carbon present in fossil fuels that are used during the process. It encourages the decrease of emissions by imposing a payment obligation on firms proportional to their emissions. Conversely, the cap-and-trade system establishes a maximum threshold or limit on the overall permissible amount of emissions. The emission permits corresponding to precise quantities of emissions are exchanged in a controlled marketplace (Chen et al., 2020). Gradually, the limit on emissions decreases, leading to a decrease in total emissions. Mission Innovation and Cement Challenge is a global effort that includes the European Commission along with 24 countries with a purpose to collectively drive progress in renewable energy technology (Oberthür et al., 2021). Its main objective is to attain significant decreases in CO_2 emissions resulting from cement manufacturing. The challenge seeks to reduce the industry's substantial impact on GHG emissions by implementing low-carbon and carbon–neutral technology. Developing nations must be made a part of such initiatives to support them in achieving low-carbon emissions.

11.2.4 Construction Industry

It is essential in the construction business to implement sustainable measures to address significant material and energy flows and their associated environmental implications. Urban areas have long been recognized as the primary sources of GHG emissions, and this trend is expected to continue growing, thus resulting in more construction. According to a projection, the urban population will exceed 70% of the total global population by the year 2050 (Kadakia & Galea, 2023). Therefore, it is viewed as the need of the hour to strategically design, construct, and manage a building structure or urban area, considering its entire lifespan, in order for it to be a valuable resource for future generations rather than a burdensome legacy.

11.2.4.1 Replicating Sustainable Design for Cities and Buildings

In order to meet the CO_2 emission reduction targets outlined in Paris agreement, just increasing the share of renewable energy sources in for energy generation it is not sufficient. It is also necessary to decrease the usage of energy by setting and attaining aggressive goals for energy efficiency. Nevertheless, the attainment of existing energy and climate goals is contingent upon the decarbonization of cities, given that cities are responsible for almost two thirds of energy consumption as well as GHG emissions (Farga-De-Cal et al., 2021). The objective of decarbonizing cities can be achieved by fostering the development of buildings and areas that produce either no or more energy than they consume, and by reproducing these models to encourage cities to meet their sustainability targets. Several projects in smart cities demonstrate innovative and pioneering approaches, often characterized by their creativity. For instance, smart city initiatives supported by the European Union may be viewed as pioneer projects. There is a REMINING-LOWEX system that utilized the heat energy generated from the mine shafts to regulate the temperature of buildings (New et al., 2023).

Similarly, the STORM project focuses on enhancing energy efficiency at the district level through the development of a district heating and cooling network controller that utilizes self-learning algorithms. Diverse methodologies are showcased in urban living laboratories, experimental sites, trial runs, and flagship initiatives which can be replicated in developing nations. The objective of this replication method must be to streamline the implementation of sustainable solutions for the cities.

The current trend places greater emphasis on the design phase of new buildings, specifically in relation to energy and GHG emissions which are referred as energy efficient buildings (Uspenskaia et al., 2021). In this initial phase of architectural development of building, it is essential to make important decisions on several aspects such as the selection of materials, arrangement of rooms, utilization of rainwater, incorporation of green areas, choice of energy sources, and the overall compactness of the building. These decisions greatly impact the future performance and the quality of the building. To achieve a comprehensive and unified approach at the building, it is necessary to combine the different needs and functions of each building component within the broader context by utilizing particular tools or tactics (Scherz et al., 2023) that are designed to address these intricate difficulties. The utilization of a systemic approach in the designing of building model will facilitate the management of complexity and minimize the need for extra effort during the initial design phase. Contractors of construction along with project managers and building owners should be provided with the tools to organize sustainability criteria, process them in a clear manner, and ensure a comprehensive supervision.

11.2.4.2 Strategic Public Procurement (SPP) in Housing Sector

The process of acquiring goods, services, or works by public entities in a deliberate and planned manner to achieve certain objectives is known as strategic public procurement (SPP). SPP can be employed as a sustainable initiative that goes beyond the conventional mode of acquiring normal goods and services for achieving policy-related and societal objectives (de Coninck et al., 2023). By implementing SPP in the former way, governments can effectively tackle this significant issue and achieve sustainable results (George et al., 2024; Patrucco et al., 2017). The housing industry is responsible for around 40% of worldwide CO_2 emissions (ICE, 2022). In the case of social housing, the owners, either individuals or local governments, should utilize SPP to actively involve local suppliers in the implementation of retrofit programs, the reduction of carbon emissions from the already constructed social houses, and the construction of environmentally friendly homes (Hawkins et al., 2023). This is achieved through various initiatives, including novel ways of competing and collaborative efforts between public and private sector organizations to develop sustainable solutions. Citizens' involvement and support are crucial for decarbonization efforts, extending beyond the participation of construction and material providers (Trischler et al., 2023).

To reduce carbon emissions in the housing sector, it would be beneficial to bring together various stakeholders, such as technicians and builders, to promote the use of innovative renewable heating technology. This approach might be particularly effective in the social housing sector, which is quite large (Hawkins et al., 2023). In addition to contractors, governments can carefully consider engaging business units, employees, functional departments, and all the involved stakeholders (George et al., 2024). Employees can greatly contribute toward achieving net zero goals in general (Bhushan & Singh, 2024). In the case of SPP, they participate in various aspects of the implementation process, including project design, auctioning and bidding, partnering with suppliers as well as end users, project execution, supervision, and control (Patrucco et al., 2023). Social housing is a substantial sector of the housing market, as social housing in the UK accounts for 15% of the country's GHG emissions, according to DBEIS 2021. Within the UK, strategic public procurement plays a crucial role in local governments due to the requirements set forth by the Public Service (Social Value) Act. This act mandates that the public sector must take sustainability factors into account when acquiring goods and services (Selviaridis et al., 2023). Similar acts may be proposed in developing nations to uplift the value of sustainable housing.

11.3 Confronting Challenges in the Way of Developing Countries

It is not easy for developing countries to achieve the net zero goals due to several reasons which are discussed as below.

11.3.1 Challenges in Energy and Heating Sector

Lack of policy support and inadequate infrastructure are the most common challenges in the way of switching to complete renewable energy sources. Financing is an additional obstacle that needs to be addressed to achieve the net zero targets. A robust policy framework that encompasses not only the legislature but also infrastructure and funding can resolve this issue. Similarly, the process of setting up a nuclear power plant is in itself very challenging, from the proposal to the approval process (Rahmanta et al., 2023). The legislative process is very cumbersome due to the high risk involved in selecting a location and the possible environmental implications. The foundation of the nuclear program can be made safe and sustainable when established through a legal framework, a special agency for managing the implementation of the entire process, and regulations governing the utilization of nuclear power. These measures would guarantee the secure establishment and functioning of a power plant. To ensure the safety of the infrastructure of such plants, there should be ongoing enhancements in multiple areas, including the decommissioning and handling of waste, to guarantee the safety of these stages and prevent any harm to the environment.

Locations with a low probability of volcanic, seismic, and tsunami disasters should be preferred to avoid contingencies.

11.3.2 Challenges in Transportation Sector

The shift of the transportation sector toward renewable energy poses significant difficulties, especially when it comes to large-scale, long-distance vehicles, and aircraft (Dominković et al., 2018). When comparing trucks with the same range, battery- or electric-powered heavy-duty trucks can transport around 40% less cargo compared to diesel-fueled trucks with internal combustion engines. A study uncovered various challenges related to electric car batteries, such as ethical dilemmas, the over-extraction of raw materials for battery production, the absence of legislation addressing emissions during manufacture, and a limited market for recycled batteries (Bonsu, 2020). Wu et al. (2021) examined the challenges and remedies for implementing vehicles running on hydrogen fuel cells across China, highlighting hurdles like inadequate infrastructure, few manufacturers, and safety issues regarding hydrogen fuel. The analysis has shown that a circular economy has the potential to reach net zero carbon emissions targets by the year 2050. Nevertheless, the concept of a circular economy should not be restricted solely to the repurposing of batteries and recycling of raw materials. Concerns such as reduction of emissions along the value chain, developing sustainable infrastructure, safeguarding the environment, and responsible utilization of natural resources should also be encompassed.

11.3.3 Challenge in Manufacturing Industry

During cement manufacturing it can be difficult to control the gas composition in cement kilns that are not completely sealed, run at extremely high temperatures, and rotate (Zheng et al., 2016). In Paris agreement nations have outlined their obligations in Nationally Determined Contributions (NDCs), to adhere to the same, governments must strategically focus on the cement industry. This entails establishing emission reduction objectives that are tailored to cement manufacturing, adopting and executing innovative and sustainable methods, and allocating resources to research and technology that minimize the environmental impact of cement production. Possible strategies could involve the implementation of storage technology, the improvement of energy efficiency, and the utilization of low-carbon alternatives in the cement production process.

11.3.4 Challenges in Construction Industry

Due to significant variations in historical, geographic, political, social, and economic attributes among metropolitan areas, replicating successful projects from developed countries to developing countries is not a straightforward task. The evidence-based methodology used in developed nations is valuable for assessing the effectiveness of outcomes, methods, tactics, and insights within a specific territorial setting. Simultaneously, it is important for the results of a pilot project to be applicable to different locations and situations. This will ensure that the project has a wider and more profound influence and will facilitate the promotion of urban transitions toward sustainability and resilience (Hawkins et al., 2023). To prevent pilot schemes from being sporadic, it is crucial to actively seek chances to expand and replicate successful pilots. In the housing sector, disagreements or conflicts among internal stakeholders on decarbonization goals and methods to achieve them may impede or disrupt the decarbonization process and diminish the generation of social value. Consequently, there is a requirement to enable stakeholders to take ownership of their actions and fully embrace sustainability. The implementation of SPP would strengthen vulnerable people who have less or no resources to make influential decisions and would increase their capacity to tackle social concerns (Patrucco et al., 2023).

11.4 Research Methodology

This research is completely based on secondary sources of data, such as journal articles, book chapters, conference papers, and internet sources, including websites and blogs. An intensive review of the extant literature is undertaken for this chapter. An in-depth review of existing research provides valuable insights into the current status of greenhouse gas emissions globally and measures being adopted by different countries to achieve net zero goals. It would help us understand the specific circumstances, target group, and scenarios in which these interventions are most effective. The challenges associated with each initiative, along with possible solutions, are also drawn from a thorough study of the literature.

11.5 Implications

Net zero targets gain greater credibility when they incorporate specific milestones, a well-defined implementation plan, and a clear declaration regarding the long-term intention of either sustaining net zero or transitioning to net negative. Omitting these elements poses the danger of inactivity, distractions, and lack of success. They serve as important benchmarks for achieving net-negative emissions targets in the future. Although sometimes overlooked, this factor determines the most effective way to achieve net zero. Those countries with the highest capability to achieve a future with a net-negative impact must make preparations for the right future for all. In light of escalating global warming, governments and companies may deem it imperative to expedite the implementation of measures. Acceleration can be promoted by establishing a consistent timetable for reviewing objectives, similar to the ones already established for the shorter-term Nationally Determined Contributions (NDCs).

This chapter highlighted the necessity for well-defined and consistent measures regarding the net zero strategy. An effective formulation of plans to achieve these goals will enhance the understanding and knowledge of the different stakeholders, such as service providers, industrialists, manufacturers, and the public at large, by highlighting the advantages. In order to ensure that measures are adopted on a large scale, the government should take the initiative. Policies must be framed at several levels with the aim of promoting the use of renewable energy sources. The public sector should first be made to function with responsibility and collaborate for energy savings. Utilizing diverse communication tools, such as concise and educational videos and graphics, can facilitate the transmission of information among all stakeholders. This ensures that the advantages of renewable electricity and heating technologies are comprehensively grasped by all stakeholders and acknowledged as a means of providing value to the end users of the service.

Similarly, in the transportation industry, it is possible to limit GHG emissions by utilizing hydrogen, bio-methane, electricity, and electro-fuels. Among these options, electricity is the most viable and convenient alternative. There is a lack of thorough study on the appropriate methods to guarantee the long-term sustainability of energy for the whole lifecycle of the transportation sector, but there are alternative measures that are feasible to improve the efficiency of the transportation industry. Implementing travel demand management strategies for decreasing both the distance and the frequency of travel can potentially enhance efficiency within the sector (Brand et al., 2020). Moreover, the expansion of sharing economies, encompassing activities like ride-sharing, parking spot sharing, and crowdsourcing information, would enhance the efficiency of the sector, leading to a reduction in carbon emissions.

The key to further adoption of low-carbon solutions for smart cities lies in sharing the best experiences, implementing sustainable initiatives, and replicating successful projects. The success stories can build trust among stakeholders and have a positive impact on the decision-making process. Local governments, as influential purchasers, have the ability to use their buying power to stimulate the market for sustainable products and services that reduce harm to the environment and enhance social value creation by promoting social inclusion. In the manufacturing sector, specifically cement, if full carbonation is guaranteed, the process of reducing emissions through carbon capture could serve as an alternate source for carbon–neutral synthetic liquid fuels. Facilitate the transfer of technology, enhance the development of skills, and secure funding from developed countries by implementing efficient and effective measures to reduce the impact of climate change on developing nations. Utilize foreign economic assistance to support developing nations in financing sustainability-oriented systems for communities and households. Unlike the traditional method of prioritizing cost-effective reductions, achieving net zero emissions involves choosing immediate steps that lay the foundation for long-term behavioral, technical, and institutional transformations necessary for all sectors to reach zero emissions or even achieve negative emissions (Bataille et al., 2020).

One must consider path dependencies and lock-in risks when commercializing new technologies. This includes the time it takes for these technologies to replace the existing ones, establish networks for supplying new types of energy, develop infrastructure for transport and new buildings, and change energy usage behaviors and land-use patterns, However, the difficulty lies in formulating country-specific approaches that effectively carry out these extensive changes while remaining in line with the unique circumstances of each nation. Every country is not required to achieve complete greenhouse gas neutrality under net zero decarbonization. But it is necessary that the emissions trajectory of each region and industry be directed by the objective of achieving carbon neutrality. While certain sectors or regions may not be able to achieve the net zero targets, this would necessitate other regions and sectors to experience a net-negative outcome in order to offset the difference. From the perspective of designing scenarios, this entails surpassing the goal of optimizing within a carbon constraint and instead concentrating on evaluating the highest achievable actions in each sector and identifying crucial country-led changes required to achieve these reductions in emissions or enhancements in carbon sinks. Critics may raise concerns about the significance of deliberating a strategy for achieving net-negative emissions while the world has not yet made progress toward attaining net zero emissions. From our perspective, these discussions are vital in molding a sustainable and enduring future for our collective planet.

11.6 Conclusions, Limitations, and Future Research Scope

Although short-term benefits hold significance, considering the cyclical nature of politics, it will be imperative to establish global and domestic backing for a vision that promotes a more sustainable, intelligent, and inclusive future. Effective management of change at both the national and local levels is crucial for providing the required assistance within countries. This would involve implementing measures at organizational levels to facilitate change, such as upgrading buildings. Additionally, it will involve promoting alternate modes of transportation such as cycling, walking, and public transport. Initially, if every country allocates resources to enhance investment and stimulate demand, all nations can reap the advantages in terms of increased demand and employment. Furthermore, robust projections of global economic expansion can stimulate and maintain investment. Furthermore, a dedication to sustainable technologies not only drives innovation but also enables cost reduction on a larger scale. Furthermore, there are several advantages that arise from the decrease in greenhouse gas emissions, which help all of us. Similar to the process of recovering and reconstructing after the Second World War, the current period is of utmost importance for global cooperation.

This chapter adds to existing research but the implications are limited to not only the four sectors chosen for study but also the sub-sectors selected for elaborated discussion. Further research is necessary to get insights into other sub-sectors as well as major sectors that could be country specific. The challenges discussed in this chapter also remained focused on specific themes; hence, there is a need to comprehend the political and economic aspects, as well as the behavioral aspects that will restrict or facilitate the shift toward achieving net zero emissions. This research should focus on generating the necessary political determination to establish durable policy frameworks that will encourage widespread sustainable innovation and investments. Additionally, it should explore methods to enhance the adoption of cleaner products and services by consumers throughout the economy rather than a country as whole. It is crucial to have a deeper understanding of how the pandemic has influenced citizens' perspectives on climate change and behavior change, as well as identifying the most effective methods for establishing agreement in the future.

References

Aktar, M. A., Harun, M. B., & Alam, M. M. (2020). Green energy and sustainable development. In W. Leal Filho, A. M. Azul, L. Brandli, A. L. Lange Salvia, & T. Wall (Eds.), *Affordable and clean energy* (pp. 1–11). Springer. 10. 1007/978-3-319-71057-0-47_1

Alhusni, H., Satria, T., Perdana, P., Purwanto, E. H., & Setyawan, H. (2023, February). Geothermal business outlook in Indonesia. In *Proceedings of the 48th Workshop on Geothermal Reservoir Engineering Stanford University, Stanford, CA, USA* (pp. 6–8).

Allen, M., Dube, O. P., Solecki, W., Aragón-Durand, F., Cramer, W., Humphreys, S., & Kainuma, M. (2018). Special report: Global warming of 1.5 °C. *Intergovernmental Panel on Climate Change (IPCC).*

Barbhuiya, S., Das, B. B., & Adak, D. (2024). Roadmap to a net-zero carbon cement sector: Strategies, innovations and policy imperatives. *Journal of Environmental Management, 359*, 121052. https://doi.org/10.1016/j.jenvman.2024.121052

Barker, D. J., Turner, S. A., Napier-Moore, P. A., Clark, M., & Davison, J. E. (2009). CO2 capture in the cement industry. *Energy Procedia, 1*(1), 87–94.

Bataille, C., Waisman, H., Briand, Y., Svensson, J., Vogt-Schilb, A., Jaramillo, M., Delgado, R., Arguello, R., Clarke, L., Wild, T., Lallana, F., & Imperio, M. (2020). Net-zero deep decarbonization pathways in Latin America: Challenges and opportunities. *Energy Strategy Reviews, 30*, 100510.

Bataille, C., Waisman, H., Colombier, M., Segafredo, L., Williams, J., & Jotzo, F. (2016). The need for national deep decarbonization pathways for effective climate policy. *Climate Policy, 16*(sup1), S7–S26.

Bhushan, M., & Singh, R. (2024). Talent management facilitates net-zero transition through employee green behavior. In R. Singh, S. Khan, A. Kumar, S. Luthra, H. Chokshi (Eds.), *Net zero economy, corporate social responsibility and sustainable value creation. CSR, sustainability, ethics & governance* (pp. 117–129). https://doi.org/10.1007/978-3-031-55779-8_7

Bonsu, N. O. (2020). Towards a circular and low-carbon economy: Insights from the transitioning to electric vehicles and net zero economy. *Journal of Cleaner Production, 256*, 120659.

Brand, C., Anable, J., Ketsopoulou, I., & Watson, J. (2020). Road to zero or road to nowhere? Disrupting transport and energy in a zero carbon world. *Energy Policy, 139*, 111334.

Carey, L., & Shepard, J. U. (2022). Congress's climate triple whammy: Innovation, investment, and industrial policy. *Rocky Mountain Institute.*

Carl, J., & Fedor, D. (2016). Tracking global carbon revenues: A survey of carbon taxes versus cap-and-trade in the real world. *Energy Policy, 96*, 50–77.

Chen, J., Zheng, W., Kong, Y., Yang, X., Liu, Z., & Xia, J. (2021). Case study on combined heat and water system for nuclear district heating in Jiaodong Peninsula. *Energy, 218*, 119546.

Chen, L., Msigwa, G., Yang, M., Osman, A. I., Fawzy, S., Rooney, D. W., & Yap, P. S. (2022). Strategies to achieve a carbon neutral society: A review. *Environmental Chemistry Letters, 20*(4), 2277–2310. https://doi.org/10.1007/s10311-022-01435-8

Chen, Y. H., Wang, C., Nie, P. Y., & Chen, Z. R. (2020). A clean innovation comparison between carbon tax and cap-and-trade system. *Energy Strategy Reviews, 29*, 100483.

Das, A., & Ghosh, A. (2023). Vision net zero: A review of decarbonisation strategies to minimise climate risks of developing countries. *Environment, Development and Sustainability*, 1–37.

De Coninck, B., Gascó-Hernández, M., Viaene, S., & Leysen, J. (2023). Determinants of open innovation adoption in public organizations: A systematic review. *Public Management Review, 25*(5), 990–1014.

Dean, C. C., Dugwell, D., & Fennell, P. S. (2011). Investigation into potential synergy between power generation, cement manufacture and CO2 abatement using the calcium looping cycle. *Energy and Environmental Science, 4*(6), 2050–2053.

Dominković, D. F., Bačeković, I., Pedersen, A. S., & Krajačić, G. (2018). The future of transportation in sustainable energy systems: Opportunities and barriers in a clean energy transition. *Renewable and Sustainable Energy Reviews, 82*, 1823–1838.

Fay, M., Hallegatte, S., Vogt-Schilb, A., Rozenberg, J., Narloch, U., & Kerr, T. (2015). *Decarbonizing development: Three steps to a zero-carbon future.* World Bank Publications.

Fraga-De Cal, B., Garrido-Marijuan, A., Eguiarte, O., Arregi, B., Romero-Amorrortu, A., Mezzasalma, G., Ferrarini, G., & Bernardi, A. (2021). Energy performance assessment of innovative building solutions coming from construction and demolition waste materials. *Materials, 14*(5), 1226.

George, G., Fewer, T. J., Lazzarini, S., McGahan, A. M., & Puranam, P. (2024). Partnering for grand challenges: A review of organizational design considerations in public–private collaborations. *Journal of Management, 50*(1), 10–40.

Ghafoori, M. S., Loubar, K., Marin-Gallego, M., & Tazerout, M. (2022). Techno-economic and sensitivity analysis of biomethane production via landfill biogas upgrading and power-to-gas technology. *Energy, 239*, 122086.

Gibb, D., Ledanois, N., Ranalder, L., Yaqoob, H., Murdock, H. E., Achury, N., Andre, T., Benachir, I., Dhar, A., Gicquel, S., Hamirwasia, V., & Levin, R. (2022). *Renewables 2022 global status report+ Renewable energy data in perspective+ Press releases+ Regional fact sheets+ Country fact sheets.*

Godin, J., Liu, W., Ren, S., & Xu, C. C. (2021). Advances in recovery and utilization of carbon dioxide: A brief review. *Journal of Environmental Chemical Engineering, 9*(4), 105644.

Guo, Z., Zhang, X., Feng, S., & Zhang, H. (2020). The impacts of reducing renewable energy subsidies on China's energy transition by using a hybrid dynamic computable general equilibrium model. *Frontiers in Energy Research, 8*, 25.

Hawkins, C. V., Krause, R. M., & Deslatte, A. (2023). Staff support and administrative capacity in strategic planning for local sustainability. *Public Management Review, 25*(5), 879–900.

Health Effects Institute. (2019). *State of global air 2019.* Special Report.

Husain, I. (2021). *Electric and hybrid vehicles: Design fundamentals.* CRC Press.

IEA, I. (2017). *Energy technology perspectives 2017. Catalysing Energy Technology Transformations.*

Institution of Civil Engineers (ICE). (2022). *State of the nation 2020: Infrastructure and the 2050 net-zero target.* https://www.ice.org.uk/news-insight/policy-and-advocacy/policy-engagement/state-of-the-nation-2020-infrastructure-and-the-2050-net-zero-target/

Iyer, G., Ledna, C., Clarke, L., Edmonds, J., McJeon, H., Kyle, P., & Williams, J. H. (2017). Measuring progress from nationally determined contributions to mid-century strategies. *Nature Climate Change, 7*(12), 871–874.

Kadakia, K. T., & Galea, S. (2023). Urbanization and the future of population health. *The Milbank Quarterly, 101*(Suppl 1), 153.

Kawajiri, K., Kobayashi, M., & Sakamoto, K. (2020). Lightweight materials equal lightweight greenhouse gas emissions? A historical analysis of greenhouse gases of vehicle material substitution. *Journal of Cleaner Production, 253*, 119805.

Leurent, M., Da Costa, P., Rämä, M., Persson, U., & Jasserand, F. (2018). Cost-benefit analysis of district heating systems using heat from nuclear plants in seven European countries. *Energy, 149*, 454–472.

Li, Z., & Hanaoka, T. (2020). Development of large-point source emission downscale model by estimating the future capacity distribution of the Chinese iron and steel industry up to 2050. *Resources, Conservation and Recycling, 161*, 104853.

Lu, B., Chen, G., Chen, D., & Yu, W. (2016). An energy intensity optimization model for production system in iron and steel industry. *Applied Thermal Engineering, 100*, 285–295.

Mertens, S. (2022). Design of wind and solar energy supply, to match energy demand. *Cleaner Engineering and Technology, 6*, 100402.

Muratori, M., Smith, S. J., Kyle, P., Link, R., Mignone, B. K., & Kheshgi, H. S. (2017). Role of the freight sector in future climate change mitigation scenarios. *Environmental Science and Technology, 51*(6), 3526–3533.

New, K., Tyler, A., Friday, A., Hazas, M., & Gormally, A. (2023). Space temperature policy towards net-zero: Recommendations from a systematic review of UK HEI heating policies. *Building Research and Information, 51*(2), 223–239.

Oberthür, S., Khandekar, G., & Wyns, T. (2021). Global governance for the decarbonization of energy-intensive industries: Great potential underexploited. *Earth System Governance, 8*, 100072.

Obobisa, E. S. (2022). Achieving 1.5 C and net-zero emissions target: The role of renewable energy and financial development. *Renewable Energy, 188*, 967–985.

Olhoff, A., & Christensen, J. M. (2020). *Emissions gap report 2020.*

Patrucco, A. S., Dimand, A. M., Neshkova, M. I., & Cevallos, M. M. (2023). How can procurement create (sustainable) public value under the Bipartisan Infrastructure Deal? *Public Administration Review, 83*(4), 960–973.

Patrucco, A. S., Luzzini, D., Ronchi, S., Essig, M., Amann, M., & Glas, A. H. (2017). Designing a public procurement strategy: Lessons from local governments. *Public Money and Management, 37*(4), 269–276.

Plocher, J., & Panesar, A. (2019). Review on design and structural optimisation in additive manufacturing: Towards next-generation lightweight structures. *Materials and Design, 183*, 108164.

Prussi, M., Padella, M., Conton, M., Postma, E. D., & Lonza, L. (2019). Review of technologies for biomethane production and assessment of Eu transport share in 2030. *Journal of Cleaner Production, 222*, 565–572.

Rahmanta, M. A., Adhi, A. C., Tambunan, H. B., Digwijaya, W., Damanik, N., & Al Hasibi, R. A. (2023). An analysis of national position, opportunity, and challenge of indonesia's nuclear program to support net-zero emissions by 2060. *Energies, 16*(24), 80–89.

Renné, D. S. (2022). Progress, opportunities and challenges of achieving net-zero emissions and 100% renewables. *Solar Compass, 1*, 100007.

Ritchie, H., Rosado, P., & Roser, M. (2024a). Sector by sector: Where do global greenhouse gas emissions come from? *Our World in data.* https://ourworldindata.org/ghg-emissions-by-sector

Ritchie, H., Rosado, P., & Roser, M. (2024b). Energy production and consumption. *Our world in data.* https://ourworldindata.org/energy-production-consumption

Rogelj, J., Geden, O., Cowie, A., & Reisinger, A. (2021). Three ways to improve net-zero emissions targets. *Nature, 591*(7850), 365–368.

Scherz, M., Hoxha, E., Maierhofer, D., Kreiner, H., & Passer, A. (2023). Strategies to improve building environmental and economic performance: An exploratory study on 37 residential building scenarios. *The International Journal of Life Cycle Assessment, 28*(7), 828–842.

Selviaridis, K., Luzzini, D., & Mena, C. (2023). How strategic public procurement creates social value: Evidence from UK anchor institutions. *Public Management Review*, 1–29.

Singh, R., Khan, S., Kumar, A., Luthra, S., & Chokshi, H. (Eds.). (2024a). *Net zero economy, corporate social responsibility and sustainable value creation. CSR, sustainability, ethics & governance.* Springer. https://doi.org/10.1007/978-3-031-55779-8_1

Singh, R., Ravesangar, K., Narayanan, S., & Singh, K. K. H. (2024b). Hrm practices in creating net zero: Supporting organizational sustainability. In R. Singh, S. Khan, A. Kumar, S. Luthra, & H. Chokshi (Eds.), *Net zero economy, corporate social responsibility and sustainable value creation. CSR, sustainability, ethics & governance* (pp. 1–18). Springer. https://doi.org/10.1007/978-3-031-55779-8_1

Soares, N., Santos, P., Gervásio, H., Costa, J. J., & Da Silva, L. S. (2017). Energy efficiency and thermal performance of lightweight steel-framed (LSF) construction: A review. *Renewable and Sustainable Energy Reviews, 78*, 194–209.

Statista. (2023). *Ember generation of electricity worldwide from 1990 to 2022.* Energy Source. Available online: https://www-statista-com.ezproxy.ugm.ac.id/statistics/273273/world-electricity-generation-by-energy-source/

Trischler, J., Svensson, P. O., Williams, H., & Wikström, F. (2023). Citizens as an innovation source in sustainability transitions—Linking the directionality of innovations with the locus of the problem in transformative innovation policy. *Public Management Review, 25*(11), 2093–2115.

Uspenskaia, D., Specht, K., Kondziella, H., & Bruckner, T. (2021). Challenges and barriers for net-zero/positive energy buildings and districts—Empirical evidence from the smart city project SPARCS. *Buildings, 11*(2), 78.

Värri, K., & Syri, S. (2019). The possible role of modular nuclear reactors in district heating: Case Helsinki region. *Energies, 12*(11), 2195.

Vercammen, S., Chalabyan, A., Ramsbottom, O., Ma, J., & Tsai, C. (2017). *The growing importance of steel scrap in China.* McKinsey Company.

Virmani, N., Agarwal, S., Raut, R. D., Paul, S. K., & Mahmood, H. (2022). Adopting net-zero in emerging economies. *Journal of Environmental Management, 321*, 115978.

Waisman, H., Bataille, C., Winkler, H., Jotzo, F., Shukla, P., Colombier, M., Buira, D., Criqui, P., Fischedick, M., Kainuma, M., La Rovere, E., & Trollip, H. (2019). A pathway design framework for national low greenhouse gas emission development strategies. *Nature Climate Change, 9*(4), 261–268.

Wei, Y. M., Chen, K., Kang, J. N., Chen, W., Wang, X. Y., & Zhang, X. (2022). Policy and management of carbon peaking and carbon neutrality: A literature review. *Engineering, 14*, 52–63. https://doi.org/10.1016/j.eng.2021.12.018

World Steel Association. (2019). *Steel statistical yearbook 2019: Concise version.* World Steel Association.

Wu, Y., Liu, F., He, J., Wu, M., & Ke, Y. (2021). Obstacle identification, analysis and solutions of hydrogen fuel cell vehicles for application in China under the carbon neutrality target. *Energy Policy, 159*, 112643.

Xi, F., Davis, S. J., Ciais, P., Crawford-Brown, D., Guan, D., Pade, C., Shi, T., Syddall, M., Lv, J., Ji, L., Bing, L., & Liu, Z. (2016). Substantial global carbon uptake by cement carbonation. *Nature Geoscience, 9*(12), 880–883.

Zeman, F., & Lackner, K. (2006). The zero-emission kiln. *International Cement Review, 56.*

Zheng, L., Hills, T. P., & Fennell, P. (2016). Phase evolution, characterisation, and performance of cement prepared in an oxy-fuel atmosphere. *Faraday Discussions, 192*, 113–124.

Chapter 12
Net Zero Policies for Sustainable Future: Issues and Challenges in the Asia Pacific Region

Mananage Shanika Hansini Rathnasiri and Dewanarayana Hiththaragedara Prasad Manuranga Gunathilaka

Abstract This study comprehensively examines the issues and challenges related to net zero policies for a sustainable future in the Asia Pacific region. Drawing upon a systematic literature review methodology, the research explores the diversity and complexity of the region's economic, geographical, and political landscape. It analyzes the current status of net zero policies in major countries such as China, India, Japan, Australia, and ASEAN nations. Key findings reveal the tension between economic growth and sustainability, the region's dependence upon fossil fuels, technological and infrastructure barriers, policy and governance issues, and the influence of social and cultural factors. The findings present case studies highlighting each country's unique approaches and challenges, from China's ambitious targets to Japan's focus on technological innovation. Implications of the research underscore the urgent need for enhanced regional cooperation, increased investment in renewable energy and technology, strengthened policy and governance frameworks, and greater public engagement. This study stands out due to its in-depth examination of the Asia Pacific region's progress towards net zero emissions. It provides crucial insights and implications for policymakers, businesses, and individuals who foster a sustainable future.

Keywords Net zero policies · Sustainable future · Issues and challenges · Asia Pacific region · Net zero emissions

---

M. S. H. Rathnasiri (✉)
Department of Marketing Management, Faculty of Management Studies, Sabaragamuwa University of Sri Lanka, Belihuloya, Sri Lanka
e-mail: shanika@mgt.sab.ac.lk

D. H. P. M. Gunathilaka
Head of Quality Assurance, Environmental, Health and Safety, International Distillers Limited, Kaduwela, Sri Lanka

R. Singh and D. Crowther (eds.), *Transition Towards a Sustainable Future*, Approaches to Global Sustainability, Markets, and Governance,
https://doi.org/10.1007/978-981-97-5756-5_12

12.1 Introduction

With over 4.5 billion inhabitants and some of the fastest-growing economies in the world, the Asia Pacific region is leading the global climate catastrophe (Shah et al., 2019; Castelo et al., 2024). According to the World Bank, this region was responsible for almost half of global carbon dioxide emissions in 2020. According to the Intergovernmental Panel on Climate Change (IPCC), regions of the Asia Pacific may see temperature increases of more than four °C by the end of the century, which would have a disastrous effect on human health, water resources, and agriculture (Canadell et al., 2023). These estimates emphasize how urgently the area must attain net zero emissions, which would mean matching the amount of greenhouse gases released into the environment with the amount removed.

Net zero emissions have become a fundamental goal in combating climate change (Yang & Lu, 2023). To achieve net zero, one must remove an equal quantity of carbon from the atmosphere through carbon capture and storage (CCS), reforestation, and other cutting-edge technologies to balance inevitable greenhouse gas emissions (Rogelj et al., 2021). By seeking to keep global warming below two °C, with a preferred objective of 1.5 °C, relative to pre-industrial levels, the Paris Agreement, agreed in 2015, established a worldwide framework to prevent catastrophic climate change (Wigley, 2018). By the middle of the century, net zero global carbon emissions are required to meet these targets. For the Asia Pacific region, achieving net zero is not just a matter of environmental responsibility but a crucial step towards ensuring sustainable development and safeguarding future generations against the severe effects of climate change in the long run.

The Asia Pacific region, which includes smaller developing nations within the ASEAN group and highly industrialized nations like South Korea and Japan, emerging economies like China, and emerging nations like India, play a critical role in the global climate context. This region substantially contributes to worldwide energy consumption and industrial production, making it a key driver of global economic growth. However, because of the quick industrialization and urbanization, significant environmental consequences have been associated with this economic success, such as increased greenhouse gas emissions and noticeable environmental degradation. As a result, the Asia Pacific area's goal of net zero emissions is essential for the region and the worldwide effort to tackle climate change (Hibino & Masui, 2024). Success in this area could serve as a potent role model for other regions, proving that environmental sustainability and economic growth are compatible.

Hence, this study explores the multifaceted issues and challenges of implementing net zero policies in Asia. The path to net zero is paved with challenges that vary significantly among nations and industries despite the growing awareness of the urgency of taking immediate action on climate change. This chapter will focus on these complexities, examining how economic growth imperatives, energy dependencies, technological capabilities, governance structures, and socio-cultural factors shape the region's path towards achieving net zero emissions. By shedding light on these challenges, this chapter provides a comprehensive understanding of the barriers that must be overcome and the strategies to facilitate this critical transition.

The chapter is structured to thoroughly explore the topic, beginning with a foundational understanding of net zero policies and their significance, followed by an in-depth analysis of the unique challenges faced by the Asia Pacific region. It will start by explaining net zero policies, highlighting their primary goals, and discussing key components such as emissions reduction, renewable energy adoption, and carbon offsetting. This will be followed by an overview of the geographical, economic, and political diversity of the Asia Pacific region, setting the stage for understanding its climate challenges and examining the current status of net zero policies in major countries, including China, India, Japan, Australia, and ASEAN nations. The chapter will then delve into the key issues and challenges, including the tension between economic growth and sustainability, reliance on fossil fuels, technological and infrastructure barriers, problems of policy and governance, and socio-cultural factors. Through detailed case studies of China, India, Japan, and ASEAN countries, the chapter will provide insights into their specific challenges and efforts. Finally, the chapter will explore potential future directions, offering policy recommendations, highlighting technological innovations, emphasizing the significance of regional cooperation, and discussing the role of private and public engagement and education. A review of the essential ideas, an outlook on the possibility of reaching net zero in the Asia Pacific area, and a call to action for interested parties to support a sustainable future will be included in the chapter's conclusion.

12.2 Methodology

A comprehensive investigation into the issues and challenges of net zero policies for a sustainable future in the Asia Pacific region was carried out through a systematic literature review (SLR). This methodological approach ensured a thorough and reliable examination by employing techniques from esteemed researchers exploring similar topics. We used several research databases and search engines, like Scopus and Google Scholar, through a broad spectrum of search terms, such as "Net Zero Policies for Sustainable Future", to cover all relevant aspects of the research.

The systematic literature review was conducted in alignment with methodologies employed by Dewasiri et al. (2022), Rathansiri and De Silva (2023), Jayawardena and Dewasiri (2023), Karunarathna et al. (2024), Dewasiri et al. (2023), and Dewasiri et al. (2024). These sources provided a robust foundation for ensuring the comprehensiveness of our review. A total of eighty-seven publications were initially identified. Rigorous inclusion and exclusion criteria were then meticulously applied to select studies most pertinent to the research topic and objectives. The review included articles published in academic research journals, magazines, and conference proceedings that concerned net zero policies and their implementation in diverse sectors.

Studies lacking adequate information or relevance to the research were systematically excluded to maintain the highest level of significance and quality. The selection process was carried out by three impartial reviewers who evaluated abstracts and titles, followed by an in-depth examination of the content based on predetermined inclusion and exclusion criteria. Ultimately, sixty-six relevant studies were included in the final review. Each study was carefully evaluated against the criteria developed for methodological rigour, sample representativeness, originality, and clarity of findings. This thorough approach ensures that this review's findings are robust and reliable. It is an essential resource for stakeholders seeking accurate and trustworthy information on net zero policies in the Asia Pacific region.

12.3 Discussion

12.3.1 Understanding Net Zero Policies

Strategies and actions to balance the removal of greenhouse gases (GHGs) from the atmosphere with their emission into it are included in net zero policies (Rogelj et al., 2021). To put it simply, reaching net zero means using technologies like carbon capture and storage (CCS), reforestation, and other cutting-edge processes to absorb an equivalent quantity of carbon dioxide from the atmosphere in order to offset any leftover GHG emissions (Adun et al., 2024). The primary objective of these policies is stabilizing and eventually lowering the amount of greenhouse gases in the atmosphere, which will lessen the effects of climate change.

International agreements and scientific consensus to keep global warming safe are the main drivers behind the worldwide targets for achieving net zero emissions (Allen et al., 2022). The Paris Agreement, which came into effect in 2015, is a significant international agreement aimed at limiting the rise in the average global temperature to significantly less than two degrees Celsius above pre-industrial levels. The goal is to cap the temperature increase at 1.5 °C (Gao et al., 2017). By the middle of the twenty-first century, net zero carbon emissions must be achieved to meet these temperature targets. This goal is crucial for preventing the worst effects of climate change, such as extreme weather, rising sea levels, and damage to ecosystems and human cultures.

12.3.1.1 Fundamental Components

Net zero policies encompass several fundamental components that work together to reduce and offset emissions (Sun et al., 2021):

1. Carbon Reduction: Substantial carbon emissions reductions in various sectors, including energy sustainability, transportation, industry, and agriculture, are the foundation of the net zero policy (Barbhuiya et al., 2024; Mishra et al., 2024). This endeavour includes increasing energy efficiency, switching to low-carbon energy sources, and adopting cleaner production techniques.

2. Renewable Energy Adoption: Broadly using renewable energy sources, such as solar, wind, hydro, and geothermal, is crucial in reducing carbon emissions (Maxmut, 2023). These sources produce little to no GHG emissions compared to fossil fuels and are essential for decarbonizing the energy sector.
3. Carbon Offsetting: Carbon offsetting mechanisms are employed for emissions that cannot be eliminated. These include afforestation and reforestation projects, soil carbon sequestration, and developing and executing carbon capture and storage (CCS) technologies (Fragkos, 2021). These methods help remove carbon dioxide from the atmosphere, offsetting residual emissions.
4. Technological Innovation: It is critical to develop and implement new technologies that reduce or capture carbon emissions (Huisingh et al., 2015). This includes advancements in energy storage, electric vehicles, and industrial processes, as well as research into new methods of carbon sequestration.
5. Policy and Regulation: Effective net zero policies require robust regulatory frameworks that set emissions targets, incentivize low-carbon technologies, and enforce compliance (Rhodes et al., 2021). This includes carbon pricing schemes that create financial incentives for reducing emissions, including carbon taxes or cap-and-trade programmes.

12.3.1.2 Global Versus Regional Policies

While the overarching goals of net zero policies are globally consistent—aiming to reduce GHG emissions and limit global warming—specific strategies and implementations can vary significantly between global and regional contexts (Satola et al., 2022).

Global Strategies:

International Agreements: Frameworks such as the Paris Agreement put forward broad objectives and principles for global climate action, encouraging countries to submit nationally determined contributions (NDCs) that outline their emissions reduction plans (Weikmans et al., 2021).

Global Initiatives: Initiatives like the Sustainable Development Goals (SDGs) integrate climate action into broader sustainability efforts, promoting worldwide cooperation and knowledge sharing (Fuso Nerini et al., 2019).

Regional Strategies in the Asia Pacific:

Economic and Development Context: The Asia Pacific region includes many economic contexts, from highly developed economies like Japan and South Korea to rapidly developing nations like China and India and smaller, less developed countries in Southeast Asia. Each country's approach to net zero is influenced by its economic development stage, industrial structure, and energy needs (Chou, 2024).

Energy Mix and Resources: The energy mix in the Asia Pacific is diverse, with some countries heavily reliant on coal and others investing significantly in renewable energy (Thavasi & Ramakrishna, 2009). Policies need to address these differences, promoting a just transition for workers in fossil fuel industries and leveraging regional renewable resources.

Cultural and Governance Factors: Social norms, cultural attitudes towards the environment, and governance structures are crucial in shaping net zero policies (Sovacool & Griffiths, 2020). Regional cooperation through entities like ASEAN can facilitate sharing best practices and joint initiatives.

For instance, China has set lofty goals to become carbon neutral by 2060 and to reach its peak carbon emissions by 2030, emphasizing increasing renewable energy and advancing CCS technology (Liu et al., 2022). India aims to achieve net zero by 2070, emphasizing renewable energy expansion and efficiency (Bhattacharyya et al., 2022). Japan targets net zero by 2050, leveraging technological innovation and energy transitions (Kainuma et al., 2024). In contrast, smaller ASEAN countries focus on integrating net zero goals with sustainable development, balancing economic growth with environmental stewardship (Nasir et al., 2019). Understanding these regional nuances is essential for designing effective net zero policies that align with the unique challenges and opportunities in the region.

12.3.2 Regional Overview: Asia Pacific

Asia Pacific is one of the world's most diverse and dynamic regions, encompassing various economic, geographical, and political landscapes. This diversity presents significant challenges and unique opportunities in pursuing net zero emissions.

1. *Economic Diversity*: The countries of China, India, Japan, and some of the smaller developing nations in Southeast Asia include some of the largest and fastest-growing economies in the world. These countries vary widely in their levels of industrialization, energy consumption patterns, and economic priorities (Sharvini et al., 2018). For instance, China is a global manufacturing hub with a significant carbon footprint, while many Pacific Island nations have minimal emissions but are highly vulnerable to climate impacts (Kumar et al., 2020).
2. *Geographical Diversity*: The geographical landscape of the Asia Pacific is incredibly varied (Hope et al., 2004), ranging from the vast deserts of Australia to the dense rainforests of Southeast Asia and the high-altitude regions of the Himalayas. This diversity influences the range of renewable energy resources available, like hydro, solar, wind, and geothermal. For example, Australia has immense solar potential, while Indonesia and the Philippines are rich in geothermal resources (Pan et al., 2019).
3. *Political Landscape*: The political landscape in the Asia Pacific region is equally diverse, encompassing democracies, monarchies, and various forms of authoritarian regimes. This diversity affects how climate policies are formulated and implemented. Countries like Japan and Australia have well-established regulatory frameworks and political stability, facilitating structured climate action (Knox-Hayes & Hayes, 2014). In contrast, some developing nations may face

political instability, limited governance capacity, and differing national priorities, which can hinder effective policy implementation.

12.3.2.1 Current Status of Net Zero Policies

China: China has set aggressive climate goals, hoping to reach carbon neutrality by 2060 and peak carbon emissions by 2030, as it is the world's top CO_2 producer (Liu et al., 2022). The government is a global leader in manufacturing and using electric vehicles and has made significant expenditures in renewable energy, notably solar and wind power. With significant funds for infrastructure development and technical innovation, China's "14th Five-Year Plan" prioritizes green development and the transition to a low-carbon economy (Shang & Lv, 2023).

India: India aims to reach net zero by 2070, balancing its developmental needs with climate action. The country focuses on expanding its renewable energy capacity, aiming to achieve 500 Gigawatts by 2030 (Singh et al., 2023). India's approach includes significant investments in solar power, enhancing energy efficiency, and promoting electric mobility. However, challenges remain regarding policy implementation, financing, and ensuring energy access for its growing population (Nandal et al., 2019).

Japan: Japan has committed to reaching net zero emissions by 2050 (Oshiro et al., 2018). The country's strategy strongly emphasizes technological innovation, energy efficiency, and adopting renewable energy sources. Japan also uses novel technologies such as hydrogen energy and CCS (Kawai et al., 2022). Policy frameworks such as the "Green Growth Strategy" outline steps to decarbonize critical sectors, including power, transportation, and industry.

Australia: Australia aims to reach net zero economy by 2050, with a target of 43% below 2005 levels by 2030 (Geroe, 2022). The country's net zero policy focuses on expanding renewable energy mechanisms, primarily solar and wind, and improving energy storage capacities. Australia also invests in carbon–neutral programmes, including reforestation and soil carbon projects (Standish & Prober, 2020). However, the nation faces many difficulties because it relies on coal exports and internal fossil fuel usage.

ASEAN Nations: The Association of Southeast Asian Nations (ASEAN) comprises a diverse group of countries at different stages of economic development. Collectively, they are working towards a low-carbon future, but each country has specific targets and strategies. For example, with international support, Indonesia aims to reach net zero by 2060 or sooner, focusing on transitioning to renewable energy (Resosudarmo et al., 2023). The Philippines aims for a 75% reduction in emissions by 2030, contingent on global financial support, and is enhancing its renewable energy infrastructure. Vietnam, Thailand, and Malaysia also increase their capacity for renewable energy and efficiency (Sahid & Sin, 2019).

Pacific Island Nations: These nations are thought to be among the most susceptible to the effects of climate change, such as increasing sea levels and harsh weather (Taylor, 2021). They pledge to achieve net zero emissions as part of the global effort to address climate change, notwithstanding their comparatively small contributions to world emissions. Pacific countries like the Marshall Islands and Fiji have set high goals for using renewable energy. They are actively involved in international climate talks, pushing for more dramatic global action.

Hence, the Asia Pacific region presents a complex and multifaceted landscape for net zero policies, with each country facing challenges and opportunities. The diversity in economic development, geographical resources, and political systems require effective strategies to reduce emissions and transition to sustainable energy systems. By understanding these nuances, policymakers can design more effective and inclusive net zero strategies that cater to each country's specific needs and contexts within the region.

12.3.3 Key Issues and Challenges

The SLR analysis explored five main themes as issues and challenges in achieving net zero policies for a sustainable future in the Asia Pacific Region. The findings are stated in Table 12.1.

12.3.3.1 *Economic Growth versus Sustainability*

The Asia Pacific region is characterized by rapid economic growth, which presents a significant challenge to achieving sustainability (Khan et al., 2022). Countries such as China, India, and Southeast Asian nations have experienced substantial economic expansion, leading to increased energy consumption and higher greenhouse gas emissions (Sharvini et al., 2018). This growth is often driven by industrialization, urbanization, and rising living standards, which are crucial for reducing poverty and improving quality of life. However, this progress comes at a cost to the environment as the demand for energy and natural resources escalates.

Balancing economic growth with sustainable practices requires a paradigm shift towards green growth strategies prioritizing environmental health and economic development. Policies incentivizing renewable energy, improving energy efficiency,

Table 12.1 Key themes identified through the SLR analysis

#	Themes
1	Economic growth versus sustainability
2	Energy dependence
3	Technological and infrastructure barriers
4	Policy and governance
5	Social and cultural factors

Source Authors' own

and promoting sustainable industrial practices are essential. For instance, transitioning to a circular economy, where waste is minimized, and resources are reused, can decouple economic growth from environmental degradation. Additionally, integrating sustainability into national development plans can help align economic objectives with climate goals, ensuring long-term resilience and prosperity.

12.3.3.2 Energy Dependence

A significant barrier to reaching net zero emissions is the Asia Pacific region's dependence on fossil fuels (Obobisa, 2022). Many countries in the region, notably China, India, and Indonesia, rely heavily on oil, coal, and natural gas for their energy requirements. This dependency is driven by the abundance of these resources, established energy infrastructures, and the relative affordability of fossil fuels compared to some renewable alternatives.

Transitioning to renewable energy is challenging due to several factors. Firstly, the existing energy infrastructure is predominantly designed for fossil fuels, requiring significant investment to retrofit or replace with renewable systems (Mutezo & Mulopo, 2021). Second, solar and wind power need advanced energy storage systems and grid management technologies to provide consistent electricity supply (Babatunde et al., 2019). Thirdly, the economic interests tied to the fossil fuel industry, including employment and revenue, create resistance to change (Trencher et al., 2019).

To overcome these challenges, policies must support the gradual phase-out of fossil fuels while scaling up renewable energy investments. This includes financial incentives for renewable energy initiatives, aids for clean energy technologies, and robust regulatory frameworks that promote renewable energy adoption. International cooperation and investment can also be vital in accelerating the region's conversion to sustainable energy systems.

12.3.3.3 Technological and Infrastructure Barriers

Technological limitations and infrastructure challenges hinder the implementation of net zero policies in the Asia Pacific region (Sovacool et al., 2023). Many countries lack advanced technologies for the efficient generation of renewable energy, CCS, and other low-carbon innovations (Dziejarski et al., 2023). Additionally, the infrastructure necessary to support these technologies, such as modernized power grids, is often inadequate.

For instance, rural and remote areas in many Asia Pacific countries face significant barriers to accessing clean energy due to the lack of infrastructure (Mehta et al., 2021). Deploying renewable energy technologies in these regions requires substantial investment in grid extension, mini-grids, and off-grid solutions. Moreover, the high upfront costs of clean technologies and the lack of technical expertise impede their widespread adoption.

Addressing these barriers requires a multifaceted approach. In order to develop clean technologies and make them more widely available and reasonably priced, the government and private sectors must undertake research and development investments. Building capacity through education and training programmes can also enhance the technical skills required for deploying and maintaining these technologies. Furthermore, international partnerships can facilitate the transfer of technology and knowledge, helping to bridge the technological gap.

12.3.3.4 Policy and Governance

Effective policy formulation, implementation, and enforcement are critical for the success of net zero initiatives (Sindhwani et al., 2022). However, the Asia Pacific region faces several governance challenges, including political will, regulatory frameworks, and corruption (Li et al., 2019). There is a lack of cohesive and comprehensive climate policies that integrate environmental, economic, and social objectives in the region. Moreover, political instability and frequent changes in government can disrupt the continuity and consistency of climate policies.

Corruption and weak regulation enforcement further undermine the effectiveness of net zero policies (Omri & Boubaker, 2024). In some cases, vested interests and lobbying by powerful industries can influence policy decisions, resulting in inadequate or biased regulations favouring short-term economic gains over long-term sustainability.

Strengthening governance requires transparent and accountable institutions that effectively design and implement climate policies (Rathnasiri et al., 2024). This involves fostering political will and leadership at all levels of government, establishing clear and enforceable regulations, and promoting stakeholder engagement in policymaking processes. Additionally, anti-corruption measures and mechanisms for monitoring and evaluating policy implementation are essential for ensuring the integrity and effectiveness of net zero initiatives.

12.3.3.5 Social and Cultural Factors

Social norms and cultural factors are significant in the acceptance and effectiveness of net zero policies (Sovacool & Griffiths, 2020). Public awareness and understanding of climate change and its impacts vary widely across the region, influencing the level of support for climate action. In some communities, there may be resistance to change due to deeply rooted cultural practices and beliefs, economic dependence on traditional industries, or a lack of information about the benefits of sustainable practices.

Education and public engagement are crucial for overcoming these barriers. Raising awareness about the significance of climate change and the benefits of net zero policies can help build public support and drive behavioural change. Community involvement in designing and implementing climate initiatives can also ensure that policies are culturally sensitive and address local needs and priorities.

Moreover, providing climate literacy at the school level and promoting environmental stewardship through media and public campaigns can foster a culture of sustainability (Jayarathne et al., 2021). Empowering local communities and providing platforms for their engagement in climate decisions can enhance the acceptance and effectiveness of net zero policies.

Hence, achieving net zero emissions in the Asia Pacific region requires addressing complex economic, technological, governance, and social challenges. By adopting a holistic approach that considers each country's unique contexts and needs, policymakers can develop effective strategies that motivate sustainable development and mitigate the impacts of climate change.

12.3.4 Case Studies

12.3.4.1 China: Ambitious Net Zero Targets and Challenges

Ambitious Targets and Measures: China, the biggest carbon dioxide emitter in the world, has set aggressive targets for the climate, including achieving carbon neutrality by 2060 and peaking carbon emissions before 2030. China is investing significantly in green energy and setting the worldwide standard for producing electric vehicles (EVs) to meet these commitments. The Chinese government has also launched several initiatives, such as the national carbon trading scheme, which began in 2021, to incentivize emissions reductions in the industrial sector (Feng et al., 2022).

Challenges: Despite these efforts, China faces significant challenges transitioning to net zero. The country's heavy reliance on coal for energy poses a significant obstacle. In 2020, coal accounted for nearly 60% of China's total energy consumption (Jia & Lin, 2021). While renewable energy capacity is rapidly expanding, the intermittency of solar and wind sources requires substantial energy storage and grid modernization investment to ensure reliability. Balancing economic growth with environmental sustainability remains critical as China prioritizes industrial output and urban development.

Key Strategies: To overcome these challenges, China focuses on technological innovation, such as advancing battery storage technologies and developing intelligent grids. The country is also investing in reforestation and other carbon offset projects. Policies to promote energy efficiency in buildings and industry, alongside stricter environmental regulations, are part of China's comprehensive approach to achieving its net zero goals.

12.3.4.2 India: Balancing Development and Sustainability

Development and Sustainability: India aims to reach net zero by 2070, balancing its developmental needs with climate action. India aims to have 500 Gigawatt by 2030, focusing on solar power, which is plentiful because of the country's location (Kumar et al., 2023). India is also investing in wind energy, hydroelectric power, and bioenergy, promoting energy efficiency across various sectors.

Challenges: India faces unique challenges in its net zero journey, including high poverty levels, a growing population, and the need for economic development. The country's energy demand is expected to grow significantly, driven by industrialization and urbanization. Transitioning from coal, which currently dominates the energy mix, to renewable sources is complex and costly. Additionally, financial constraints and limited access to clean technologies pose significant hurdles.

Key Strategies: India is seeking international financial support and technological cooperation to address these challenges. The government is implementing laws to motivate private investment in green energy projects and improve the regulatory environment to facilitate clean energy transitions. Initiatives like the International Solar Alliance, which India co-founded, aim to foster collaboration and resource sharing among solar-rich countries.

12.3.4.3 Japan: Technology and Innovation for Net Zero

Japan has committed to achieving net zero emissions by 2050. The country's strategy emphasizes technological innovation and energy efficiency. Japan is a global leader in developing advanced technologies like hydrogen energy, CCS, and next-generation nuclear reactors. The government's "Green Growth Strategy" outlines measures to decarbonize critical sectors, including power, transportation, and industry.

Challenges: Japan faces challenges in balancing its energy needs with climate goals. Following the 2011 Fukushima nuclear disaster, the country relied heavily on fossil fuel imports, particularly liquefied natural gas (LNG), which reduced atomic power usage (Hayashi & Hughes, 2013). Ensuring energy security while phasing out fossil fuels is a significant concern. Additionally, Japan's ageing population and low economic growth challenge sustaining investment in green technologies.

Key Strategies: To address these issues, Japan is increasing its investment in renewable energy and energy storage solutions. The country is also promoting energy conservation and efficiency measures across all sectors. International partnerships and collaborations on green technology development are essential to Japan's approach to achieving its net zero targets.

12.3.4.4 Australia: Navigating Fossil Fuel Dependency and Renewable Potential

Ambitious Targets and Measures: Australia aims to become a net zero economy by 2050, emphasizing shifting from its predominant reliance on natural green energy sources. Due to its abundance of green energy potential, especially solar and wind, Australia is investing in large-scale renewable energy projects, battery storage systems, and hydrogen production (Milani et al., 2020).

Challenges: The country's significant fossil fuel exports present a unique challenge, as coal and natural gas exports are significant economic contributors. Transitioning the domestic energy market while managing the financial implications of reduced fossil fuel exports requires careful policy planning. Additionally, the country faces political challenges and varying levels of commitment to climate action across different states and territories.

Key Strategies: Australia is building an advanced clean energy sector using abundant renewable energy resources. Among the initiatives are the Clean Energy Finance Corporation (CEFC), which makes investments in energy efficiency and renewable energy, and the Australian Renewable Energy Agency (ARENA), which offers financial aid for innovative green energy initiatives. Policies to aid the growth of the green hydrogen industry are also being implemented to harness Australia's renewable energy potential for domestic use and export.

12.3.4.5 ASEAN Countries: Diverse Efforts and Challenges

Efforts and Initiatives: The Association of Southeast Asian Nations (ASEAN) comprises nations with diverse economic and energy profiles, each facing unique challenges and opportunities in their net zero journey. For instance, Indonesia has committed to achieving net zero by 2060, focusing on transitioning from coal to renewable energy sources like geothermal and hydropower. The Philippines aims to cut emissions by 75% by 2030, contingent on international support, and is enhancing its renewable energy infrastructure (La-Viña et al., 2018).

Challenges: Common challenges across ASEAN countries include heavy reliance on fossil fuels, limited financial resources, and technological constraints. Many ASEAN nations face significant barriers to accessing advanced renewable energy technologies and infrastructure. Additionally, political instability and varying levels of governance capacity can impede the implementation of cohesive and effective climate policies.

Key Strategies: ASEAN countries emphasize regional cooperation and resource sharing to overcome these challenges. Initiatives like the ASEAN Plan of Action for Energy Cooperation (APAEC) seek to improve energy safety and encourage the use of green energy throughout the region. International aid and investment are critical for supporting these efforts. Countries also focus on capacity building and knowledge transfer to improve their technical and regulatory capabilities for implementing net zero policies.

Accordingly, the case studies of China, India, Japan, Australia, and ASEAN countries illustrate the diverse approaches and challenges nations in the Asia Pacific region face in their quest for net zero emissions. Each country's strategy is shaped by its unique economic, geographical, and political context, highlighting the need for tailored solutions and international cooperation to achieve global climate goals.

12.4 Conclusion and Future Directions

12.4.1 Conclusion

The Asia Pacific area is vital to the worldwide fight against climate change. With over 4.5 billion populations and among the world's leading economies, this region contributes significantly to global carbon emissions and has the potential to be a climate action leader. As detailed in the preceding sections, the region's road to net zero is dynamic and holistic, shaped by a unique blend of economic, geographical, and political factors.

Understanding Net Zero Policies: The foundational concepts of net zero policies, including emissions reduction, renewable energy adoption, and carbon offsetting, are essential for mitigating climate change. The global targets set by agreements like the Paris Agreement provide a framework within which the Asia Pacific countries must operate. However, regional adaptations and strategies are necessary to address specific local challenges.

Regional Overview: The Asia Pacific region's diversity, encompassing significant economies like China, India, and Japan, as well as smaller ASEAN nations, requires a nuanced approach to policy formulation and implementation. Each country's current status in adopting net zero policies reflects its economic priorities, resource availability, and political will.

Key Issues and Challenges: The tension between rapid development and the need for sustainable practices is a recurring theme. The region's reliance on fossil fuels, technological and infrastructure barriers, policy and governance issues, and socio-cultural factors pose significant hurdles. Addressing these challenges requires a balanced approach that promotes sustainable development without compromising economic growth.

Case Studies: The experiences of China, India, Japan, Australia, and ASEAN countries illustrate the diverse strategies and obstacles in achieving net zero. These case studies highlight the importance of technological innovation, international cooperation, and tailored policy frameworks to meet each country's unique needs and circumstances.

12.4.2 Future Directions

Enhancing Regional Cooperation: The Asia Pacific region can benefit significantly from enhanced regional cooperation. Collaborative initiatives, such as sharing best practices, technological innovations, and financial resources, can help bridge the gap among the region's countries. Regional organizations and frameworks, such as ASEAN, can be important in promoting this collaboration.

Investing in Renewable Energy and Technology: It is essential to continue funding green energy initiatives such as hydroelectric, solar, and wind power. Furthermore, cutting-edge technologies like energy storage systems, green hydrogen, and CCS can significantly eliminate emissions. Governments should encourage research and development to spur the development of innovative clean technology.

Strengthening Policy and Governance: Effective policy formulation and robust governance structures are essential for the successful implementation of net zero policies. This includes creating clear regulatory frameworks, ensuring transparency, and combating corruption. Engaging multiple stakeholders is critical for fostering a collective effort towards sustainability.

Promoting Sustainable Economic Growth: Balancing economic growth with environmental sustainability remains a key challenge. Governments need to integrate sustainability into their financial planning and development strategies. This includes promoting green industries, supporting sustainable agriculture and forestry practices, and encouraging energy efficiency across all sectors.

Addressing Social and Cultural Factors: Introducing public awareness and education campaigns to change social norms and behaviours towards more sustainable practices is essential. Engaging communities and promoting grassroots initiatives can enhance the acceptance and effectiveness of net zero policies. Policy design should consider cultural factors to ensure they are culturally sensitive and widely accepted.

Mobilizing Financial Resources: Achieving net zero will require significant financial investment. International financial institutions, development banks, and private sector investors should be mobilized to provide the necessary funding. Innovative financial products, like green bonds and climate funds, can also facilitate the shift.

Monitoring and Accountability: Robust monitoring and accountability mechanisms are essential to tracking progress and ensuring compliance with net zero targets. This includes setting up independent oversight bodies, regular reporting, and using data-driven approaches to measure and verify emissions reductions.

In conclusion, the Asia Pacific region faces formidable challenges and possesses immense potential in reaching net zero emissions. The region can succeed towards a sustainable future by leveraging its diversity, fostering regional cooperation, investing in innovation, and strengthening governance. Policymakers, businesses, and individuals must collaborate and act decisively to turn this potential into reality, ensuring a healthier planet for future generations.

12.4.3 Call to Action

The challenge of achieving net zero emissions in the Asia Pacific region requires collective action and shared responsibility. As we stand at a significant juncture in the combat against climate change, policymakers, businesses, and individuals must step up and contribute to building a sustainable future for future generations.

1. *For Policymakers*: We urge policymakers to prioritize climate action and implement ambitious net zero policies. This includes setting clear targets, enacting supportive regulatory frameworks, and fostering international cooperation. Policymakers must lead by example, demonstrating political willingness and commitment to driving the conversion to a net zero region.
2. *For Businesses*: Businesses drive innovation and shape market dynamics. We call upon businesses to embrace sustainability as a core principle of their operations. This involves adopting green technologies, reducing emissions throughout the supply chain, and investing in renewable energy solutions. By integrating sustainability into business strategies, companies can mitigate climate risks and seize opportunities for growth and innovation.
3. *For Individuals*: Everyone has a part in transitioning to a sustainable future. We encourage individuals to make informed decisions to eliminate carbon emissions, such as opting for sustainable transportation, conserving energy and water, and supporting eco-friendly products and initiatives. Raising awareness and advocating for climate action within communities can amplify collective efforts towards achieving net zero emissions.

Together, through collaborative and concerted action, we can overcome the challenges and issues of climate change and build a more resilient and prosperous future for all. Let us seize this opportunity to safeguard our planet and ensure a sustainable legacy for future generations. The time to act is now.

References

Adun, H., Ampah, J. D., Bamisile, O., Ozsahin, D. U., & Staffell, I. (2024). Sustainability implications of different carbon dioxide removal technologies in the context of Europe's climate neutrality goal. *Sustainable Production and Consumption, 47*, 598–616.

Allen, M. R., Friedlingstein, P., Girardin, C. A., Jenkins, S., Malhi, Y., Mitchell-Larson, E., Peters, G. P., & Rajamani, L. (2022). Net zero: Science, origins, and implications. *Annual Review of Environment and Resources, 47*, 849–887.

Babatunde, O. M., Munda, J. L., & Hamam, Y. (2019). A comprehensive state-of-the-art survey on power generation expansion planning with intermittent renewable energy sources and energy storage. *International Journal of Energy Research, 43*(12), 6078–6107.

Barbhuiya, S., Das, B. B., & Adak, D. (2024). Roadmap to a net-zero carbon cement sector: Strategies, innovations and policy imperatives. *Journal of Environmental Management, 359*, 121052.

Bhattacharyya, R., Singh, K. K., Grover, R. B., & Bhanja, K. (2022). Estimating minimum energy requirement for transitioning to a net-zero, developed India in 2070. *Current Science*, 517–527.

Canadell, J. G., Monteiro, P. M., Costa, M. H., Cotrim da Cunha, L., Cox, P. M., Eliseev, A. V., & Zickfeld, K. (2023). Intergovernmental panel on climate change (IPCC). Global carbon and other biogeochemical cycles and feedbacks. In *Climate change 2021: The physical science basis. Contribution of working group I to the sixth assessment report of the intergovernmental panel on climate change* (pp. 673–816). Cambridge University Press.

Castelo, S., Antunes, L., & Ashrafuzzaman, M. (2024). The impact of the climate crisis on gender inequality. Looking to the frontlines in search of priorities for policy. *Frontiers in Sustainable Cities, 6*, 1304535.

Chou, K. T. (2024). Developmental netzeroism. *Renewable and Sustainable Energy Reviews, 197*, 114319.

Dewasiri, N. J., Baker, H. K., Banda, Y. W., & Rathnasiri, M. S. H. (2022). The dividend decision model: A possible solution for the dividend puzzle. In *Exploring the latest trends in management literature* (pp. 249–267). Emerald Publishing Limited. https://doi.org/10.1108/978-1-80262-277-520231021

Dewasiri, N. J., Dharmarathna, D. G., & Choudhary, M. (2024). Leveraging artificial intelligence for enhanced risk management in banking: A systematic literature review. In R. Singh, S. Khan, A. Kumar, & V. Kumar (Eds.), *Artificial intelligence enabled management: An emerging economy perspective* (pp. 197–213). De Gruyter. https://doi.org/10.1515/9783111172408

Dewasiri, N. J., Karunarathne, K. S. S. N., Menon, S., Jayarathne, P. G. S. A., & Rathnasiri, M. S. H. (2023). Fusion of artificial intelligence and Blockchain in the banking industry: Current application, adoption, and future challenges. In A. Saini, & V. Garg (Eds.), *Transformation for sustainable business and management practices: Exploring the spectrum of industry 5.0* (pp. 293–307). Emerald Publishing Limited. https://doi.org/10.1108/978-1-80262-277-520231021

Dziejarski, B., Krzyżyńska, R., & Andersson, K. (2023). Current status of carbon capture, utilization, and storage technologies in the global economy: A survey of technical assessment. *Fuel, 342*, 127776.

Feng, R., Lin, P., Hou, C., & Jia, S. (2022). Study the effect of China's emissions trading scheme on promoting regional industrial carbon emission reduction. *Frontiers in Environmental Science, 10*, 947925.

Fragkos, P. (2021). Assessing the role of carbon capture and storage in mitigation pathways of developing economies. *Energies, 14*(7), 1879.

Fuso Nerini, F., Sovacool, B., Hughes, N., Cozzi, L., Cosgrave, E., Howells, M., Tavoni, M., Tomei, J., Zerriffi, H., & Milligan, B. (2019). Connecting climate action with other sustainable development goals. *Nature Sustainability, 2*(8), 674–680.

Gao, Y., Gao, X., & Zhang, X. (2017). The 2 °C global temperature target and the evolution of the long-term goal of addressing climate change—From the United Nations framework convention on climate change to the Paris agreement. *Engineering, 3*(2), 272–278.

Geroe, S. (2022). 'Technology not taxes': A viable Australian path to net zero emissions? *Energy Policy, 165*, 112945.

Hayashi, M., & Hughes, L. (2013). The Fukushima nuclear accident and its effect on global energy security. *Energy Policy, 59*, 102–111.

Hibino, G., & Masui, T. (2024). Development of AIM (Asia–Pacific Integrated Model) and its contribution to policy-making for realising decarbonised societies in Asia. *Sustainability Science, 19*(1), 223–239.

Hope, G., Kershaw, A. P., van der Kaars, S., Xiangjun, S., Liew, P. M., Heusser, L. E., Takahara, H., McGlone, M., Miyoshi, N., & Moss, P. T. (2004). History of vegetation and habitat change in the Austral-Asian region. *Quaternary International, 118*, 103–126

Huisingh, D., Zhang, Z., Moore, J. C., Qiao, Q., & Li, Q. (2015). Recent advances in carbon emissions reduction: Policies, technologies, monitoring, assessment and modelling. *Journal of Cleaner Production, 103*, 1–12.

Jayarathne, P. G. S. A., Sachithra, V., & Dewasiri, N. J. (2021). Sustainable practices through competitive capabilities in the apparel industry: Sri Lankan experience. *FIIB Business Review*. https://doi.org/10.1177/23197145211060398

Jayawardena, N. S., & Dewasiri, N. J. (2023). Food acquisition and consumption issues of South Asian countries: A systematic literature review and future research agenda. *FIIB Business Review*. https://doi.org/10.1177/23197145231194113

Jia, Z., & Lin, B. (2021). How to achieve the first step of the carbon-neutrality 2060 target in China: The coal substitution perspective. *Energy, 233*, 121179.

Karunarathna, K. S. S. N., Dewasiri, N. J., Singh, R., & Rathnasiri, M. S. H. (2024). What does artificial intelligence–Powered ChatGPT bring to academia? A review. In R. Singh, S. Khan, A. Kumar, & V. Kumar (Eds.), *Artificial intelligence enabled management: An emerging economy perspective* (pp. 3–13). De Gruyter. https://doi.org/10.1515/9783111172408

Kainuma, M., Gross, R., Hourcade, J. C., La Motta, S., Lechtenböhmer, S., & Masui, T. (2024). Accelerating actions for leveraging a climate-neutral sustainable society. *Sustainability Science, 19*(1), 1–6.

Kawai, E., Ozawa, A., & Leibowicz, B. D. (2022). Role of carbon capture and utilization (CCU) for decarbonization of industrial sector: A case study of Japan. *Applied Energy, 328*, 120183.

Khan, I., Hou, F., Zakari, A., Irfan, M., & Ahmad, M. (2022). Links among energy intensity, non-linear financial development, and environmental sustainability: New evidence from Asia Pacific Economic Cooperation countries. *Journal of Cleaner Production, 330*, 129747.

Knox-Hayes, J., & Hayes, J. (2014). Technocratic norms, political culture and climate change governance. *Geografiska Annaler: Series b, Human Geography, 96*(3), 261–276.

Kumar, C. M. S., Singh, S., Gupta, M. K., Nimdeo, Y. M., Raushan, R., Deorankar, A. V., Kumar, T. A., Rout, P. K., Chanotiya, C. S., Pakhale, V. D., & Nannaware, A. D. (2023). Solar energy: A promising renewable source for meeting energy demand in Indian agriculture applications. *Sustainable Energy Technologies and Assessments, 55*, 102905.

Kumar, L., Jayasinghe, S., Gopalakrishnan, T., & Nunn, P. D. (2020). Climate change and the Pacific Islands. *Climate change and impacts in the Pacific* (pp. 1–31).

La-Viña, A. G., Tan, J. M., Guanzon, T. I. M., Caleda, M. J., & Ang, L. (2018). Navigating a trilemma: Energy security, equity, and sustainability in the Philippines' low-carbon transition. *Energy Research and Social Science, 35*, 37–47.

Li, S., Park, S. H., & Bao, R. S. (2019). The transition from relation-based to rule-based governance in East Asia: Theories, evidence, and challenges. *International Journal of Emerging Markets, 14*(1), 171–186.

Liu, Z., Deng, Z., He, G., Wang, H., Zhang, X., Lin, J., Qi, Y., & Liang, X. (2022). Challenges and opportunities for carbon neutrality in China. *Nature Reviews Earth and Environment, 3*(2), 141–155.

Maxmut O'g'li, X. F. (2023). Renewable energy sources: Advancements, challenges, and prospects. *International Journal of Advance Scientific Research, 3*(08), 14–25.

Mehta, K., Ehrenwirth, M., Trinkl, C., Zörner, W., & Greenough, R. (2021). The energy situation in Central Asia: A comprehensive energy review focusing on rural areas. *Energies, 14*(10), 2805.

Milani, D., Kiani, A., & McNaughton, R. (2020). Renewable-powered hydrogen economy from Australia's perspective. *International Journal of Hydrogen Energy, 45*(46), 24125–24145.

Mishra, U., Rathnasiri, M. S. H., Dewasiri, N. J. (2024). Net-zero economy and energy sustainability: A Nepalese perspective. In R. Singh, S. Khan, A. Kumar, S. Luthra, & H. Chokshi (Eds.), *Net zero economy, corporate social responsibility and sustainable value creation. CSR, sustainability, ethics & governance* (pp. 145–161). Springer. https://doi.org/10.1007/978-3-031-55779-8_9

Mutezo, G., & Mulopo, J. (2021). A review of Africa's transition from fossil fuels to renewable energy using circular economy principles. *Renewable and Sustainable Energy Reviews, 137*, 110609.

Nandal, V., Kumar, R., & Singh, S. K. (2019). Barriers identification and analysis of solar power implementation in Indian thermal power plants: An Interpretative Structural Modeling approach. *Renewable and Sustainable Energy Reviews, 114*, 109330.

Nasir, M. A., Huynh, T. L. D., & Tram, H. T. X. (2019). Role of financial development, economic growth & foreign direct investment in driving climate change: A case of emerging ASEAN. *Journal of Environmental Management, 242*, 131–141.

Obobisa, E. S. (2022). Achieving 1.5 °C and net-zero emissions target: The role of renewable energy and financial development. *Renewable Energy, 188*, 967–985.

Omri, A., & Boubaker, S. (2024). When do climate change legislation and clean energy policies matter for net-zero emissions? *Journal of Environmental Management, 354*, 120275.

Oshiro, K., Masui, T., & Kainuma, M. (2018). Transformation of Japan's energy system to attain net-zero emission by 2050. *Carbon Management, 9*(5), 493–501.

Pan, S. Y., Gao, M., Shah, K. J., Zheng, J., Pei, S. L., & Chiang, P. C. (2019). Establishment of enhanced geothermal energy utilization plans: Barriers and strategies. *Renewable Energy, 132*, 19–32.

Rathnasiri, M. S. H., & De Silva, V. (2023). The role of shopping values, cognitive and affective evaluations on the relationship between store environment and store patronage intention: A future research agenda, Sri Lanka. *Journal of Management Studies, 5*(1), 100–120.

Rathnasiri, M. S. H., Dewasiri, N. J., & Kumar, A. A. (2024). Policy framework and implementation strategies for Sri Lanka's transition to a net-zero economy. In R. Singh, S. Khan, A. Kumar, S. Luthra, & H. Chokshi (Eds.), *Net zero economy, corporate social responsibility and sustainable value creation. CSR, sustainability, ethics & governance* (pp. 43–60). Springer. https://doi.org/10.1007/978-3-031-55779-8_3

Resosudarmo, B. P., Rezki, J. F., & Effendi, Y. (2023). Prospects of energy transition in Indonesia. *Bulletin of Indonesian Economic Studies, 59*(2), 149–177.

Rhodes, E., Scott, W. A., & Jaccard, M. (2021). Designing flexible regulations to mitigate climate change: A cross-country comparative policy analysis. *Energy Policy, 156*, 112419.

Rogelj, J., Geden, O., Cowie, A., & Reisinger, A. (2021). Three ways to improve net-zero emissions targets. *Nature, 591*(7850), 365–368.

Sahid, E. J. M., & Sin, T. C. (2019). Energy security in ASEAN region: A case study of Malaysia energy security performance with renewable energy implementation. *Journal of Advanced Research in Fluid Mechanics and Thermal Sciences, 61*(2), 190–201.

Satola, D., Wiberg, A. H., Singh, M., Babu, S., James, B., Dixit, M., Sharston, R., Grynberg, Y., & Gustavsen, A. (2022). Comparative review of international approaches to net-zero buildings: Knowledge-sharing initiative to develop design strategies for greenhouse gas emissions reduction. *Energy for Sustainable Development, 71*, 291–306.

Shah, S., Abbas, G., Hanif, M., Anees-Ur-Rehman, Zaman, M., Riaz, N., Altaf, A., Hassan, S. U., Saleem, U., & Shah, A. (2019). Increased burden of disease and role of health economics: Asia-Pacific region. *Expert Review of Pharmacoeconomics and Outcomes Research, 19*(5), 517–528.

Shang, W. L., & Lv, Z. (2023). Low carbon technology for carbon neutrality in sustainable cities: A survey. *Sustainable Cities and Society, 92*, 104489.

Sharvini, S. R., Noor, Z. Z., Chong, C. S., Stringer, L. C., & Yusuf, R. O. (2018). Energy consumption trends and their linkages with renewable energy policies in East and Southeast Asian countries: Challenges and opportunities. *Sustainable Environment Research, 28*(6), 257–266.

Sindhwani, R., Singh, P. L., Behl, A., Afridi, M. S., Sammanit, D., & Tiwari, A. K. (2022). Modeling the critical success factors of implementing net zero emission (NZE) and promoting resilience and social value creation. *Technological Forecasting and Social Change, 181*, 121759.

Singh, K., Meena, R. S., Kumar, S., Dhyani, S., Sheoran, S., Singh, H. M., Pathak, V. V., Khalid, Z., Singh, A., Chopra, K., & Bajar, S. (2023). India's renewable energy research and policies to phase down coal: Success after Paris agreement and possibilities post-Glasgow Climate Pact. *Biomass and Bioenergy, 177*, 106944.

Sovacool, B. K., Del Rio, D. F., & Zhang, W. (2023). The political economy of net-zero transitions: Policy drivers, barriers, and justice benefits to decarbonization in eight carbon-neutral countries. *Journal of Environmental Management, 347*, 119154.

Sovacool, B. K., & Griffiths, S. (2020). The cultural barriers to a low-carbon future: A review of six mobility and energy transitions across 28 countries. *Renewable and Sustainable Energy Reviews, 119*, 109569.

Standish, R. J., & Prober, S. M. (2020). Potential benefits of biodiversity to Australian vegetation projects registered with the Emissions Reduction Fund—is there a carbon-biodiversity trade-off? *Ecological Management and Restoration, 21*(3), 165–172.

Sun, T., Ocko, I. B., Sturcken, E., & Hamburg, S. P. (2021). The path to net zero is critical to climate change. *Scientific Reports, 11*(1), 22173.

Taylor, S. (2021). The vulnerability of health infrastructure to the impacts of climate change and sea level rise in small island countries in the South Pacific. *Health Services Insights, 14*, 11786329211020856.

Thavasi, V., & Ramakrishna, S. (2009). Asia energy mixes from socio-economic and environmental perspectives. *Energy Policy, 37*(11), 4240–4250.

Trencher, G., Healy, N., Hasegawa, K., & Asuka, J. (2019). Discursive resistance to phasing out coal-fired electricity: Narratives in Japan's coal regime. *Energy Policy, 132*, 782–796.

Weikmans, R., van Asselt, H., & Roberts, J. T. (2021). Transparency requirements under the Paris Agreement and their (un) likely impact on strengthening the ambition of nationally determined contributions (NDCs). In *Making climate action more effective* (pp. 107–122). Routledge.

Wigley, T. M. L. (2018). The Paris warming targets: Emissions requirements and sea level consequences. *Climatic Change, 147*, 31–45.

Yang, W. C., & Lu, W. M. (2023). Achieving net zero—An illustration of carbon emissions reduction with a new meta-inverse DEA Approach. *International Journal of Environmental Research and Public Health, 20*(5), 4044.

Chapter 13
Understanding Factors Responsible for Framework to Design Sustainable Net Zero Policies—An Empirical Analysis on Indian Small and Medium Enterprises (SMEs)

D. M. Arvind Mallik

Abstract Background: Today, SMEs play a decisive part in the developing economy; due to limited resources they frequently encounter challenges in implementing sustainability initiatives. Addressing the escalating concerns surrounding climate change demands urgent action in crafting sustainable net-zero policies tailored specifically for Small and Medium Enterprises (SMEs). **Objectives**: This study purposes to create a specialized framework for Indian SMEs to promote sustainable net zero policies, evaluating primary factors impacting policies, pinpointing obstacles to SMEs' sustainability efforts, and suggesting a conceptual framework for designing sustainable net zero policies. **Design/Methodology**: This research utilizes a quantitative analysis approach. Primary data collection involves distributing 300 questionnaires across 30 different companies, evenly distributed among Manufacturing, Technology, Transportation, Retailers, Energy, and Hospitality and Tourism. SEM analysis is performed on 265 completed questionnaires, testing 8 hypotheses. Additionally, secondary data is collected to assess existing policies and frameworks. **Findings**: The study identifies key factors influencing the development of net zero policies for Indian SMEs, such as organizational empowerment, technological innovation, and alignment with Sustainable Development Goals. Barriers to SMEs' adoption of sustainable practices include financial constraints, lack of awareness, and regulatory challenges. **Results**: Proposed conceptual framework will eventually support SMEs in transitioning toward designing net Zero sustainable policies. Partnership between government, industry, and civil society is emphasized to drive collective action toward sustainable net-zero policies. **Originality**: This study offers original insights by addressing the specific factors essential to Indian SMEs in

---

D. M. Arvind Mallik (✉)
Associate Professor, Department of MBA, PES Institute of Technology and Management, Shivamogga 577204, Karnataka, India
e-mail: aravind.mallik@gmail.com

R. Singh and D. Crowther (eds.), *Transition Towards a Sustainable Future*, Approaches to Global Sustainability, Markets, and Governance,
https://doi.org/10.1007/978-981-97-5756-5_13

adopting sustainable practices and crafting tailored policy interventions. The framework takes a holistic approach, considering the unique socioeconomic context of SMEs must align with global sustainability objectives. **Social Implications**: This study recognizes the factors influencing to design Net Zero policy, offering potential for positive social impact. By promoting environmental stewardship, creating green jobs, and strengthening SMEs against climate risks, it supports India's Paris Agreement commitments and encourages inclusive growth through SME involvement in sustainability initiatives.

Keywords Sustainability · Net zero policies · Small and medium enterprises (SMEs) · India · Conceptual framework · Policy design · Climate change · Structural equantional modeling (SEM)

13.1 Introduction

Among mounting global distresses about climate change and the domineering to decrease carbon releases, the objective of attaining net-zero emissions has become paramount worldwide. This goal, underscored by international agreements like the Paris Agreement, has transformed sustainability from a mere aspiration into a critical necessity (Rogeli et al., 2018). The journey toward achieving "net zero" in combating climate change stands as a complex challenge, intertwining technology, economics, and politics. "Unlike the conventional approach of merely mitigating climate change or slashing greenhouse gas emissions by predefined percentages, attaining net zero demands comprehensive decarbonization across all economic sectors while significantly curbing emissions. While certain sectors, such as electricity or buildings, may boast clear pathways to net zero, they remain exceptions rather than the norm (Davis et al., 2018; Dubois et al., 2019; International Energy Agency, 2021)".

Predictions indicated that by 2050, oil and gas production will decline by 55% and 70%, respectively, compared to current levels. Coal production is expected to nearly cease altogether by 2050 (Krishnan et al., 2022) conversely, demand for alternative energy sources is projected to increase tenfold by 2050 and deviations in regulations, technologies, consumer and investor preferences have triggered significant modifications in the demand for various goods and services (IEA, 2015). The report also suggests that certain sectors may experience long-term cost savings due to initial capital investments required for achieving net-zero transformations (Krishnan et al., 2022). The pressing need to confront climate change has now reached a pivotal juncture, compelling a profound overhaul of both corporate practices and governmental policies. Net-zero emissions, where greenhouse gas emissions are counterbalanced by removal or offsetting measures, have emerged as a pivotal strategy in this endeavor (Rogeli et al., 2018).

Global Trends and Initiatives:

The changeover toward low-carbon energy and the commitment to achieving net-zero emissions have become central themes in global efforts to combat climate change (IPCC, 2021, 2018). Governments and private firms worldwide are increasingly pledging to reduce greenhouse gas emissions to net-zero, marking a significant shift toward sustainability (Grubler et al., 2018). "Over 120 countries, representing 61% of global greenhouse gas emissions, have announced commitments to reach net-zero by midcentury or soon after, including major emitters like the United States, China, the European Union, the United Kingdom, Canada, and Japan (ECIU, 2021a, 2021b). While the notion of identifying entire countries that have achieved net-zero status may seem daunting 8 nations have already accomplished this feat across various regions located across Asia (Bhutan), South America (Suriname, Panama, and Guyana), Africa (Comoros, Gabon, and Madagascar), and the Pacific Islands (Niue)" (Nick, 2022).

Policy Frameworks and Challenges:

To achieve net-zero emissions, it is domineering to apply indicative and comprehensive policies that embrace sustainable approaches encompassing emission reduction, enhancing carbon sequestration, and adopting resilient practices to mitigate climate change impacts (Davis et al., 2010). A growing number of nations are enacting such policies, including carbon pricing mechanisms and energy efficiency measures, while also actively promoting research, development, and deployment of low-carbon technologies (Le Quéré et al., 2020).

> Despite the momentum towards a net-zero economy, progress in decarbonization has faced a recent slowdown (Mishra et al., 2022a, 2022b). Despite commitments from 137 nations to reduce greenhouse gas emissions by up to 25%, none of these countries rank among the top five polluters. Additionally, there's a concerning lack of disclosure regarding carbon emission data from companies, and public awareness of climate change remains limited (Mishra et al., 2022a, 2022b). Moreover, achieving decarbonization in the supply chain (SC) encounters various challenges, including operational, policy and regulatory, technological, economic, and market-related obstacles (Mishra et al., 2022a, 2022b). This is particularly evident in emerging economies like India, where small and medium enterprises (SMEs) play a pivotal role in advancing sustainable practices.

Empirical Analysis and Policy Implications:

This study aims to progress a framework for understanding how net-zero policies are shaped for Indian SMEs on their path to sustainability. By analyzing this sector, it provides insights for policymakers, industry, and academia to create effective sustainability policies, empowering SMEs to prioritize sustainability.

13.1.1 Net Zero

Net-zero emissions involve maintaining equilibrium between greenhouse gas emissions and removals, effectively cancelling out overall emissions through strategies like carbon capture, reforestation, and renewable energy investments. According to

Table 13.1 Definitions of net zero by various autho

Author	Definitions on net zero
Davis et al. (2018)	"Achieving net zero means decarbonizing all parts of the economy in addition to reducing emissions significantly. It encompasses not only reducing emissions but also enhancing carbon sequestration and adopting resilient practices to withstand the impacts of climate change."
International Energy Agency (2015)	"Net zero involves reducing emissions significantly across all sectors of the economy, with a particular emphasis on decarbonizing "difficult-to-decarbonize" industries such as aviation, shipping, and cement production"
ECIU (2021a, b)	"Net zero implies reducing emissions to the extent that any remaining emissions are balanced by removal or offsetting measures, aiming to achieve a state where the net emissions are zero"
IPCC (2021)	"Net-zero emissions occur when anthropogenic emissions of greenhouse gases to the atmosphere are balanced by anthropogenic removals over a specified period"

the UK government, "net zero occurs when greenhouse gas emissions are entirely offset by removal or negative emissions, with a target for all economic sectors to achieve this by 2050". While variations in definitions may exist (Howarth et al., 2020), the overarching aim remains consistent: to balance emissions and removals in the combat against environmental variation influencing raise in global temperature (Table 13.1).

13.1.2 Importance and Impact on Net Zero

Over the past decade, public concern regarding climate change has steadily risen, with significant majorities in many countries expressing apprehension about the issue (IPSOS, 2023; Poortinga et al., 2019). There's an increasing recognition of the urgency to combat climate change and the imperative for substantial changes (Steentjes et al., 2021). However, environmental action predominantly revolves around small-scale lifestyle adjustments such as recycling (IPSOS, 2023; Whitmarsh, 2009).

Humanity faces alarming climate change records, with 2023 projected to be the warmest year on record. Greenhouse gas emissions reached unprecedented levels in 2022, leading to a 1.8 °C rise in global average temperatures above pre-industrial levels by September 2023. The Emissions Gap Report underscores the urgent need for action, suggesting that fully implementing mitigation efforts outlined in the Paris Agreement could limit temperature rise to 2.9 °C by 2100. However, even with these measures, a concerning 2.5 °C rise is projected, highlighting the critical need for immediate and significant action (IPCC, 2022).

Progress since the Paris Agreement (Smith et al., 2016a, 2016b) shows a shift: while initial projections expected a 16% upsurge in greenhouse gas emissions by 2030, today's estimate is only 3% (Smith et al., 2016a, 2016b). However, to meet the Paris agreement's goals of limiting global warming to 2 °C or 1.5 °C, emissions must decrease by 28% and 42% respectively by 2030 (Smith et al., 2016a, 2016b). This requires swift economy-wide transitions toward low-carbon development, particularly in energy, with high-emitting countries taking more ambitious actions and supporting developing nations (Smith et al., 2016a, 2016b). Low- and middle-income countries, responsible for over 2/3rd of emissions, should pursue low-emissions growth to meet development needs while providing universal energy access (Smith et al., 2016a, 2016b). "The upcoming COP 28 will set new NDCs for 2035, crucial for aligning with climate goals. Public support for net-zero policies can create the political mandate needed for ambitious climate action (Höppner & Whitmarsh, 2012; Howarth et al., 2020; Stern & Dietz, 2008)".

To drastically "reduce greenhouse gas emissions, more impactful measures are necessary, including minimizing air travel and driving while adopting plant-based diets (Ivanova et al., 2020; Wynes & Nicholas, 2017). These actions not only mitigate emissions but also offer health, economic, and environmental benefits. For example, transitioning to plant-based diets can cut food-related greenhouse gas emissions by up to 70% and decrease global mortality rates by up to 10% (Springmann et al., 2016). Similarly, embracing lower-carbon transportation modes like electric vehicles, public transit, and active travel not only alleviates congestion, air, and noise pollution but also reduces road accidents (Smith et al., 2016a, 2016b). Furthermore, walking and cycling contribute to increased physical activity, leading to additional health advantages (Shaw et al., 2014)".

13.1.3 India and SME (Small and Medium Enterprise)

India, a rapidly developing nation, grapples with the task of balancing economic growth with carbon footprint reduction. Small and Medium Enterprises (SMEs), integral to India's industrial landscape, play a vital role in this challenge (Salvioni et al., 2021). Despite their scale, SMEs significantly contribute to the country's industrial output and employment. However, amidst the pressing need to address climate change, there's an increasing call for SMEs to embrace sustainable practices to mitigate environmental impact.

India's developmental journey requires economic growth to align with environmental sustainability, with net zero for Indian SMEs signifying the minimization of carbon emissions to achieve overall carbon neutrality (Biswas et al. 2018). The global prominence of achieving net-zero emissions reflects nations' pledges to vindicating the effects of climate change, with Indian SMEs transitioning toward sustainable practices to contribute to India's climate goals (Singh et al., 2023). At its essence, net-zero entails balancing the amount of greenhouse gases emitted into the atmosphere with removal or offsetting measures, thereby achieving carbon neutrality (Sindhwani et al., 2022).

Indian Small and Medium Enterprises (SMEs) wield significant influence in global trade, driving India's economic growth and international market presence. However, recent climate negotiations at COP28 highlight the imperative for these enterprises to devise emission reduction strategies to thrive in global value chains (Riahi et al., 2017). Challenges such as limited access to finance, technology, and awareness hinder their adoption of sustainable practices (Hardey et al., 2023). Achieving net-zero emissions for Indian SMEs necessitates balancing emissions with removal or offset measures (Miller et al., 2021), with governance and institutional capacity playing crucial roles in policy implementation (Modak et al., 2023).

India's net-zero policy should prioritize inclusive growth to ensure that vulnerable communities benefit from the transition to a low-carbon economy (Kumar et al., 2023). Empowering SMEs to embrace sustainability can drive innovation, efficiency gains, and competitive advantage, fostering both economic growth and environmental stewardship. As highlighted by Virmani et al. (2022), Integrating climate action with broader development goals, India can leverage targeted policy interventions and stakeholder engagement to harness the transformative potential of its SME sector for a more sustainable and inclusive future North (2016). For India, a nation experiencing rapid industrialization and urbanization, achieving net-zero emissions is imperative not only to mitigate its carbon footprint but also to ensure a sustainable pathway to development. This study aims to tackle this urgent matter by exploring which major factors impacts are designing framework for designing sustainable net zero policies specifically tailored to Indian SMEs.

SMEs are vital to India's shift to a low-carbon economy, contributing significantly to industrial output and employment. However, they face unique challenges in adopting sustainable practices, including limited access to finance and technological capabilities (Kumar et al., 2023). Tailored frameworks and policies are urgently needed to support SMEs in achieving net-zero emissions, addressing their specific barriers while leveraging their strengths (Kumar et al., 2023) In this research work, researcher have identified 6 different possible ways in getting Net Zero in near future would suggests varying levels of engagement across sectors, Manufacturing, Technology, Transportation, Retailers, Energy, Financial Institutions, Hospitality and Tourism potentially affecting business positively as they are and we have used some of these sectors in designing our conceptual framework too.

13.2 Conceptual Framework and Literature Review

This subdivision examines two interconnected strands of investigation, "Exploring the Role of Human Values in Small and Medium Enterprises (SMEs)": Investigates how human values influence decision-making in SMEs, crucial for promoting pro-environmental practices. "The Significance of Intermediaries in Facilitating SME Engagement" (Sivropoulos-Valero, 2021) Focuses on the role of intermediaries in aiding SMEs' participation in sustainability efforts, essential for overcoming barriers and providing resources. We also explore the intersection of SMEs, intermediaries, and human values, considering how alignment in these areas can promote sustainability. This understanding informs the progress of a conceptual framework for designing net-zero policies, incorporating factors like values, intermediary roles, and SME engagement in pro-environmental practices.

13.2.1 Exploring the Role of Human Values in Small and Medium Enterprises (SMEs)

In Small and Medium Enterprises (SMEs), human values significantly shape organizational culture and behavior. These values, encompassing principles, beliefs, and ethical standards, serve as guiding principles for actions and decisions (Gehman et al., 2013). Due to flat organizational structures and emphasis on interpersonal relationships, the influence of human values is particularly pronounced in SMEs. This impact is evident in strong cultural norms and shared values among employees (Littlewood et al., 2018; Williams & Preston, 2018). For instance, a commitment to social responsibility and environmental sustainability influences decisions regarding product development, supply chain management, and community engagement (Dieleman & Koning, 2020; García-Álvarez & López-Sintas, 2001; Hemingway, 2005). Misalignment between individual and organizational values can lead to conflicts, disengagement, and reduced productivity (Hammann et al., 2009; Jamali et al., 2017). Moreover, human values extend to external relationships and stakeholder interactions, with SMEs prioritizing building robust relationships based on shared values and mutual respect. These relationships bolster the SME's reputation, influencing customer loyalty, brand perception, and long-term sustainability.

13.2.2 The Significance of Intermediaries in Facilitating SMEs Engagement

In business ecosystems, Small and Medium Enterprises (SMEs) are acknowledged as pivotal backers to economic growth and innovation (BEIS, 2016; Egels-Zandén, 2017; Mole et al., 2017). Nonetheless, their limited size and resource constraints frequently pose challenges in accessing markets (Schaefer et al., 2020), resources, and opportunities. Intermediaries play a crucial role in this context, serving as vital facilitators that bridge the gap between SMEs and stakeholders such as customers, suppliers, investors, and government agencies. Intermediaries provide support services such as market research, access to financing, and business development assistance, serving as advocates for SME interests (Eadson, 2014; Egels-Zandén, 2017; Revell, 2007). Despite their significance, SME owners and managers often overlook intermediary functions and impacts due to their diverse forms, including industry associations and specialized service providers (Mole et al., 2017; Revell & Blackburn, 2007). Initiatives to support SMEs in transitioning to Net Zero face challenges in adoption due to a lack of knowledge and expertise (Blundel & Hampton, 2021; Johnstone, 2020). Communication is essential in addressing these issues, with social media providing opportunities for SMEs to exchange ideas and promote Net Zero initiatives (Fletcher & Cruz, 2020). Intermediaries facilitate this exchange through networking events, workshops, and peer-to-peer learning programs, fostering a sense of community and collective growth within the entrepreneurial ecosystem (Carbon Trust, 2020).

13.2.3 SMEs, Intermediary and Values

The environmental business support policy encounters a fundamental contradiction within Small and Medium Enterprises (SMEs), operating in intricate ecosystems where intermediary actors significantly influence organizational values and behaviors (Ahn et al., 2011). While these intermediaries play a crucial role in shaping SMEs' decision-making processes, they often prioritize economic and environmental benefits over values-based approaches (Hampton, 2018; Mukherjee, 2018). Integrating Net Zero policy changes is vital for both UK and for other global businesses, with European countries intensifying efforts since the establishment of the Kyoto protocol in 1997 (Lombardi & Liserre, 2022). Recent research extensively explores Net Zero emissions and innovation for SMEs, highlighting the importance of analyzing social media content and networks to influence SMEs in embracing net zero business goals (Kivimaa et al., 2020). Despite environmental support initiatives, SMEs often miss potential energy and resource savings (Bose, 2013). Intermediary actors play a crucial role in shaping SME values by guiding ethical and sustainable business practices, yet many lack the skills needed to foster meaningful relationships with SMEs (Ghosh & Roy, 2011; Rajesh & Vinayagamurthy, 2013; Hampton, 2018). Transitioning to net-zero emissions is imperative for SMEs in India's industrial landscape to combat climate change.

13.2.4 Conceptual Framework

The study's conceptual framework outlines key factors like organizational empowerment, sectorial approaches, financial performance, socioeconomic factors, environmental performance, public awareness and engagement, technological innovation, and alignment with Sustainable Development Goals (SDGs) that influence nets zero policies. It aims to understand these interconnections and their impact on climate change mitigation and future generations. This framework informs policy development for holistic, inclusive, and equitable sustainability transitions.

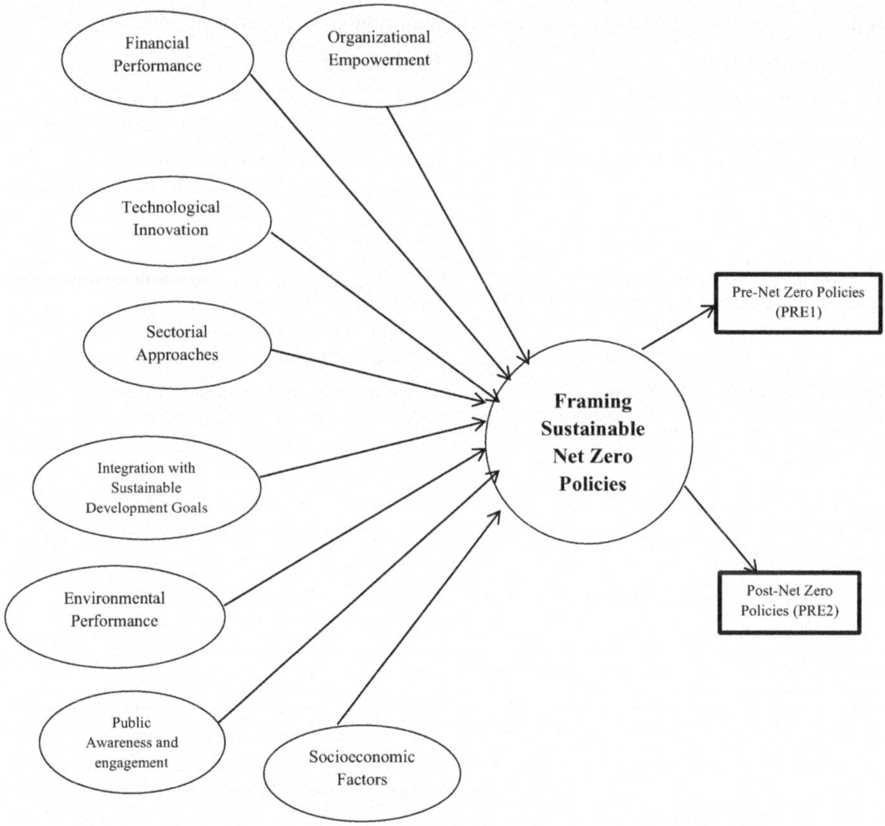

Fig. 13.1 Proposed conceptual framework to design sustainable net zero policies

13.2.5 Explaining Constructs

The policies' effectiveness hinges on their breadth, rigor, and harmony with broader climate objectives, underscoring the need for robust strategies to tackle climate change and foster sustainable development. Each variable—scope, stringency, and alignment—uniquely shapes policy effectiveness and future outcomes (Table 13.3). Through empirical analysis, researchers can unravel these dynamics, informing policy for sustainable and fair outcomes.

Table 13.2 Exogenous and endogenous constructs for framing sustainable net zero policies

Exogenous constructs	**Explanation**
Organizational empowerment (OE)	Empowering SMEs with sustainability tools, knowledge, and support for implementation
Financial performance (FIP)	Emphasizing the importance of financial viability for sustainability initiatives, including cost-effective solutions, financial incentives, and green financing options
Technological innovation (TI)	Encouraging SMEs to embrace innovative technologies for sustainable operations through funding, access facilitation, and knowledge exchange
Sectorial approaches (SA)	Tailoring policies and support measures to address the specific challenges and opportunities within different industries where SMEs operate
Integration with sustainable development goals (SDGs)	Aligning net zero policies with broader SDGs to ensure holistic sustainability and societal well-being
Environmental sustainability (ENP)	Focusing on reducing greenhouse gas emissions, conserving resources, and minimizing pollution through targets, regulations, and monitoring
Public awareness and engagement (PAE)	Promoting understanding of climate change, low-carbon benefits, and fostering sustainability culture through awareness campaigns and SME empowerment
Socioeconomic factors (SEF)	Considering equity, affordability, and job creation to ensure inclusive and equitable net zero policies that support all SMEs
Endogenous construct	**Description**
Framing sustainable net zero policies (FSNP)	Development of comprehensive strategies aimed at achieving carbon neutrality while encouraging sustainable development, assimilating environmental, social, and economic considerations
– Pre-net zero policies (PRE1)	Evaluation of the proportion of energy generated from renewable sources to reduce reliance on fossil fuels and mitigate greenhouse gas emissions before reaching net zero
– Post-net zero policies (PRE2)	Indicating progress toward cleaner and more sustainable energy sources by increasing the percentage of energy from renewables after achieving net zero, facilitating the transition to a low-carbon economy

Table 13.3 Author own explanation for each construct (Hair et al., 2017; Diamantopoulos & Siguaw, 2006; Tenenhaus et al., 2005; Kline, 2015)

Construct	Item	Explanation
Organizational empowerment (OE)	OE1	Assessing the organization's ability to delegate authority, encourage initiative, and foster innovation and accountability
	OE2	Evaluating the degree to which employees are involved in decision-making processes and granted autonomy in their roles
	OE3	Examining the organization's efforts in providing training and development opportunities to enhance employees's kills and capabilities
	OE4	Assessing the effectiveness of communication channels and feedback mechanisms to facilitate transparency and employee engagement
	OE5	Measuring the organization's dedication to diversity, equity, and inclusion, promoting equal opportunities and a supportive work environment
	OE6	Evaluating the organization's responsiveness to change and its ability to adapt to evolving market conditions and technological advancements
	OE7	Assessing leadership effectiveness and its role in inspiring and motivating employees toward common goals and objectives
	OE8	Examining the organization's investment in employee well-being and initiatives promoting work-life balance to enhance job satisfaction and retention rates
Financial performance (FIP)	FIP1	Measures the financial performance improvement resulting from net zero advantages, including enhancements in ROI, market share, and total profit from products/services
	FIP2	Assesses the organization's market competitiveness and its share of the industry's total revenue after implementing net zero initiatives
	FIP3	Evaluates the organization's overall profitability and financial gains derived from products/services developed under net zero initiatives
Environmental performance (ENP)	ENP1	Examines reductions in hazardous materials in product/service provision to indicate improved environmental performance
	ENP2	Assesses reductions in electricity consumption as a measure of improved energy efficiency and reduced environmental impact
	ENP3	Evaluates reductions in total fuel consumption in product/service transportation, indicating lower carbon emissions and environmental impact
	ENP4	Measures reductions in total raw material usage in manufacturing processes, reflecting improved resource efficiency and reduced environmental footprint
	ENP5	Assesses reductions in total packaging materials used, indicating progress toward sustainable packaging practices and waste reduction efforts
	ENP6	Examines reductions in air emissions resulting from cleaner production processes and emissions control technologies

(continued)

Table 13.3 (continued)

Construct	Item	Explanation
	ENP7	Evaluates reductions in solid waste disposal, indicating improved waste management practices and reduced environmental pollution
Socioeconomic factors (SEF)	SOF1	Assesses socioeconomic factors like income levels, education, employment rates, and social demographics influencing the organization's adoption of sustainable practices and support for net zero policies
	SOF2	Evaluates the organization's efforts in promoting social equity and inclusion through employment opportunities and community development initiatives
	SOF3	Assesses the organization's engagement with local communities and stakeholders to discourse societal challenges and promote sustainable development goals
	SOF4	Measures the organization's contribution to economic growth and stability through job creation, income generation, and investment in local economies
	SOF5	Evaluates the organization's adherence to ethical and responsible business practices, including transparency, accountability, and regulatory compliance
Technological innovation (TI)	TE1	Adoption of energy-efficient technologies to reduce energy consumption and carbon emissions
	TE2	Integration of renewable energy sources (e.g., solar, wind) to generate clean energy onsite
	TE3	Use of digital technologies and data analytics to optimize resource use and minimize environmental impact
	TE4	Implementation of low-carbon transportation solutions (e.g., electric vehicles) to reduce emissions from SMEs' transportation activities
	TE5	Adoption of innovative waste management practices to minimize waste generation and promote recycling
	TE6	Leveraging digital platforms and tools to facilitate collaboration, knowledge sharing, and innovation among SMEs toward sustainability goals
Sectorial approaches (SEA)	SEA1	Tailoring net zero policies to address the specific challenges and opportunities within a particular sector of SMEs
	SEA2	Implementing sector-specific incentives and support mechanisms to encourage adoption of net zero practices
	SEA3	Collaborating with industry associations and stakeholders to develop sector-specific sustainability standards and guidelines
	SEA4	Facilitating knowledge sharing and best practice exchange among SMEs operating within the same sector to accelerate the adoption of net zero policies
Integration with sustainable development goals (SDGs)	SDG1	Ensuring that efforts toward carbon neutrality contribute to poverty eradication and inclusive economic growth
	SDG2	Promoting sustainable agriculture and food systems while reducing emissions

(continued)

Table 13.3 (continued)

Construct	Item	Explanation
	SDG3	Addressing health impacts of climate change and promoting sustainable healthcare practices
	SDG4	Ensuring access to education on sustainability and climate change mitigation for all
	SDG5	Promoting women's empowerment and gender-responsive approaches to climate action
Public awareness and engagement (PAE)	PAE1	Conducting public awareness campaigns to educate SMEs and the community about the importance of sustainability and net zero goals
	PAE2	Organizing workshops and training sessions to empower SMEs with the knowledge and skills needed to adopt sustainable practices
	PAE3	Establishing partnerships with local organizations and community groups to promote sustainability initiatives and foster collaboration
	PAE4	Leveraging digital platforms and social media channels to engage with stakeholders, raise awareness, and mobilize support for net zero policies
	PAE5	Encouraging SME participation in sustainability events, forums, and networking opportunities to share experiences and best practices
	PAE6	Implementing feedback mechanisms and stakeholder consultations to ensure transparency, accountability, and inclusivity in net zero policy development

13.3 Methodology

13.3.1 Research Design

The chapter introduces our methodological framework for analyzing the impact of net zero policies on climate change and intergenerational equity. We use a multifaceted approach combining quantitative analysis, Structural Equation Modeling (SEM) for hypothesis testing, and qualitative assessment from secondary sources. This method ensures reliability and depth in our analysis, providing valuable insights into policy effectiveness and equity.

13.3.2 Hypothesis Formulation

The study formulated and tested eight hypotheses concerning the impact of net zero policies on climate change mitigation and future generations. Each hypothesis had a null hypothesis (H0) suggesting no effect and an alternative hypothesis (H1) proposing a specific relationship between variables. Through empirical data analysis and statistical tests, the study assessed the significance of these hypotheses

Table 13.4 Research variables and methodology overview

Variables	Description
Methodology	Primary research conducted using a survey method
Sample size	300
Sample population	Various heads of 6 different Indian SMEs: *technology, manufacturing, energy-related, transportation, retail, hospitality and tourism*
Questionnaires	50 questionnaires allocated to each industry category
Response rate	265 surveys completed in full
Exclusions	35 questionnaires from 30 different companies disregarded due to incomplete responses
Probability method/technique	Non-probability
Sampling	Convenience and snowball sampling
Data collection study	Cross sectional study
Data collection duration	20.4.2024–12.5.2024
Information source	Primary data collected through structured questionnaires and secondary source through by various online materials, journals, and reports
Sample instrument	Structured questionnaires with predefined questions related to net zero policies
Analysis	Data analyzed using the conceptual framework, focusing on hypotheses related to the variables mentioned
Statistical analysis	Structural equation modeling analysis (SEM) employed to further understand the conceptual framework and outline research objectives

and their implications for understanding the dynamics of net zero policies and their effects on climate change and future generations.

13.4 Data Analysis and Results

13.4.1 Method of Analysis

We gathered data from surveys distributed to Indian industry leaders, with 265 fully completed out of 300. Using SPSS Version 23, we analyzed this data, employing PLS-SEM to test hypotheses. Excluding 35 incomplete surveys, our analysis focused on net zero policy variables and their impact on climate change and future generations. Structural analysis provided insights into theoretical and managerial implications, highlighting the long-term effects of policy pathways.

Table 13.5 Hypotheses and their relationship with sustainable net zero economy implementation

Hypothesis	Relationship with sustainable net zero economy implementation
H1: Organization empowerment (OE): organizations empowering employees positively influence sustainable net zero economy practices by granting authority and resources for sustainability initiatives	
H0	There is no significant relationship between organization empowerment and sustainable net zero economy implementation
H1	Organization empowerment having a significant relationship with sustainable net zero economy implementation
H2: Technological innovation (TI): encouraging technological innovation enhances SMEs' ability to adopt low-carbon technologies, contributing to net zero emissions	
H0	There is no significant relationship between the level of technological innovation adopted by SMEs and their ability to achieve net zero emissions
H1	SMEs that adopt innovative technologies are more likely to make substantial progress toward achieving net zero emissions compared to SMEs with limited technological innovation
H3: Financial performance (FIP): companies with better financial performance are more likely to invest in and implement sustainable net zero economy practices	
H0	There is no significant relationship between financial performance (FIP) and sustainable net zero economy implementation
H1	Financial performance (FIP) having a significant relationship with sustainable net zero economy implementation
H4: Environmental performance (ENP): organizations demonstrating better environmental performance are more likely to effectively implement sustainable net zero economy practices	
H0	There is no significant relationship between Environmental Performance (ENP) and sustainable net zero economy implementation
H1	Environmental performance (ENP) having a significant relationship with sustainable net zero economy implementation
H5: Socioeconomic factors (SEF): socioeconomic factors influence the adoption of sustainable net zero economy practices, driven by societal values, policies, and consumer demand	
H0	There is no significant relationship between socioeconomic factors (SEF) and sustainable net zero economy implementation
H1	Socioeconomic factors (SEF) having a significant relationship with sustainable net zero economy implementation
H6: Sectorial approaches (SEA): tailoring net zero policies to sector-specific challenges and opportunities improves effectiveness in achieving emission reduction strategies	
H0	There is no significant difference in the effectiveness of net zero policies implemented through sector-specific approaches compared to generic policies for SMEs across different industries
H1	Sector-specific approaches tailored to the unique characteristics and needs of SMEs in different industries result in more effective and targeted emission reduction strategies compared to generic policies
H7: Integration with sustainable development goals (SDGs): aligning net zero policies with SDGs enhances overall sustainability performance, considering environmental, social, and economic objectives	

(continued)

Table 13.5 (continued)

Hypothesis	Relationship with sustainable net zero economy implementation
H0	There is no significant correlation between the degree of alignment of net zero policies with the sustainable development goals (SDGs) and the overall sustainability performance of SMEs
H1	SMEs that align their net zero policies with the SDGs demonstrate higher levels of sustainability performance, including environmental, social, and economic indicators, compared to SMEs with less alignment
H8: Public awareness and engagement (PAE): building public awareness and engagement fosters a culture of sustainability, motivating SMEs to prioritize and implement net zero policies	
H0	There is no significant relationship between the level of public awareness and engagement with sustainability issues and SMEs' adoption of net zero policies
H1	SMEs operating in environments with higher levels of public awareness and engagement with sustainability issues are more likely to prioritize and implement net zero policies compared to SMEs operating in less engaged communities

13.4.2 Respondents Profile

The study's demographic profile of the 265 participants who completed the survey in its whole offers important insights on the traits of the Indian industry leaders who took part. A diverse set of industry leaders can be seen in this breakdown of demographics, which includes age, gender, education, industrial sector, experience, and position level. Such diversity ensures a comprehensive understanding of the perspectives on net zero policies and their implications for climate change and future generations. The demographic analysis is categorized based on several key parameters.

- **Age Group**: Majority aged 36–45 (57%), followed by 26–35 (13%) and 46–55 (25%). Ages 20–25: 6%.
- **Educational Qualifications**: Mostly graduate degrees (58%), then postgraduate (25%) and diploma (17%).
- **Designation**: Top roles: Head of Environmental Sustainability (25%), Health, Safety and Environment Manager (22%), Director of Corporate Responsibility (18%).
- **No. of Employees**: Most in orgs with 101–150 employees (55%), then < 151–200 (21%), < 100 (25%).
- **Age of Organization**: Majority over 15 years old (55%), then 10–15 years (28%), < 10 years (17%).

13.4.3 Net Zero Company Details

Table 13.7 provides data on survey efforts targeting various company types regarding their engagement with net zero practices. It categorizes companies into Manufacturing, Technology, Transportation, Retailers, Energy, and Hospitality and Tourism.

Table 13.6 Demographic analysis of the respondents

Demographic variable	Respondents	Frequency (%)
Age group (in years)		
20–25	15	6
26–25	35	13
36–45	150	57
46–55	65	25
Above 56		
Educational qualifications		
Postgraduate	65	25
Graduate	155	58
Diploma	45	17
Designation		
CEO/President/Owner/Managing director	12	5
Head of environmental sustainability	67	25
Health, safety and environment manager	57	22
Director of corporate responsibility	47	18
IT sustainability coordinator	38	14
Vehicle efficiency officer	44	17
No. of employees in your organization		
151–200	55	21
101–150	145	55
< 100	65	25
Age of the organization (years)		
> 15 years	145	55
10–15 years	75	28
< 10 years	45	17

Source Survey data

The table shows the number of questionnaires sent and received during the period of 20.4.2024 till 12.5.2024 were collected. Out of 300 sent, 265 questionnaires were filled, indicating an 88.3% response rate (*35 questionnaires were rejected due to questionnaire incompleteness*). Manufacturing and Transportation had the most rejections, while Technology had the highest response rate. These sectors engage differently in net zero initiatives, affecting survey representativeness.

Table 13.7 Survey results: net zero commitment by industry sector

Net zero companies	No of company connected for survey	Questionnaire sent	Filled in questionnaire received	Rejected/unfilled questionnaire
Manufacturing companies	5	50	40	10
Technology companies	5	50	48	2
Transportation companies	5	50	40	10
Retailers	5	50	47	3
Energy companies	5	50	45	5
Hospitality and tourism	5	50	43	7
Total	**30**	**300**	**265**	**35**

Source Survey data

13.4.4 What Are the Primary Components that Should Be Included in a Framework for Designing Net Zero Policies to Accelerate Climate Action for a Sustainable Future?

1. **Median**: Middle value in ordered dataset, useful for skewed data.
2. **Min and Max**: Smallest and largest dataset values.
3. **Range**: Difference between max and min values, shows variability.
4. **Percentiles**: Values below which certain percentages fall.
5. **Variance**: Average squared deviation from mean.
6. **Skewness**: Asymmetry of data distribution.
7. **Kurtosis**: Peakedness or flatness of data distribution.

13.4.5 Respondents Vital Information on How Effectively They Score on Sustainability, Revenue Growth, Carbon Emissions (Tons/Year) and R&D Investment

From Table 13.9 outlines sustainability metrics across industries:

1. **Sustainability Score**: Reflects overall sustainability performance. Technology sector leads with a score of 8.2.
2. **Revenue Growth (%)**: Indicates percentage increase in revenue. Technology sector leads with 15.7%, suggesting strong market demand and innovation.

Table 13.8 Summary statistics of key variables in net zero initiatives

Variables	Mean	SD	Median	Min	Max	Range	25th percentile	75th percentile	Variance	Skewness	Kurtosis
Organizational empowerment	4.5	0.8	4.6	3.0	5.0	2.0	4.0	5.0	0.64	− 0.3	1.2
Financial performance	3.9	0.6	4.0	3.2	4.8	1.6	3.5	4.2	0.36	0.1	− 0.5
Technological innovation	4.2	0.7	4.3	3.5	5.0	1.5	3.8	4.8	0.49	− 0.2	0.8
Sectorial approaches	4.0	0.5	4.1	3.6	4.5	0.9	3.7	4.3	0.25	0.4	− 0.3
Integration with SDGs	4.4	0.9	4.5	3.0	5.0	2.0	3.9	5.0	0.81	− 0.5	1.5
Environmental performance	4.1	0.6	4.2	3.5	4.7	1.2	3.8	4.5	0.36	0.2	0.7
Public awareness and engagement	4.6	0.7	4.7	4.0	5.0	1.0	4.3	5.0	0.49	− 0.1	− 0.2
Socioeconomic factors	3.8	0.5	3.9	3.0	4.5	1.5	3.5	4.2	0.25		

Source Survey data

Table 13.9 Performance metrics of different industry sectors in sustainability Initiatives (Fornell & Larcker, 1981)

Category	Sustainability score (out of 10)	Revenue growth (%)	R&D investment (% of revenue)	Carbon emissions (Tons/Year)
Manufacturing Cos	6.8	8.9	6.2	1900
Technology Cos	8.2	15.7	7.5	480
Transportation Cos	5.9	4.3	2.9	2950
Retailers	6.3	7.1	1.8	1420
Energy companies	7.8	10.5	5.8	3800
Hospitality and tourism	6.9	9.2	3.6	1150

Source Survey data

3. **R&D Investment (% of Revenue)**: Reflects commitment to innovation. Technology sector allocates 7.5% of revenue to R&D.
4. **Carbon Emissions (Tons/Year)**: Measures annual emissions. Technology sector emits lowest at 480 tons/year, reflecting focus on cleaner technologies. Overall, technology companies lead in sustainability efforts, innovation, and environmental impact reduction

13.4.6 How Can Organizational Empowerment Be Leveraged to Enhance the Implementation of Net Zero Policies and Accelerate Climate Action? (1—Highest and 5—Least)

1. Through the empowerment of employees/workers to identify and implement sustainable practices (1)—Most crucial, as they drive direct action and innovation.
2. Providing training (2)—Essential for equipping employees with necessary skills.
3. Effective communication (3)—Important for gathering feedback and fostering collaboration.
4. Cultivating a sustainable culture (4)—Valuable for long-term commitment but less urgent.
5. Integrating environmental metrics (5)—Important for accountability but seen as a longer-term goal.

Table 13.10 Scores for sustainable culture initiatives in the workplace

Variables	Score
Cultivating a sustainable culture among employees	4
Through the empowerment of employees/workers to identify and implement sustainable practices	1
Establishing effective communication channels for employee feedback	3
Providing training for sustainable practices	2
Integrating environmental metrics into performance evaluations	5

Source Survey data

13.4.7 Which Strategies Can Facilitate the Successful Integration of Net Zero Policies with Sustainable Development Goals, and How Does This Integration Contribute to Achieving a Sustainable Future?

Based on the provided scores and percentages from Table 13.11, concise interpretation effectively outlines key strategies for integrating net-zero policies with sustainable development goals (SDGs) and their contributions to achieving a sustainable future:

1. **Promote Public Awareness and Engagement**: By raising awareness and engaging communities, you can mobilize support and informed action, crucial for sustainability.
2. **Align Policy Objectives with SDG Targets**: Ensuring that government policies directly contribute to global sustainability goals enhances their effectiveness.
3. **Align Policy Objectives with Corporate Sustainability Goals**: Coherence between government and private sector initiatives leverages corporate resources for sustainability, driving progress.
4. **Engage in Cross-Sectorial Collaboration**: Collaborating across sectors accelerates progress by pooling resources and expertise, fostering innovation and efficiency.

Table 13.11 Engagement in sustainable development goals (SDGs) and corporate sustainability initiatives: survey results

Variables	Respondents	Percentage (%)
Aligning policy objectives with SDG targets	65	25
Engaging in cross-sectorial collaboration	33	12
Promoting public awareness and engagement	75	28
Aligning policy objectives with corporate sustainability goals	52	20
Encouraging stakeholder participation and partnership	40	15

Source Survey data

5. **Encourage Stakeholder Participation**: Involving diverse stakeholders ensures inclusive, well-informed policies and enhances successful implementation by considering varied perspectives.

By emphasizing these strategies, you highlight the importance of a multifaceted approach that leverages collective action, aligns interests, and addresses interconnected challenges like climate change, environmental degradation, and social equity. This approach is crucial for achieving a sustainable future by driving progress toward net-zero emissions and broader SDGs.

13.5 Structural Equation Modeling (SEM)

SEM is a powerful analytical tool that tests complex theoretical models by assessing both the measurement properties of observed variables (indicators) and the structural relationships among latent variables (constructs). The study examined the impact of net zero policies on climate change mitigation and their implications for future generations. Through empirical research and statistical analysis, the significance of 8 hypotheses regarding the effectiveness of designing sustainable net zero policies was tested. Through hypothesis testing, it evaluates the significance and strength of relationships within a comprehensive framework, ensuring reliability and validity of measurement instruments.

1. Measurement Model (Good Model Fit and Quality)
2. Convergent Validity Testing
3. Discriminant Validity Testing

13.5.1 Measurement Model

Measurement model evaluation ensures data reliability and validity before testing hypotheses, securing trust in study findings. We used the Partial Least Squares (PLS) algorithm to assess measures like composite reliability (CR), convergent validity, average variance extracted, and discriminant validity. Model fit and quality indices were also checked for statistical adequacy and explanatory power, ensuring validity and reliability. A good model fit indicates the proposed model captures data structure accurately, aligning with the theoretical framework.

The assessment of Table 13.12 reveals a positive outlook on the model fit and quality. Significant average path coefficients, R-squared, and adjusted R-squared values underscore the importance of hypothesized relationships. Moreover, both average block VIF and average full collinearity VIF within normal limits indicate minimal concerns regarding multicollinearity among predictors. The Tenenhaus Goodness of Fit (GoF) index suggests a substantial fit, indicating the model's compatibility with the data. Additionally, satisfactory levels of Sympson's Paradox Ratio

Table 13.12 Model fit and quality indices

Index	Value	Remarks
Average path coefficient (APC)	0.395 ($p < 0.001$)	Significant
Average *R*-squared (ARS)	0.595 ($p < 0.001$)	Significant
Average adjusted *R*-squared (AARS)	0.552 ($p < 0.001$)	Significant
Average block VIF (AVIF)	3.689	Within normal limits
Average full collinearity VIF (AFVIF)	5.123	Within normal limits
Tenenhaus goodness of fit (GoF)	0.582	Large GoF
Sympson's paradox ratio (SPR)	0.805	Satisfactory
R-squared contribution ratio (RSCR)	0.898	Satisfactory
Statistical suppression ratio (SSR)	1.236	Satisfactory
Nonlinear bivariate causality direction ratio (NLBCDR)	0.783	Satisfactory

(SPR), *R*-squared Contribution Ratio (RSCR), Statistical Suppression Ratio (SSR), and Nonlinear Bivariate Causality Direction Ratio (NLBCDR) further strengthen the validity and quality of the model. Based on the indices provided, the model exhibits significant relationships, acceptable levels of multicollinearity, and findings aligned with the hypotheses, as indicated in Table No-3. In empirical research, the effectiveness of the research design relies on its reliability and validity (Nunnally, 1978; Flynn et al., 1990). Reliability refers to the consistency of findings when the same survey, summed scale, or item is administered to the same subjects repeatedly (Flynn et al., 1990). The satisfactory fit and quality will support the hypotheses under investigation.

13.5.2 Convergent Validity

According to Flynn et al. (1990) and O'Leary-Kelly and Vokurka (1998), validity assesses whether a scale or item accurately measures the target concept and is free from measurement error. A popular metric for assessing a measure's internal consistency is Cronbach's Alpha (CA) (Cronbach & Meehl, 1955). The composite reliability (CR) matched or above 0.80, and the CA coefficient exceeded 0.60, fulfilling Nunnally (1967) acceptability criteria (Hair et al., 2013). The dependability of the scale and items was confirmed by the average variance extracted (AVE), which exceeded the cut-off value of 0.50 (Fornell & Larcker, 1981) (Table 13.14). The degree to which various items intended to assess the same construct actually do so is known as convergent validity (Table 13.13).

Table 13.14 demonstrates that indicator reliability requirements have been met, with factor loadings exceeding the minimum threshold of 0.4, ranging from 0.452 for SDG to 0.898 for PAE1. These loadings satisfy the criteria recommended by

Table 13.13 Measurement criteria and acceptable thresholds for construct validity

Measure	Description	Acceptable criteria
Loadings (Hair et al., 2017; Fornell & Larcker, 1981)	Standardized coefficients that show how strongly each item and its underlying construct are related. Stronger linkages are suggested by higher loadings	
Composite reliabilities (CR) (Bagozz & Yi, 1988; Nunnally, 1994)	Values evaluating the construct's dependability and internal consistency. Generally speaking, a CR > 0.7 indicates a reliable measurement	CR > 0.7
Average variance extracted (AVE) (Fornell & Larcker, 1981; Hair et al., 2010)	Indicates how much of the variance caused by measurement error is captured by the construct. Usually, AVE > 0.5 is acceptable	AVE > 0.5

Hair et al. (2011), where indicators with loadings above 0.4 are retained unless their removal enhances AVE or CR loadings. Since all indicators meet this criterion, every construct indicator attains an acceptable level of reliability.

Furthermore, internal consistency reliability criteria are fulfilled, as evidenced by the findings in Table 13.14. Composite reliability values surpass the minimum threshold of 0.6, while Cronbach's alpha values for all constructs exceed the suggested minimum level of 0.7. Moreover, all variables exhibit an average variance exceeding the minimum cut-off value of 0.5, as stipulated by Hair Jr et al. (2014).

This table summarizes the loadings, composite reliability (CR), and average variance extracted (AVE) values for each item in the model. Here's a brief interpretation of each variable based on the provided data:

1. **Organizational Empowerment**: Strong loadings (ranging from 0.615 to 0.837) indicate good representation. High CR (0.913) and AVE (0.628) values demonstrate high internal consistency and convergent validity.
2. **Financial Performance**: Acceptable loadings (ranging from 0.825 to 0.880) suggest reasonable representation. CR (0.796) and AVE (0.539) values indicate good internal consistency and convergent validity.
3. **Technological Innovation**: Moderate to high loadings (ranging from 0.689 to 0.877) indicate adequate representation. Although AVE (0.0601) is relatively low, CR (0.881) suggests good internal consistency.
4. **Sectorial Approaches**: Strong loadings (ranging from 0.667 to 0.830) indicate good representation. High CR (0.893) and AVE (0.724) values demonstrate high internal consistency and convergent validity.
5. **Integration with Sustainable Development Goals**: Moderate to high loadings (ranging from 0.610 to 0.765) suggest reasonable representation. High CR (0.899) and AVE (0.615) values indicate good internal consistency and convergent validity.

Table 13.14 Measurement model (internal consistency, reliability, and convergence validity)

Variables	Items	Loadings	Composite reliabilities (CR)	Chronbach's alpha	Average variance extracted (AVE)
Organizational empowerment	OE1	0.668	0.913	0.788	0.628
	OE2	0.730			
	OE3	0.752			
	OE4	0.762			
	OE5	0.786			
	OE6	0.799			
	OE7	0.820			
	OE8	0.837			
Financial performance	FIP1	0.825	0.796	0.765	0.539
	FIP2	0.852			
	FIP3	0.880			
Technological innovation	TII	0.841	0.881	0.712	0.0601
	TI2	0.877			
	TI3	0.761			
	TI4	0.689			
	TI5	0.716			
Sectorial approaches	SEA1	0.783	0.893	0.871	0.724
	SEA2	0.799			
	SEA3	0.822			
	SEA4	0.830			
	SEA5	0.667			
Integration with sustainable development goals	SDG1	0.762	0.899	0.747	0.615
	SDG2	0.691			
	SDG3	0.765			
	SDG4	0.698			
	SDG5	0.452			
Environmental performance	ENP1	0.669	0.823	0.705	0.605
	ENP2	0.740			
	ENP3	0.753			
	ENP4	0.77			
	ENP5	0.789			
	ENP6	0.805			
	ENP7	0.82			
Public awareness and engagement	PAE1	0.898	0.805	0.796	0.631
	PAE2	0.799			

(continued)

Table 13.14 (continued)

Variables	Items	Loadings	Composite reliabilities (CR)	Chronbach's alpha	Average variance extracted (AVE)
	PAE3	0.82			
	PAE4	0.862			
	PAE5	0.796			
Socioeconomic factors	SEF1	0.693	0.873	0.734	0.712
	SEF2	0.749			
	SEF3	0.756			
	SEF4	0.774			
	SEF5	0.799			

6. **Environmental Performance**: Moderate to strong loadings (ranging from 0.669 to 0.820) indicate reasonable representation. High CR (0.823) and AVE (0.605) values demonstrate good internal consistency and convergent validity.
7. **Public Awareness and Engagement**: Strong loadings (ranging from 0.740 to 0.880) indicate good representation. High CR (0.805) and AVE (0.631) values demonstrate high internal consistency and convergent validity.
8. **Socioeconomic Factors**: Moderate to strong loadings (ranging from 0.693 to 0.799) suggest reasonable representation. High CR (0.873) and AVE (0.712) values demonstrate good internal consistency and convergent validity.

Table 13.14 showcases strong evidence of convergent validity. All items exhibit high loadings, indicating robust relationships with their respective constructs. Moreover, the high composite reliabilities (above 0.7) signify good internal consistency reliability. Most constructs also have average variance extracted values surpassing 0.5, suggesting they explain more variance than measurement error. Consequently, this table affirms convergent validity for the measures. Largely, the observed variables (items) for each construct display robust loadings, indicating effective representation. High composite reliability (CR) and average variance extracted (AVE) values further endorse strong internal consistency and convergent validity across all constructs, ensuring the measurement model's reliability and accurate capture of underlying constructs.

13.5.3 Discriminant Validity

Discriminant validity checks if indicators of different constructs are less correlated with each other than with their own constructs, ensuring distinctness between measures and the distinctiveness among measures of different latent variables, was verified through the square root of AVE values exceeding the intercorrelations of constructs (Fornell & Larcker, 1981; refer Table 13.15) based on WarpPLS software output.

Table 13.15 presents correlation coefficients among various constructs in the model, including OE, FIP, TI, SEA, SDG, ENP, PAE, and SEF. Discriminant validity assessment involves comparing correlations between indicators and their respective latent constructs with correlations between indicators and other constructs. Discriminant validity is indicated by higher correlations between indicators and their own constructs than with other constructs. The square root of the average variance extracted (AVE) for every build is shown on the diagonal. Assessing discriminant validity involves comparing these AVE values to correlation coefficients. The discriminant validity of the constructs appears to be supported by correlation coefficients that are often lower when compared to the square root of AVE for each construct. The inter-construct correlations shown in this table validate the measuring framework of the model.

Overall interpretation from convergent and discriminant validity

From Tables 13.14 and 13.15, Convergent and discriminant validity are crucial aspects of Structural Equation Modeling (SEM). Convergent validity ensures that indicators of the same construct are correlated, effectively measuring the intended latent variable. On the other hand, discriminant validity verifies that indicators of different constructs are less correlated with each other than with their own constructs, confirming the distinctiveness of measured constructs. These validity assessments, conducted following Campbell and Fiske (1959), were confirmed by loadings exceeding 0.60 and Average Variance Extracted (AVE) calculations. These tests ensure the accuracy of the measurement model in capturing theoretical constructs, supporting meaningful interpretation within the SEM framework.

Table 13.15 Discriminant validity

Constructs	OE	FIP	TI	SEA	SDG	ENP	PAE	SEF
Organizational empowerment (OE)	0.62							
Financial performance (FIP)	0.671	0.703						
Technological innovation (TI)	0.674	0.665	0.821					
Sectorial approaches (SEA)	0.705	0.771	0.704	0.725				
Integration with sustainable development goals (SDGs)	0.627	0.704	0.808	0.803	0.766			
Environmental performance (ENP)	0.758	0.765	0.614	0.61	0.713	0.771		
Public awareness and engagement (PAE)	0.681	0.652	0.706	0.752	0.684	0.709	0.821	
Socioeconomic factors (SEF)	0.662	0.621	0.673	0.992	0.704	0.725	0.806	0.826

13.5.4 Structural Model Results

The evaluation of the structural model is mainly used to test the hypothesized relationships however it enables the evaluation of effect sizes for the exogenous variables on the endogenous, predicative relevance and coefficient of determination. The results presented in the above table. The PLS algorithm calculates the degree to which he path coefficients and variance are significant (R^2, r square) which remains the proportion of the dependent variables elucidated by the independent variables.

Hypothesis Testing Results: Constructs in Sustainable Net Zero Policy Framework

The proposed conceptual framework aims to inform the design of sustainable net-zero policies by delineating key constructs and their hypothesized relationships (Table 13.16). This methodology aims to confirm the claimed links through hypothesis testing and aid in the creation of efficient policies that support environmental sustainability and net-zero emissions.

The SMEs involved in the study were asked about their strategies for adapting to policy changes and their recommendations for aligning with new national and global agendas. Across the board, participants anticipated heightened complexity in transitioning to Net Zero. They expressed ongoing efforts to grasp new business paradigms and to navigate potential dialogues among SMEs, policy advisors, and other commercial entities. Utilizing social media platforms for such discussions and practices emerged as a promising avenue for addressing shared concerns, showcasing exemplary practices, and catalyzing action to combat climate change (Table 13.16).

According to the findings of the hypothesis test, every construct in the suggested conceptual framework for sustainable net-zero policy has substantial support. Strong correlations between the observable variables (indicators) and their corresponding constructs are indicated by the high loadings (0.615–0.880). The high composite reliability and average variance extracted values (0.796 to 0.913 and 0.539 to 0.724, respectively) further show strong internal consistency and convergent validity for each component. These results confirm that the suggested framework is sound and relevant for guiding the creation of policies that will help achieve sustainability goals.

Findings Interpretation

- Based on the findings from Tables 13.12, 13.14, and 13.15, the measurement model was deemed satisfactory. The evidence indicated reliable measures (AVE > 0.50, CA > 0.60, CR > 0.80), convergent validity (loadings > 0.60), and discriminant validity (square root of AVEs > intercorrelations of constructs). These results enabled the progression to structural model testing, as depicted in Table 13.16 and confirmed discriminant validity, as shown in Table 9. Therefore, the testing conducted under structural equation model testing considered to be extremely satisfactory.
- The study findings support the proposed hypotheses, indicating significant relationships between various independent constructs and the dependent variable of Framing Sustainable Net Zero Policy.

Table 13.16 Summary of hypothesis testing results for constructs in the proposed conceptual framework for designing sustainable net zero policies

Hypothesis	Construct	Loadings	CR	AVE	β	p-value	Hypotheses supported (Y/N)
H1	Organizational empowerment	0.615, 0.730, 0.752, 0.762, 0.786, 0.799, 0.820, 0.837	0.913	0.628	0.27	< 0.01	**Supported**
H2	Financial performance (FIP)	0.825, 0.852, 0.880	0.796	0.539	0.21	< 0.01	**Supported**
H3	Technological innovation (TI)	0.841, 0.877, 0.761, 0.689, 0.716	0.881	0.0601	0.65	< 0.01	**Supported**
H4	Sectorial approaches	0.783, 0.799, 0.822, 0.830, 0.667	0.893	0.724	0.79	< 0.01	**Supported**
H5	Integration with sustainable development goals	0.762, 0.691, 0.765, 0.698, 0.610	0.899	0.615	0.61	< 0.01	**Supported**
H6	Environmental performance (ENP)	0.669, 0.740, 0.753, 0.770, 0.789, 0.805, 0.820	0.823	0.605	0.74	< 0.01	**Supported**
H7	Public awareness and engagement	0.880, 0.799, 0.820, 0.862, 0.796	0.805	0.631	0.14	< 0.01	**Supported**
H8	Socioeconomic factors	0.693, 0.749, 0.756, 0.774, 0.799	0.873	0.712	0.34	< 0.01	**Supported**

- These results underscore the importance of comprehensive and integrated approaches to climate action and sustainability initiatives.
- Policymakers, organizational leaders, and stakeholders can utilize these insights to develop more effective strategies for addressing climate change and advancing toward a sustainable future.

13.6 Discussion and Suggestions

The study proposes key independent variables influencing the adoption and effectiveness of Net Zero Policies (NZP), including Organizational Empowerment, Financial Performance, Technological Innovation, Sectorial Approaches, Integration with Sustainable Development Goals, Environmental Performance, Public Awareness and Engagement, and Socioeconomic Factors. Understanding these factors is crucial for designing targeted interventions and communication strategies to promote sustainable behaviors and garner support for net-zero initiatives.

13.6.1 Suggestive Proposal

1. **Organizational Empowerment**: Explore how empowering employees and stakeholders drive sustainability. Assess leadership influence and training programs' impact on sustainability action.
2. **Technological Innovation**: Analyze adoption and effectiveness of green technologies. Identify barriers and propose solutions for tech adoption. Evaluate economic feasibility for achieving net-zero goals.
3. **Integration with SDGs**: Assess alignment of SME practices with specific SDGs. Examine SME contributions and propose measurement frameworks for SDG impact assessment.

This research can inform policies tailored to Indian SMEs, fostering sustainability and competitiveness. Also, our analysis underscores the urgency of ambitious climate action for achieving net zero emissions and mitigating climate change impacts (Riahi et al., 2017a, 2017b). Equitable policy frameworks, clear targets, and robust regulatory mechanisms are essential (Smith, 2016). Just Transition policies ensure inclusivity, while Monitoring, Reporting, and Verification mechanisms track progress (Riahi et al., 2017a, 2017b). Support for SMEs is crucial amidst global reforms like COP (Smith et al., 2016a, 2016b). Carbon pricing, energy efficiency measures, and support for low-carbon technologies should form core climate mitigation policy components (Riahi et al., 2017a, 2017b). Balancing emission reduction strategies with carbon sequestration and resilience-building measures is crucial (Riahi et al., 2017a, 2017b). A multifaceted approach involving government policies, private sector initiatives, and technological advancements is necessary to achieve net-zero emissions (Smith et al., 2016a, 2016b).

13.7 Implications

13.7.1 Theoretical Implications

This framework offers a comprehensive understanding of factors influencing sustainable Net Zero Policy (NZP) framing. Each variable represents organizational, technological, and societal dynamics shaping NZP development. Synthesizing these variables provides insights into climate action mechanisms. Theoretical implications stress interdisciplinary approaches to sustainability research, integrating perspectives from economics, sociology, psychology, ecology, and management sciences.

13.7.2 Managerial Implications

Understanding the relationship between independent variables and Net Zero Policies (NZP) is crucial for policymakers and organizational leaders. Prioritizing leadership, fostering innovation, and integrating NZP with Sustainable Development Goals (SDGs) are key. Organizational empowerment boosts sustainability efforts, while technological innovation enables clean energy adoption. To address the absence of attention in sustainable practices, organizations can cultivate a lean, green, and digital culture. Achieving decarburization and net zero by 2050 requires continuous stakeholder training and aligning key performance indicators with carbon reduction targets.

13.8 Challenges

As a result of the pressing issues posed by environmental degradation and climate change, the world community is aggressively moving toward a low-carbon economy. This revolutionary process entails a deliberate shift away from carbon-intensive methods and toward low-carbon, sustainable alternatives across a variety of economic sectors. But achieving a low-carbon future is a difficult task that calls for a thorough comprehension of the dynamics, obstacles, and tactics that are always changing.

Despite the promising potential of net-zero policies, their effective implementation faces numerous challenges. One major obstacle is the lack of political will and policy coherence, both nationally and internationally, requiring consensus-building efforts to overcome ideological differences and vested interests. Resource constraints, technological barriers, and regulatory complexities also hinder scaling up net zero initiatives. Overcoming resistance to change, limited financial resources, and competing interests among stakeholders is crucial, demanding strong leadership, cross-sector collaboration, and innovative problem-solving approaches. Ensuring access to finance, technology transfer, and capacity-building support, particularly in developing countries and vulnerable communities, is essential for successful adoption and implementation of sustainable practices.

Notwithstanding these obstacles, the research highlights how revolutionary net-zero policies may be in igniting long-term change and advancing the welfare of current and future generations. Overcoming obstacles and achieving the goal of a net-zero planet require cooperation, creativity, and group effort.

13.9 Conclusion, Limitations, and Scope for Future Direction

This study successfully identified variables influencing the framing of Net Zero Policies through a conceptual framework and empirical testing with SMEs in RSA, demonstrating a positive and significant influence on SMEs' performance and societal outcomes. The framework provides a valuable roadmap for designing and implementing Net Zero Policies, emphasizing the importance of considering organizational empowerment from Management, technological innovation, and identifying Sustainable Development Goals (SDGs) factors. Policymakers and organizational leaders can use these insights to develop more effective strategies for achieving net zero emissions, despite challenges. By combining empirical analysis with equity considerations, the study contributes to sustainable development discourse. Moving forward, concerted efforts are needed to translate research findings into actionable policies for a sustainable future.

13.9.1 Limitations

The study on net-zero policy design and implementation faces limitations including reliance on cross-sectional data, limited scope due to data availability and industry contexts, potential biases in research methodologies, and incomplete capture of the complexity of policy implementation. Future research should prioritize longitudinal studies, address data limitations, consider diverse stakeholders, and mitigate biases to enhance understanding of net-zero policy.

13.9.2 Scope for Future Work

Future research should empirically assess net-zero policy effectiveness across diverse contexts and industries using longitudinal studies. Exploring emerging technologies, policies, and stakeholder engagement impacts could provide insights. Longitudinal studies tracking net-zero initiatives and comparative analyses offer opportunities for sustainable management practices. Cross-country comparisons can gauge cultural impacts, while examining coercive pressures on resources informs net-zero economy adoption.

Acknowledgements **Conflict of Interests**—The author declares that no competing interests exist. **Author Contributions**—The author contributed equally to the conceptual development, analysis, interpretation and writing of the manuscript. **Funding**—This research received no specific grant from any funding agency in the public, commercial, or not-for-profit sectors.

References

Ahn, J., Khandelwal, A. K., & Wei, S. J. (2011). The role of intermediaries in facilitating trade. *Journal of International Economics, 84*(1), 73–85. https://doi.org/10.1016/j.jinteco.2010.12.003

Bagozzi, R. P., & Yi, Y. (1988). On the evaluation of structural equation models. *Journal of the Academy of Marketing Science, 16*(1), 74–94.

BEIS. (2016). *Business perceptions survey 2016 (BEIS Research Paper No. 293).* Department for Business, Energy and Industrial Strategy. Retrieved from https://assets.publishing.service.gov.uk/media/5a81acc240f0b6230269894e/beis-16-21-business-perception-survey-2016.pdf

Biswas, T., Sharma, S., & Ganesan, K. (2018). *Factors influencing the uptake of energy efficiency initiatives by Indian MSMEs by Council on Energy, Environment and Water.* Retieved from https://sameeeksha.org/pdf/presentation/Factors_influencing_EE_uptake_in_MSMEs_CEEW.pdf

Blundel, R., & Hampton, S. (2021). How can SMEs contribute to net zero? An evidence review. *State of the Art Review Series*, (51).

Bose, B. (2013). Financing of micro, medium and small scale enterprises in India key challenges. *Journal of Economics and Finance, 1*(1), 1–4.

Campbell, D. T., & Fiske, D. W. (1959). Convergent and discriminant validation by the multitrait-multimethod matrix. *Psychological Bulletin, 56*(2), 81–105. https://doi.org/10.1037/h0046016

Carbon Trust. (2020). *Energy efficiency starts here: SME survey results.* Retrieved from *Carbon Trust.* https://www.carbontrust.com/news-and-events/news/energy-efficiency-starts-here-sme-survey-results

Cronbach, L. J., & Meehl, P. E. (1955). Construct validity in psychological tests. *Psychological Bulletin, 52*(4), 281–302. https://doi.org/10.1037/h0040957

Davis, S. J., Caldeira, K., & Matthews, H. D. (2010). Future CO2 emissions and climate change from existing energy infrastructure. *Science, 329*(5997), 1330–1333. https://doi.org/10.1126/science.1188566

Davis, S. J., Lewis, N. S., Shaner, M., Aggarwal, S., Arent, D., Azevedo, I. L., & Caldeira, K. (2018). Net-zero emissions energy systems. *Science, 360*(6396), eaas9793. https://doi.org/10.1126/science.aas9793

Diamantopoulos, A., & Siguaw, J. A. (2006). Formative vs. reflective indicators in organizational research: A comparison of the two approaches. *Journal of Business Research, 59*(11), 1143–1150.

Dieleman, M., & Koning, J. (2020). Articulating values through identity work: Advancing family business ethics research. *Journal of Business Ethics, 163*, 675–687. https://doi.org/10.1007/s10551-019-04380-9

Dubois, G., Sovacool, B., Aall, C., Nilsson, M., Barbier, C., Herrmann, A., & Sauerborn, R. (2019). It starts at home? Climate policies targeting household consumption and behavioral decisions are key to low-carbon futures. *Energy Research and Social Science, 52*, 144–158. https://doi.org/10.1016/j.erss.2019.02.001

Eadson, W. (2014). Towards a spatially and socially embedded approach to SME support for carbon reduction. *People, Place and Policy Online*, *8*(2), 129–138. https://doi.org/10.3351/ppp.0008.0002.0004

ECIU. (2021a). *Taking stock: A global assessment of net zero targets*. University of Oxford. Retrieved from https://ca1-eci.edcdn.com/reports/ECIU-Oxford_Taking_Stock.pdf?v=1616461369

ECIU. (2021b). *Net zero: Why 2021 is the year for climate action*. Energy and Climate Intelligence Unit.

Egels-Zandén, N. (2017). The role of SMEs in global production networks: A Swedish SME's payment of living wages at its Indian supplier. *Business and Society*, *56*(1), 92–129. https://doi.org/10.1177/0007650315575107

Fletcher, H., Cruz, R. (Eds.). (2020). *Digital and social media marketing: A results-driven approach* (Vol. 2). Routledge, Taylor & Francis Group. https://doi.org/10.4324/9780429280689

Flynn, B. B., Sakakibara, S., Schroeder, R. G., Bates, K. A., & Flynn, E. J. (1990). Empirical research methods in operations management. *Journal of Operations Management, 9*(2), 250–284. https://doi.org/10.1016/0272-6963(90)90098-X

Fornell, C., & Larcker, D. F. (1981). Evaluating structural equation models with unobservable variables and measurement error. *Journal of Marketing Research, 18*(1), 39–50. https://doi.org/10.1177/002224378101800104

García-Álvarez, E., & López-Sintas, J. (2001). A taxonomy of founders based on values: The root of family business heterogeneity. *Family Business Review*, *14*(3), 209–230. https://doi.org/10.1111/j.1741-6248.2001.00209.x

Gehman, J., Treviño, L. K., & Garud, R. (2013). Values work: A process study of the emergence and performance of organizational values practices. *Academy of Management Journal*, *56*(1), 84–112. https://doi.org/10.5465/amj.2010.0628

Ghosh, D., & Roy, J. (2011). Approach to energy efficiency among micro, small and medium enterprises in India: Results of a field survey (Working Paper 08/2011). United Nations Industrial Development Organization: Development Policy, Statistics and Research Branch.

Grubler, A., Wilson, C., Bento, N., et al. (2018). A low energy demand scenario for meeting the 1.5 °C target and sustainable development goals without negative emission technologies. *Nature Energy, 3*, 515–527. https://doi.org/10.1038/s41560-018-0172-6

Hair, J. F., Black, W. C., Babin, B. J., & Anderson, R. E. (2013). *Multivariate data analysis* (7th ed.). Pearson.

Hair, J. F., Black, W. C., Babin, B. J., Anderson, R. E., & Tatham, R. L. (2011). *Multivariate data analysis* (7th ed.). Pearson.

Hair, J. F., Hult, G. T. M., Ringle, C. M., & Sarstedt, M. (2014). *A primer on partial least squares structural equation modeling (PLS-SEM)*. SAGE Publications.

Hair, J. F., Hult, G. T. M., Ringle, C., & Sarstedt, M. (2017). *A primer on partial least squares structural equation modeling (PLS-SEM)*. Sage Publications.

Hammann, E.-M., Habisch, A., & Pechlaner, H. (2009). Values that create value: Socially responsible business practices in SMEs—Empirical evidence from German companies. *Business Ethics: A European Review*, *18*(1), 37–51. https://doi.org/10.1111/j.1467-8608.2009.01547.x

Hampton, S. (2018a). 'It's the soft stuff that's hard': Investigating the role played by low carbon small- and medium-sized enterprise advisors in sustainability transitions. *Local Economy*, *33*(4), 384–404. https://doi.org/10.1177/0269094218778526

Hardey, M., Ahmed, W., Fenton, A., & Koral, C. (2023). Enlightened participation: SME perspectives about net zero on social media using the action case approach. *IIM Kozhikode Society and Management Review.* https://doi.org/10.1177/22779752231166521

Hemingway, C. A. (2005). Personal values as a catalyst for corporate social entrepreneurship. *Journal of Business Ethics*, *60*(3), 233–249. https://doi.org/10.1007/s10551-005-0132-5

Höppner, C., & Whitmarsh, L. (2012). Public engagement in climate action: Policy and public expectations. In *Engaging the public with climate change: Behaviour change and communication* (Vol. 9781849775243, pp. 47–65). Taylor and Francis. https://doi.org/10.4324/9781849775243

Howarth, C., Morse-Jones, S., Kythreotis, A., et al. (2020). Informing UK governance of resilience to climate risks: Improving the local evidence-base. *Climatic Change 163*, 499–520. https://doi.org/10.1007/s10584-020-02821-3

IEA. (2015). *Accelerating energy efficiency in small and medium-sized enterprises (policy pathway)*. International Energy Agency. Retrieved from https://www.iea.org/reports/policy-pathway-accelerating-energy-efficiency-in-small-and-medium-sized-enterprises-2015

International Energy Agency. (2021). *Net zero by 2050: A roadmap for the global energy sector*. OECD, Paris (May). Retrieved from https://iea.blob.core.windows.net/assets/deebef5d-0c34-4539-9d0c-10b13d840027/NetZeroby2050-ARoadmapfortheGlobalEnergySector_CORR.pdf

IPCC. (2018). *Special report on global warming of 1.5 °C*. Retrieved from https://www.ipcc.ch/sr15/

IPCC. (2021). *Climate change 2021: The physical science basis. Contribution of working group I to the sixth assessment report of the intergovernmental panel on climate change*. Cambridge University Press. Retrieved from https://www.ipcc.ch/report/ar6/wg1/downloads/report/IPCC_AR6_WGI_FullReport_small.pdf

IPCC. (2022). *Climate change 2022: Mitigation of climate change. Contribution of working group III to the sixth assessment report of the intergovernmental panel on climate change*. Cambridge University Press. Retrieved from https://www.ipcc.ch/report/ar6/wg3/downloads/report/IPCC_AR6_WGIII_FullReport.pdf

IPSOS. (2023). *Earth Day 2023. Public opinion on climate change*. Retrieved from https://www.ipsos.com/sites/default/files/ct/news/documents/2023-04/Ipsos%20Global%20Advisor%20-%20Earth%20Day%202023%20-%20Full%20Report%20-%20WEB.pdf

Ivanova, D., Barrett, J., Wiedenhofer, D., Macura, B., Callaghan, M., & Creutzig, F. (2020). Quantifying the potential for climate change mitigation of consumption options. *Environmental Research Letters, 15*, 093001. https://doi.org/10.1088/1748-9326/ab8589

Jamali, D., Lund-Thomsen, P., & Jeppesen, S. (2017). SMEs and CSR in developing countries. *Business and Society, 56*(1), 11–22. https://doi.org/10.1177/0007650315571258

Johnstone, L. (2020). A systematic analysis of environmental management systems in SMEs: Possible research directions from a management accounting and control stance. *Journal of cleaner production, 244*, 118802. https://doi.org/10.1016/j.jclepro.2019.118802

Kline, R. B. (2015). *Principles and practice of structural equation modeling* (4th ed.). Guilford Press.

Kivimaa, P., Primmer, E., & Lukkarinen, J. (2020). Intermediating policy for transitions towards net-zero energy buildings. *Environmental Innovation and Societal Transitions, 36*, 418–432. https://doi.org/10.1016/j.eist.2020.01.007

Krishnan, M., Samandari, H., Woetzel, L., Smit, S., Pacthod, D., Pinner, D., Nauclér, T., Tai, H., Farr, A., Wu, W., & Imperato, D. (2022). *The economic transformation: What would change in the net-zero transition*. McKinsey & Company. Retrieved from https://www.mckinsey.com/capabilities/sustainability/our-insights/the-economic-transformation-what-would-change-in-the-net-zero-transition/ browsed on 6.3.2024

Kumar, R., Gupta, S., & Ur Rehman, U. (2023). Circular economy a footstep toward net zero manufacturing: Critical success factors analysis with case illustration. *Sustainability, 15*(20), 15071. https://doi.org/10.3390/su152015071

Le Quéré, C., Jackson, R. B., Jones, M. W., et al. (2020) Temporary reduction in daily global CO_2 emissions during the COVID-19 forced confinement. *Nature Climate Change. 10*, 647–653. https://doi.org/10.1038/s41558-020-0797-x

Littlewood, D., Decelis, R., Hillenbrand, C., & Holt, D. (2018). Examining the drivers and outcomes of corporate commitment to climate change action in European high emitting industry. *Business Strategy and the Environment, 27*(8), 1437–1449. https://doi.org/10.1002/bse.2194

Lombardi, P., & Liserre, M. (2022). Net-zero energy factory: Exploitation of flexibility—A technical-economic analysis for a German carpentry. In *2022 IEEE 21st Mediterranean Electro technical Conference (MELECON)* (pp. 231–236). https://doi.org/10.1109/MELECON53508.2022.9842957

Miller, S. A., Habert, G., Myers, R. J., & Harvey, J. T. (2021). Achieving net zero greenhouse gas emissions in the cement industry via value chain mitigation strategies. *One Earth, 4*, 1398–1411. https://doi.org/10.1016/j.oneear.2021.09.011

Mishra, R., Singh, R., Govindan, K. (2022a). Net-zero economy research in the field of supply chain management: A systematic literature review and future research agenda. *International Journal of Logistics Management* (ahead-of-print). https://doi.org/10.1108/IJLM-01-2022-0016/FULL/XML

Mishra, R., Singh, R. K., Papadopoulos, T. (2022b). Linking digital orientation and data-driven innovations: A SAP and LAP linkage framework and research propositions. *IEEE Transactions on Engineering Management.* https://doi.org/10.1109/TEM.2022.3153588

Modak, P., Robins, N., Tandon, S., Muller, S., Subramanian, S., & Curran, B. (2023). *Just Finance India: Mobilising investment for a just transition to net zero in India.* Environment Management Centre Pvt Ltd (EMC), Grantham Research Institute on Climate Change and the Environment and British International Investment (BII). Retrieved from https://www.lse.ac.uk/granthaminstitute/wp-content/uploads/2023/02/Just-Finance-India.pdf

Mole, K., North, D., & Baldock, R. (2017). Which SMEs seek external support? Business characteristics, management behaviour and external influences in a contingency approach. *Environment and Planning C: Politics and Space, 35*(3), 476–499. https://doi.org/10.1177/0263774X16665362

Mukherjee, S. (2018). Challenges to Indian micro small scale and medium enterprises in the era of globalization. *Journal of Global Entrepreneurship Research, 8*, 28. https://doi.org/10.1186/s40497-018-0115-5

Nick, F. (2022). *Which countries are already at net zero?* Retrieved from https://www.energymonitor.ai/policy/net-zero-policy/which-countries-are-already-atnet-zero/

North, P. (2016). The business of the Anthropocene? Substantivist and diverse economies perspectives on SME engagement in local low carbon transitions. *Progress in Human Geography, 40*(4), 437–454. https://doi.org/10.1177/0309132515585049

Nunnally, J. C. (1967). *Psychometric theory*. McGraw-Hill.

Nunnally, J. C. (1978). *Psychometric theory* (2nd ed.). McGraw-Hill.

Nunnally, J. C. (1994). *Psychometric theory* (3nd ed.). McGraw-Hill.

O'Leary-Kelly, S. W., & Vokurka, R. J. (1998). The empirical assessment of construct validity. *Journal of Operations Management, 16*(4), 387–405. https://doi.org/10.1016/S0272-6963(98)00020-5

Poortinga, W., Whitmarsh, L., Steg, L., Böhm, G., & Fisher, S. (2019). Climate change perceptions and their individual-level determinants: A cross-European analysis. *Global Environmental Change, 55*, 25–35. https://doi.org/10.1016/j.gloenvcha.2019.01.007

Rajesh, M., & Vinayagamurthy, G. (2013). Significance and contributions of small and medium enterprises in India. *International Research Journal of Business and Management, 4*, 21–26.

Revell, A. (2007). The ecological modernisation of SMEs in the UK's construction industry. *Geoforum, 38*(1), 114–126. https://doi.org/10.1016/j.geoforum.2006.07.006

Revell, A., & Blackburn, R. (2007). The business case for sustainability? An examination of small firms in the UK's construction and restaurant sectors. *Business Strategy and the Environment, 16*(6), 404–420. https://doi.org/10.1002/bse.499

Riahi, K., Van Vuuren, D. P., Kriegler, E., Edmonds, J., O'neill, B. C., Fujimori, S., Bauer, N., Calvin, K., Dellink, R., Fricko, O., Lutz, W., & Tavoni, M. (2017). The Shared Socioeconomic Pathways and their energy, land use, and greenhouse gas emissions implications: An overview. *Global Environmental Change, 42*, 153–168. https://doi.org/10.1016/j.gloenvcha.2016.05.009

Riahi, K., van Vuuren, D. P., Kriegler, E., Edmonds, J., O'Neill, B. C., Fujimori, S., Bauer, N., Calvin, K., Dellink, R., Fricko, O., Lutz, W., Popp, A., Cuaresma, J. C., KC, S., Leimbach, M.,

Jiang, L., Kram, T., Rao, S., Emmerling, J., & Tavoni, M. (2017). The shared socioeconomic pathways and their energy, land use, and greenhouse gas emissions implications: An overview. *Global Environmental Change, 42*, 153–168. https://doi.org/10.1016/j.gloenvcha.2016.05.009

Rogeli, J., Shindell, D., Jiang, K., Fifita, S., Forster, P., Ginzburg, V., Handa, C., Kheshgi, H., Kobayashi, S., Kriegler, E., Mundaca, L., & Zickfeld, K. (2018). Mitigation pathways compatible with 1.5 °C in the context of sustainable development. In *Global warming of 1.5 °C* (pp. 93–174). Intergovernmental Panel on Climate Change.

Salvioni D. M., Bosetti L., & Fornasari T. (2021). Implementing and monitoring circular business models: An analysis of Italian SMEs. *Sustainability: Science Practice and Policy, 14*(1), 270. https://doi.org/10.3390/su14010270

Schaefer, A., Williams, S., & Blundel, R. (2020). Individual values and SME environmental engagement. *Business and Society, 59*(4), 642–675. https://doi.org/10.1177/0007650317750134

Shaw, C., Hales, S., Howden-Chapman, P., & Edwards, R. (2014). Health co-benefits of climate change mitigation policies in the transport sector. *Nature Clinical Practice Endocrinology and Metabolism, 4*, 427–433. https://doi.org/10.1038/nclimate2247

Sindhwani, R., Singh, P. L., Behl, A., Afridi, M. S., Sammanit, D., & Tiwari, A. K. (2022). Modeling the critical success factors of implementing net zero emission (NZE) and promoting resilience and social value creation. *Technological Forecasting and Social Change, 181*, 121759. https://doi.org/10.1016/j.techfore.2022.121759

Singh, J., Pandey, K. K., Kumar, A., Naz, F., & Luthra, S. (2023). Drivers, barriers and practices of net zero economy: An exploratory knowledge based supply chain multi-stakeholder perspective framework. *Operations Management Research, 16*(3), 1059–1090. https://doi.org/10.1007/s12063-022-00255-x

Sivropoulos-Valero, A. V. (2021). Budget 2021: While important steps have been taken, bolder and better coordinated action is needed for a sustainable recovery. *British Politics and Policy at LSE*. Retrieved from https://eprints.lse.ac.uk/110098/1/politicsandpolicy_budget_2021.pdf

Smith, A. C., Holland, M., Korkeala, O., Warmington, J., Forster, D., ApSimon, H., et al. (2016a). Health and environmental co-benefits and conflicts of actions to meet UK carbon targets. *Climate Policy, 16*, 253–283. https://doi.org/10.1080/14693062.2014.980212

Smith, P. (2016). Soil carbon sequestration and biochar as negative emission technologies. *Global Change Biology, 22*(3), 1315–1324. https://doi.org/10.1111/gcb.13178

Smith, P., Bustamante, M., Ahammad, H., Clark, H., Dong, H., Elsiddig, E. A., Haberl, H., Harper, R., House, J., Jafari, M., Masera, O., Mbow, C., Ravindranath, N. H., Rice, C. W., Robledo Abad, C., Romanovskaya, A., Sperling, F., & Tubiello, F. (2016). Agriculture, forestry and other land use (AFOLU). In B. Metz, O. R. Davidson, P. R. Bosch, R. Dave, & L. A. Meyer (Eds.), *Climate change 2014: Mitigation of climate change. Contribution of working group III to the fifth assessment report of the intergovernmental panel on climate change* (pp. 811–922). Cambridge University Press.

Springmann, M., Godfray, H. C. J., Rayner, M., & Scarborough, P. (2016). Analysis and valuation of the health and climate change cobenefits of dietary change. *Proceedings of the National Academy of Sciences, 113*, 4146–4151. https://doi.org/10.1073/pnas.1523119113

Steentjes, K., Poortinga, W., Demski, C., & Whitmarsh, L. (2021). *UK perceptions of climate change and lifestyle changes*. CAST Briefing Paper 08. Retrieved from https://cast.ac.uk/wp-content/uploads/2021/03/CAST-Briefing-08.pdf

Stern, P. C., & Dietz, T. (Eds.). (2008). *Public participation in environmental assessment and decision making*. National Academies Press. Retrieved from https://nap.nationalacademies.org/read/12434/chapter/1

Tenenhaus, M., Vinzi, V. E., Chatelin, Y.-M., & Lauro, C. (2005). PLS path modeling. *Computational Statistics & Data Analysis, 48*(1), 159–205.

Virmani, N., Agarwal, S., Raut, R. D., Paul, S. K., & Mahmood, H. (2022). Adopting net-zero in emerging economies. *Journal of Environmental Management, 321*, 115978. https://doi.org/10.1016/j.jenvman.2022.115978

Whitmarsh, L. (2009). What's in a name? Commonalities and differences in public understanding of "climate change" and "global warming". *Public Understanding of Science, 18*(4), 401–420. https://doi.org/10.1177/0963662506073088

Williams, S., & Preston, D. (2018). Working with values: An alternative approach to win-win. *International Journal of Corporate Strategy and Social Responsibility, 1*(4), 302–319. https://doi.org/10.1504/IJCSSR.2018.099445

Wynes, S., & Nicholas, K. A. (2017). The climate mitigation gap: Education and government recommendations miss the most effective individual actions. *Environmental Research Letters, 12*, 074024. https://doi.org/10.1088/1748-9326/aa7541

Chapter 14
Behavioral Changes and Public Engagement in Net-Zero Policies: Issues and Challenges

A. Arun Kumar, T. Sowmyya, Smitha Sambrani, and Sameer Shekhar

Abstract The study examines the complex economic, social, cultural, political, and technological relationships affecting sustainable behavior. It highlights the critical need for targeted interventions and the significance of equity and inclusion in formulating effective strategies to effectively bridge socioeconomic disparities. Promoting behavioral changes and public participation in Net-Zero policies, which are firmly based on extensive research, constitute a sophisticated approach with significant long-term ramifications. This study also tried to co-relate and synchronize cultural views with socio-communal policies. Additionally, it argues that policy innovation, comprehensive education reform, and increased corporate engagement are necessary to make significant progress towards reaching Net-Zero goals. By leveraging these opportunities, policymakers, community leaders, and stakeholders can accelerate the transition to a more resilient, sustainable, and environmentally conscious future.

Keywords Behavioral changes · Public engagement · Net Zero · Sustainable · Policies

---

A. Arun Kumar
ICFAI Law School, ICFAI Foundation for Higher Education (Deemed to Be University Under Sec 3 of UGC Act, 1956), Hyderabad, Telangana 501203, India
e-mail: arunkumar@ifheindia.org

T. Sowmyya (✉)
Forensic Science, Department of Chemistry, University College of Science, Osmania University, Hyderabad, Telangana 500007, India
e-mail: dr.sowmyya@osmania.ac.in

S. Sambrani
Department of Business Management, University College of Commerce and Business Management, Osmania University, Hyderabad, Telangana 500007, India

S. Shekhar
Department of Commerce (KSEC), Kalinga Institute of Industrial Technology (KIIT), Deemed to be University, Bhubaneshwar, Odisha, India

R. Singh and D. Crowther (eds.), *Transition Towards a Sustainable Future*, Approaches to Global Sustainability, Markets, and Governance,
https://doi.org/10.1007/978-981-97-5756-5_14

14.1 Introduction

Globally, climate change presents an unprecedented challenge that calls for quick action and coordinated efforts from all spheres of society. Pursuing Net-Zero greenhouse gas (GHG) emissions is central to the most ambitious attempt to address this crisis. Technology breakthroughs and policy changes are unquestionably significant. Still, it is also essential to recognize that behavioral changes and public participation are equally crucial to achieving these goals (Becker et al., 2023). The main objective will be to examine the approaches and programs used to promote sustainable behavior. It highlights challenges faced, encourages public involvement in environmental policies, and provides recommendations for improving their effectiveness. Global CO_2 emissions must reach Net Zero by roughly 2050 to stall the temperature increase. Any breach of this barrier would have catastrophic and possibly irreversible effects on human health, ecosystems, and global economies. To achieve this lofty goal, radical changes must be made in every area, including waste management, industry, transportation, energy, and agriculture. A crucial component in the reduction of GHG emissions is behavioral changes. We can help ourselves even more by reducing energy use and adopting sustainable modes (Attari et al., 2010).

However, bringing about and maintaining these shifts require more than awareness; it involves understanding and addressing the psychological, social, and economic factors that shape behavior. Effective public participation is identified as a critical component of climate policy. Participating communities in implementing Net-Zero policies ensure that these regulations are appropriate from a social, economic, and cultural standpoint. Participation from the general public can increase the credibility of climate policies, foster a sense of ownership, and motivate people to adopt sustainable habits (Singh et al., 2024). Engaged citizens can also push for more comprehensive and successful climate policies by holding corporations and governmental entities responsible. Even though public engagement and behavioral changes are essential, several issues remain. These challenges include low knowledge and understanding of climate-related problems, resistance to change because of perceived inconvenience or financial consequences, and limited availability of sustainable alternatives (Yuanxiong Lin, 2023). Moreover, differences in socioeconomic status may hinder certain groups' ability to participate in and benefit from climate policies, thereby exacerbating already-existing inequalities. Through identifying critical obstacles and presenting effective strategies, the research is well-positioned to provide practical advice that will influence the development and implementation of future Net-Zero regulations (Rathnasiri et al., 2024). Climate change is one of the twenty-first century's most significant and complex issues, necessitating quick global action. Of all the tactics developed to combat climate change, achieving Net-Zero emissions has received much attention. Ultimately, this research aims to support international efforts to slow climate change and advance a sustainable future for all.

The balance between greenhouse gas (GHG) emissions and atmospheric removals is Net Zero. To achieve Net Zero, greenhouse gas emissions must be as low as possible, with the remaining emissions being offset by carbon removal technologies or natural sinks like forests (Midha, 2024). This paper explores the vital role that Net-Zero policies play in reducing climate change and offers a thorough overview of them. This study analyzes the effectiveness of various strategies and initiatives to promote behavioral changes and public engagement in Net-Zero policies. It identifies the challenges and barriers to achieving widespread public participation and sustainable behavior adoption.

14.2 The Science Behind Net Zero

The Intergovernmental Panel on Climate Change (IPCC) is adamant on limiting to 1.5 °C above pre-industrial levels. Global CO_2 emissions must reach Net Zero by roughly 2050 to accomplish this goal. The logic behind this target is based on a 'carbon budget,' or the total amount of CO_2 emissions that can be released with a reasonable chance of warming to 1.5 °C. Thus, achieving Net-Zero emissions is essential to stay within this constraint and prevent irreversible damage to human societies and ecosystems. It clarifies the fundamental ideas of biology, chemistry, and physics that govern climate change and shows how human activity drives greenhouse gas emissions. It investigates possible decarbonization routes and the contributions of various industries, including energy, transportation, industry, agriculture, and land use, to emissions. Additionally, the study explores the scientific aspects of using natural carbon sinks like forests and oceans to offset emissions and carbon capture, utilization, and storage (CCUS). By developing a thorough scientific understanding of the challenges and opportunities associated with achieving Net-Zero emissions, scientists, policymakers, and stakeholders can collaborate to develop evidence-based plans and technologies to tackle climate change effectively (Flores & Moore, 2024).

14.3 Policy Frameworks

A wide range of actions, from market-based and regulatory tools to voluntary initiatives and international agreements, are essential to reduce emissions. These regulations usually entail establishing voluntary or mandatory goals to reduce greenhouse gas (GHG) emissions, usually about a designated reference year. These objectives could be intensity-based, like lowering emissions per GDP or production output unit, or absolute, like cutting emissions by a specific percentage. Economic incentives for reducing emissions are established through implementing carbon taxes or cap-and-trade systems. A low-carbon economy is promoted by carbon pricing, which internalizes the external costs of greenhouse gas emissions and incentivizes consumers and businesses to adopt sustainable practices and cleaner technologies (Verfuerth et al., 2023).

Furthermore, tax breaks, regulatory guidelines, and subsidies encourage renewable energy. The transition to renewable energy is essential to decarbonizing the power industry, a significant contributor to worldwide emissions. Furthermore, it is necessary to implement standards and laws to improve the energy efficiency of industrial processes, appliances, and buildings. Enhanced energy efficiency is frequently more affordable than alternative mitigation techniques and lowers energy demand and associated emissions. Investing in the research, development, and deployment of low-carbon technologies, negative emission technologies (NETs), and advanced renewable energy systems is essential. Developing innovative strategies to achieve significant decarbonization requires creativity (Atul Rawat et al., 2024).

Specialized regulations for high-emission industries such as transportation (fuel economy requirements, electric car subsidies), agriculture (sustainable farming methods, methane reduction), and industry (process improvements, carbon capture) are crucial. In addition, improving soil carbon sequestration, afforestation, and reforestation can improve biodiversity and resilience while significantly reducing carbon emissions. Signing international treaties and collaborations guarantee a coordinated worldwide effort. One prominent instance is the 2015 Paris Agreement, which mandates that countries to minimize greenhouse gas emissions must update their Nationally Determined Contributions (NDCs) (Rice & Reed, 2019).

Achieving Net-Zero emissions is essential for maintaining global temperature stability and reducing the harmful consequences of climate change. The shift to a low-carbon economy has the potential to stimulate job creation and economic growth. Energy efficiency, sustainable infrastructure, and renewable energy investments can boost competitiveness, foster innovation, and develop new markets and industries (Ioannis Dafnomilis, 2023).

In addition to this, lowering GHG emissions also reduces air pollution, which is directly linked to diseases of the heart and lungs. Improved public health, lower healthcare costs, and a higher standard of living can all be achieved with cleaner air. Reducing reliance on imported fossil fuels can improve energy security through diversifying energy sources and boosting the use of renewable energy. By doing this, energy resilience can be increased and vulnerability to price swings and geopolitical unpredictability is reduced. Biodiversity conservation is supported by incorporating nature-based solutions into Net-Zero strategies. Healthy ecosystems, such as soil fertility, flood control, and water purification, provide essential functions necessary for human well-being. Social equity can also be advanced by implementing Net-Zero policies (Wenguang Tang, 2023).

14.4 Challenges in Implementing Net-Zero Policies

Reaching Net Zero requires firm, unwavering resolve of Governments and building robust institutions that create, carry out, and maintain policies. Progress in this endeavor may be impeded by political resistance, a lack of unity, and entrenched interests within the fossil fuel industry. High upfront costs are associated with the shift to a low-carbon economy, which may impede development (Ordonez et al., 2023). Getting enough money together and offering sustainable project financing options at a reasonable price are one of the biggest challenges to overcome; this task is especially difficult for developing countries. Many low-carbon technologies are already on the market, but some are still being developed, such as advanced CCS and certain NETs. Therefore, expediting the pace of technological innovation and deployment is of utmost importance in this context.

The public must be heavily involved and educated to change societal norms and behaviors towards sustainability. One of the biggest challenges on this journey is overcoming inertia and resistance to change, particularly about consumption habits and lifestyle choices. A fair shift necessitates carefully analyzing the disparate effects of climate policies on different groups within society. Policies must be comprehensive and assist marginalized communities to stop the current disparities from worsening. To ensure accountability and transparency in pursuing Net-Zero goals, it is essential to establish effective mechanisms for tracking, reporting, and validating reductions in emissions. Furthermore, developing solid data collection and analysis skills is critical for monitoring advancement and implementing necessary policy changes (Ma et al., 2024).

14.5 Case Studies

Denmark: Renewable Energy and Public Engagement

Denmark is a well-known example of how to implement Net-Zero policies efficiently, particularly regarding renewable energy. About half of Denmark's electricity comes from wind power, the country's most well-known source. The government has actively supported wind energy by passing laws advantageous to wind energy, providing subsidies, and encouraging partnerships between the public and private sectors.

The participation and engagement of the public have dramatically influenced Denmark's achievements. The Danish government has actively involved local communities in renewable energy projects by offering financial incentives to promote local ownership of wind turbines. Additionally, Denmark has significantly reduced energy consumption and emissions by enforcing strict energy efficiency standards for appliances and buildings (State of Green Summit, 2024).

Japan: Energy Conservation and Innovation

On the other hand, Japan's approach to reaching Net Zero primarily emphasizes technological advancements and energy conservation. The country faces unique challenges because of its small natural resources and dense population. Japan has conducted widespread national campaigns to encourage energy-saving practices, especially after the Fukushima nuclear accident. Initiatives such as 'Cool Biz' are designed to motivate people to reduce daily energy. Japan is a central Research,

Development, and Demonstration (RD&D) investor for cutting-edge low-carbon technologies like energy storage, smart grids, and hydrogen energy. Japan has implemented sector-specific policies, like fuel efficiency standards and vehicle regulations, to improve industrial energy efficiency, which has led to significant emission reductions in important sectors (Japan Energy Summit, 2024).

Rwanda: Nature-Based Solutions and Community Involvement

Rwanda is a prime example of how Net-Zero strategies can successfully incorporate natural solutions. The nation has prioritized reforestation and sustainable land management practices to support its efforts at sequestering carbon dioxide. Rwanda has initiated ambitious reforestation projects to expand its forest cover and restore degraded lands. International collaborations and national policies have supported these efforts. Rwanda's reforestation efforts are zoned on local communities, and the government works closely with them to adopt beneficial, socially inclusive, and sustainable land management techniques. Rwanda employs nature-based solutions to improve ecosystem resilience, encourage biodiversity, and sequester carbon. Sustainable development is supported by essential services provided by healthy ecosystems, such as soil fertility and water regulation (IUCN, 2024).

14.6 Results and Discussion

Successful climate action requires significant changes in individual and societal behavior, policy changes, and technological advancements. Implementing policies that support public involvement and behavioral modifications is crucial to reach Net-Zero emissions. Using case studies and empirical data to highlight the successes and difficulties faced in these endeavors, this critical analysis explores the effectiveness of various strategies and initiatives to promote these crucial changes. Public awareness campaigns are essential to this issue because they inform people about the severity of climate change and the value of taking personal responsibility for their actions. Initiatives such as the UK's 'Act on CO_2' and Germany's 'Klimaschutzkampagne' have effectively raised public awareness of and influenced attitudes toward sustainability and energy use (Perdan et al., 2017).

However, converting knowledge into long-term behavioral change is difficult because awareness does not always translate into action, particularly when people face financial or practical obstacles. Norway has the highest per capita EV ownership in the world, for example, thanks to its generous EV incentives. However, financial limitations may limit the effectiveness of economic incentives, particularly in lower-income households. Moreover, there is a chance that wealthier individuals may unintentionally benefit, worsening inequality.

Emissions can be significantly reduced by regulatory measures like forbidding specific pollutants and enforcing mandatory energy efficiency standards. For example, the strict CO_2 emission regulations set by the European Union for automobiles have encouraged automakers to develop new technologies and reduce emissions.

Regulators, however, frequently face opposition from businesses and consumers who view them as unduly costly or restrictive. Another issue is effective enforcement, especially in areas with weak regulatory capacity. Initiatives that are community-based and involve local stakeholders in climate action typically have high levels of effectiveness. Communities are empowered by initiatives like community solar projects and Transition Towns, which promote increased participation and a long-term dedication to sustainable practices. However, maintaining strong leadership and coordination is necessary for these initiatives, which can be difficult. Moreover, the impact of community-based initiatives may be constrained in scale compared to broader policy measures (Tianyu Li, 2024).

Initiatives for Enhancing Public Engagement

Since the effects of educational programs can take time to become apparent, it can be difficult to gauge their direct impact on behavioral changes. Evaluating these programs' immediate effects on behavior change can be challenging because many external and intervening variables could affect the results. Educational programs must undergo periodic content updates to incorporate the latest scientific findings and policy developments. Regular updates are required to keep the information distributed to the intended audience accurate and relevant. Government agencies, corporations, and civil society organizations working together can significantly increase campaigns' public engagement and efficacy to promote sustainable behaviors. By pooling resources and expertise from different sectors, these collaborations can synergize their efforts and create a more impactful and comprehensive approach to promoting sustainability (Malone et al., 2010).

A noteworthy endeavor in this context is the RE 100, which requires businesses to source all their energy from renewable sources. Participating businesses lower their carbon emissions through this program, spread the word about the value of adopting renewable energy sources, and set standards that other companies can follow. Nonetheless, it can be challenging to keep the varied stakeholders in these partnerships' interests aligned. Within these cooperative frameworks, a fine line exists between furthering corporate interests and preserving more general social and environmental goals (Ray, 2019).

Public engagement initiatives have seen a radical change thanks to digital tools and platforms, which have opened new channels for communication and engagement. Social media campaigns, carbon footprint tracking mobile applications, and online community forums have significantly increased the accessibility and reach of sustainability initiatives. Initiatives such as the UN's Act Now campaign, which uses digital tools, have inspired people to take proactive steps toward sustainability. However, differences in digital literacy levels and access to technology among various populations may compromise the efficacy of digital engagement strategies. Furthermore, user motivation and long-term engagement are critical to the success of these tools (Sharp et al., 2009).

Encouraging public participation in Net-Zero policies and behavioral changes ensures a smooth and long-lasting transition to a low-carbon future. While several initiatives and strategies have shown promise in this area, their effectiveness depends on several variables, such as technological infrastructure, regulatory frameworks, community engagement levels, and economic constraints. Adopting integrated strategies that integrate public involvement, regulatory measures, incentive programs, and awareness-raising campaigns is essential to overcoming these obstacles and optimizing the effectiveness of sustainability initiatives. Policymakers and stakeholders who want to drive meaningful progress toward a sustainable future must adopt critical practices such as learning from successful case studies and customizing strategies to fit local contexts (Curry, 2004).

Challenges and Barriers to Achieving Public Participation and Sustainable Behavior

Achieving broad public engagement and acceptance of sustainable behavior is essential to the success of Net-Zero initiatives. However, there are a lot of barriers and difficulties that could make these attempts difficult. Many people and households may be discouraged by these costs. Long-term adoption of these sustainable measures may be impeded by operation and maintenance costs, even with subsidies for initial expenses.

Lower-income communities often struggle to obtain the funds to invest in environmentally friendly technologies or participate in specific programs, exacerbating already-existing disparities. It may be more challenging to win over people to Net-Zero policies if workers in high-carbon industries oppose transitioning because they fear job insecurity and unstable economic conditions. Many people are reluctant to change due to their deeply embedded routines and habits, even when they know the benefits to the environment that come with these changes (Gregor Vulturius, 2022).

One recurring problem frequently noted is people's ignorance of their actions' impact on climate change and the benefits of adopting sustainable practices. Implementing sustainable practices is commonly hampered by misconceptions about their efficacy and viability. Many people view climate change as an abstract or far-off issue, which makes adopting sustainable practices less urgent. Some people may put off taking responsibility for climate change until future generations because they think they will not be directly impacted by its effects. The attitudes and behaviors of one's peers have a significant impact on one's actions. People tend not to engage in sustainable practices when not commonplace in their community or social group (Inhwan Ko, 2022).

Taking environmentally friendly actions, like taking public transportation rather than owning a private vehicle, may encounter social stigma or lack of acceptance. Reluctance to change can result from cultural traditions and values that counter sustainable practices. Moreover, societal values prioritizing economic growth and consumption may impede the adoption of sustainable behaviors. Policies that are unclear or inconsistent can cause misunderstandings and reduce public participation. Establishing stable, unambiguous, and comprehensive policies is essential for directing public behavior (Sovacool et al., 2020).

Adopting sustainable behaviors can be hampered by a lack of infrastructure, such as inadequate recycling facilities, few renewable energy sources, or limited options for public transportation. Policies that fail to recognize the value of accessibility and convenience may not be able to promote public involvement effectively. Insufficient access to technology and internet connectivity can hinder the spread of knowledge and the uptake of digital tools that encourage environmentally friendly practices. Adoption opportunities may be limited in some areas due to the difficulty of accessing sustainable technologies and solutions. People who lack technological literacy and comprehension may find it challenging to use sustainable technologies and benefit from them (Flanagan et al., 2003).

To overcome these challenges and barriers, the following strategies can be implemented. Financial incentives and subsidies could be provided to reduce sustainable technologies' upfront and ongoing costs. Improving their affordability and accessibility for a broader range of stakeholders is the aim. By promoting investments in environmentally friendly solutions, this kind of approach is essential to accelerating the transition to a low-carbon economy. Creating economic transition plans is also necessary to support workers in high-carbon industries, guaranteeing a smooth transition to more sustainable sectors, offering opportunities for retraining, and ensuring job security. These plans can promote a fair and inclusive green transition by reducing the possibility of social and economic upheavals from the phase-out of carbon-intensive operations (Stephenson, 2018).

To further educate the public about the range of impacts of climate change and the numerous benefits of adopting sustainable practices, focused awareness campaigns could be started. These campaigns aim to bridge knowledge gaps and empower individuals to make informed decisions that support environmental conservation by dispelling misconceptions and providing clear, helpful information. Furthermore, social norms and peer influence can be effectively tapped to promote sustainable behaviors through social marketing techniques. This approach recognizes social networks' influential role in people's attitudes and actions, making them vital for promoting constructive change in communities (United Nations, 2024).

As these influential people have the power to mobilize support and promote coordinated action, engaging with communities and cultural leaders is equally crucial to ensure sustainable practices align with cultural values and beliefs. Governments and organizations can foster an environment that supports sustainable development by formulating consistent, long-term policies that offer a stable framework for sustainable behaviors. Another essential component of encouraging sustainable practices is investing in infrastructure upgrades, such as recycling centers, renewable energy sources, and public transportation. These investments improve community sustainability and have a minor negative environmental impact (Bożena Gajdzik, 2023).

Expanding access to technology and boosting digital connectivity can help spread knowledge and effectively use digital engagement tools to encourage sustainable behaviors. To ensure that people can use sustainable technologies effectively and reap the maximum benefits, it is imperative that training and resources can be made available to improve technological literacy among people. The successful implementation of Net-Zero policies can be supported by encouraging widespread public participation and the adoption of sustainable behaviors by addressing these barriers through a comprehensive and integrated approach.

Promoting Behavioral Changes and Public Engagement in Net-Zero Policies

Reaching Net-Zero emissions is a difficult task that calls for the proactive participation of people, groups, businesses, and governmental organizations. While many tactics have been used to encourage public participation and behavioral changes, focused recommendations can significantly increase efficacy. With the support of relevant statistical data, this critical analysis offers policymakers, community leaders, and stakeholders' practical recommendations. Improving tax breaks and subsidies for buying electric cars, renewable energy systems, and energy-efficient appliances could effectively promote the uptake of sustainable technologies.

To promote fairness and accessibility and address socioeconomic differences in adopting sustainable technologies, these incentives may specifically target lower-income households. Furthermore, enabling a fair transition requires developing job training and reskilling programs for workers moving from high-carbon industries to the green economy. According to the UK's Green Jobs Taskforce, up to 2 million new green jobs could be created by 2030, highlighting the possibility of job creation through reskilling initiatives (IEA 50, 2023).

Various media, such as social media, television, and community events, can successfully engage audiences with tailored messages emphasizing sustainable practices' benefits, raising awareness, and promoting behavioral change. One noteworthy example of how focused awareness campaigns can result in quantifiable changes in energy consumption behaviors is Germany's 'Klimaschutzkampagne,' which was quite successful. Including a lot of climate education in adult education and school curricula can raise awareness of environmental issues and promote sustainable practices in the community (Frankfurt.de, 2024).

Eco-schools, a program in 67 countries, has demonstrated efficacy in educating students about environmental sustainability and engaging communities in eco-friendly endeavors. Social media marketing can help create new social norms around sustainability, encouraging people to incorporate eco-friendly practices into their daily lives. The Transition Towns movement has demonstrated the effectiveness of peer influence in promoting positive environmental outcomes by utilizing social norms to implement community-led sustainability projects (Eco-schools, 2024).

Community buy-in and support for sustainable practices can be enhanced by collaborating with religious and cultural leaders to match sustainability initiatives with cultural values and practices. The success of initiatives like the Green Faith Alliance in involving religious communities in environmental stewardship highlights the significance of cultural congruence in sustainability endeavors. Climate policies must remain stable and consistent to create a clear framework for individual and corporate actions that support long-term sustainability and stimulate investment in green technologies (Chambers et al., 2022).

Making investments in renewable energy grids, recycling facilities, and public transportation are examples of infrastructure that promotes sustainable practices. High rates of environmentally friendly commuting result from significant investments in bicycle infrastructure and public transit by cities like Amsterdam and Copenhagen. Use of digital engagement tools and platforms increases digital connectivity and access to promote sustainable initiatives through online platforms. Increased involvement in online sustainability initiatives has been observed in programs that improve digital connectivity in rural areas (Avelino, 2017). Implementation of training programs to increase technological literacy to ensure that people use and benefit from sustainable technologies effectively is another initiative towards sustainable practices. Research has indicated that focused technology training initiatives can significantly boost low-income communities' adoption and efficient use of sustainable technologies.

14.7 Limitations and Future Scope of the Study

Case studies and data from particular regions are the study's main areas of interest. This restriction might ignore the difficulties and solutions found elsewhere, especially in developing nations. The case studies and data cited are found on comparatively recent advancements. Understanding sustained impacts may be limited by fully capturing long-term effects and trends. The study discusses cultural barriers in general, but it might not capture all the subtle variations within and between cultures that significantly impact behavior and acceptance of policies. The study makes assumptions about political stability and that might not apply to all areas, which could affect how well-suggested strategies are implemented.

More varied case studies from different geographical areas, including developing nations, should be included in future research to comprehend various difficulties and practical solutions in a broader context. Extensive research conducted in settings can offer a more profound understanding of how social norms and cultural values impact the adoption of sustainable behaviors and public participation. More thorough information on sustainability and behavior modification can be obtained by conducting longitudinal studies to evaluate the long-term efficacy of different initiatives and strategies. Long-term trends in sustainable behavior adoption and public engagement can be monitored to determine which methods work best over time.

14.8 Policy Implications of the Study

The literature on encouraging public participation and behavioral changes in support of Net-Zero policies provides crucial information that can direct the development and implementation of effective climate policies. To guarantee equitable involvement in sustainable practices, policymakers must create incentives and subsidies that are easily affordable for lower-income households. Targeted subsidies for installing renewable energy, energy-efficient appliances, and electric vehicles should be prioritized, emphasizing giving impoverished communities financial support. To lessen resistance and promote economic stability, assisting workers moving from high-carbon industries to green jobs is critical. To enable a smooth transition to the green economy, extensive job training programs and reskilling initiatives must be developed in partnership with industry stakeholders. The dissemination of information through multi-platform campaigns aimed at increasing public awareness can significantly drive large-scale behavioral changes. Investments in extensive public awareness campaigns utilizing social media, television, and community gatherings can effectively educate the public about the advantages of sustainable behaviors and the pressing need for climate action.

Social norms and community engagement can be used to increase the efficacy of sustainability projects. It is crucial to support local leaders and influencers in their advocacy for sustainable practices and to support community-led sustainability projects. Connecting sustainability initiatives with cultural values can boost participation and acceptance rates. Incorporating sustainability into culturally significant practices and events can be achieved through collaboration with religious and cultural entities.

Encouraging investments in sustainable technologies and bringing about long-term behavioral changes depend heavily on the continuity and stability of climate policies, ensuring that climate policies are stable, consistent, and have long-term goals. It is best to avoid frequent policy changes because they can lead to uncertainty and make long-term planning more difficult. Expanding access to digital technologies can improve public participation and adoption of sustainable practices. Expanding digital infrastructure requires investments, especially in underserved and rural areas, to ensure that all people have equitable access to information and digital engagement tools.

Changes in education that integrate sustainability into core curricula have the potential to create long-lasting behavioral changes. To incorporate sustainability principles into the educational framework, educational policies should include comprehensive sustainability education from early childhood education through higher education.

14.9 Societal Implications of the Study

Research encouraging public participation in Net-Zero policies and behavioral changes has revealed significant societal ramifications, including social, political, cultural, and economic. The implications highlight the noteworthy capacity of efficacious sustainability tactics to revolutionize society. The research highlights the importance of providing financial incentives and subsidies to lower-income households, implying that economic inequality may decrease within the population. With this approach, it is possible to close the gap between different socioeconomic groups and provide long-term advantages to all people, regardless of their financial situation. Additionally, the study emphasizes the necessity of developing green jobs and assisting workers to leave high-carbon industries to mitigate job displacement and support economic stability in the face of environmental challenges. This economic shift is crucial for safeguarding people's livelihoods while advancing sustainability objectives in the long run.

Furthermore, normalizing sustainable behaviors and incorporating them into daily life may require actively involving communities and utilizing social norms. This normalization process has the potential to impact a broad spectrum of people in society by bringing about a collective shift in attitudes toward sustainability. The public's support and involvement in sustainability initiatives can grow due to social desirability and acceptance of sustainable practices. Additionally, raising public awareness and incorporating climate education into curricula can contribute to developing a better-informed society and being conscious of its environmental impact. Such initiatives can foster a long-lasting, generation-spanning culture of sustainability, ensuring a more sustainable future for all.

These strategies may result in a more robust, educated, and equitable society that can achieve and sustain Net-Zero emissions over time. Legislators, community leaders, and stakeholders must collaborate to implement these recommendations and realize the full benefits of a shift towards sustainability for society.

14.10 Conclusion

A shift to Net-Zero greenhouse gas emissions is imperative to mitigate climate change and prevent its severe and potentially irreversible effects on ecosystems, human health, and global economies. Technological developments, legislative actions, and significant behavioral shifts on both an individual and a group level are all necessary to achieve Net Zero. Strategies that promote sustainable behaviors and increase public participation are essential for effective climate action. Initiatives from the community, financial incentives, laws, and public awareness campaigns have all shown varied degrees of success in promoting sustainable practices.

To make climate policies more socially and economically viable and culturally aligned, public engagement is essential to improving their overall efficacy. Economic and logistical obstacles make it challenging to translate knowledge into long-lasting behavioral change, even with awareness campaigns. Financial limitations may unintentionally benefit wealthier people by limiting economic incentives, especially in lower-income households. Effective enforcement mechanisms are necessary because industry and consumers frequently oppose regulatory measures. Community-based initiatives can have limited scope and require strong leadership and coordination, even though they are effective locally. Policy innovation, reform in education, and proactive corporate involvement are significant forces behind the advancement of Net-Zero objectives. The conclusions drawn from this most recent study offer a valuable framework for advancing international efforts to move towards a sustainable and low-carbon future.

References

Attari, S. Z., DeKay, M. L., Davidson, C. I., & De Bruin, W. B. (2010). Public perceptions of energy consumption and savings. *Proceedings of the National Academy of Sciences of the United States of America, 107*(37), 16054–16059.

Atul Rawat, C. P. (2024). Analysis of the key hydrogen fuel vehicles adoption barriers to reduce carbon emissions under net zero target in emerging markets. *Energy Policy*. https://doi.org/10.1016/j.enpol.2023.113847

Avelino, F. (2017). Power in sustainability transitions: Analysing power and (dis)empowerment in transformative change towards sustainability. *Environmental Policy Governance, 27*(2017), 505–520. https://doi.org/10.1002/eet.1777

Becker, S., Demski, C., Smith, W., & Pidgeon, N. (2023). Public perceptions of heat decarbonization in Great Britain. *Wiley Interdisciplinary Reviews: Energy and Environment, 12*(6), e492. https://doi.org/10.1002/wene.492

Bożena Gajdzik, R. W. (2023). Process of transformation to net zero steelmaking: Decarbonisation scenarios based on the analysis of the polish steel industry. *Energies*. https://doi.org/10.3390/en16083384

Chambers, J. M., Wyborn, C., Klenk, N. L., Ryan, M., Serban, A., Bennett, N. J., Brennan, R., Charli-Joseph, L., Fernández-Giménez, M. E., Galvin, K. A., Goldstein, B. E., Haller, T., Hill, R., Munera, C., Nel, J. L., Österblom, H., Reid, R. S., Riechers, M., Spierenburg, M., … Rondeau, R. (2022). Co-productive agility and four collaborative pathways to sustainability transformations. *Global Environmental Change, 72*, 102422. https://doi.org/10.1016/j.gloenvcha.2021.102422

Climate Change Mitigation. (2004). Massachusetts Institute of Technology. Retrieved from https://sequestration.mit.edu/pdf/Tom_Curry_Thesis_June2004.pdf

Climate Change Mitigation. (2024). *MIT on climate change, massachusetts institute of technology*. Retrieved from https://sequestration.mit.edu/pdf/Tom_Curry_Thesis_June2004.pdf

Curry, T. E. (2004). *Public awareness of carbon capture and storage: A survey of attitudes toward climate change mitigation*. Massachusetts Institute of Technology. Retrieved from https://sequestration.mit.edu/pdf/Tom_Curry_Thesis_June2004.pdf

Dafnomilis, I. (2023). Achieving net-zero emissions targets: An analysis of long-term scenarios using an integrated assessment model. *Annals of the New York Academy of Sciences*. https://doi.org/10.1111/nyas.14970

Eco-schools. (2024). *Engaging the youth of today to protect the planet of tomorrow*. Retrieved from https://www.ecoschools.global/

Frankfurt.de (2024). Climate-Neutral Cities by 2030 Kick-off for NetZero Cities Project, Retrieved from https://www.frankfurtschool.de/en/home/newsroom/news/2021/Oktober/net-zero-cities

Flanagan, S. C., & Lee, A.-R. (2003). The new politics, culture wars, and the authoritarian-libertarian value change in advanced industrial democracies. *Comparative Political Studies, 36*(3), 235–270.

Flores, R. M., & Moore, T. A. (2024). Coal and coalbed gas: Future directions and opportunities. *Coal and Coalbed Gas: Future Directions and Opportunities*, 1–748. https://doi.org/10.1016/C2019-0-05012-1

Gregor Vulturius, A. M. (2022). Sustainability-linked bonds—Their potential to promote issuers' transition to net-zero emissions and future research directions. *Journal of Sustainable Finance and Investment*. https://doi.org/10.1080/20430795.2022.2040943

IEA 50. (2023). *UK Green jobs taskforce and North Sea transition deal*. Retrieved from https://www.iea.org/policies/17856-uk-green-jobs-taskforce-and-north-sea-transition-deal

Inhwan Ko, N. D. (2022). Have renewable energy leaders announced aggressive emission reduction goals? Examining variations in the stringency of country-level net-zero emission pledges. *Plos Climate*. https://doi.org/10.1371/journal.pclm.0000094

International Union for Conservation of Nature. (2024). IUCN Rwanda. Retrieved from https://www.iucn.org/sites/default/files/2023-09/iucn-rwanda-country-office-brochure-lowres_april-2023_compressed.pdf

Japan Energy Summit. (2024). *Accelerating Japan's energy transition through innovation and global connections.* Retrieved from https://www.japanenergyevent.com/

Ma, Z., Pu, D., Liang, H. (2024). Financing net-zero energy integration in smart cities with green bonds and public-private partnerships. *Sustainable Energy Technologies and Assessments, 64*(103708). https://doi.org/10.1016/j.seta.2024.103708

Madsen, K. (2022). Eco-schools: Engaging the youth of today to protect the planet of tomorrow. *Childhood Education, 98*(5), 6–17. https://doi.org/10.1080/00094056.2022.2115791

Malone, E. L., Dooley, J. J., & Bradbury, J. A. (2010). Moving from misinformation derived from public attitude surveys on carbon dioxide capture and storage towards realistic stakeholder involvement. *International Journal of Greenhouse Gas Control, 4*(2), 419–425. https://doi.org/10.1016/j.ijggc.2009.09.004

Malone, T. W. (2010). The future of work. Boston: Harvard Business School Press.

Midha, A. (2024). Towards achieving net zero emissions in India by 2070. *Renewable Power for Sustainable Growth.* https://doi.org/10.1007/978-981-99-6749-0_66

Ordoñez, G.L. (2023). Macroeconomics and financial crises: Bound together by information dynamics. Princeton: Princeton University Press. https://doi.org/10.1515/9780691227023

Ordonez, J. A. (2023). India's just energy transition: Political economy challenges across states and regions. *Energy Policy, 179*(113621). https://doi.org/10.1016/j.enpol.2023.113621

Perdan, S., Jones, C. R., & Azelaic, A. (2017). Public awareness and acceptance of carbon capture and utilization in the UK. *Sustainable Production and Consumption, 10*, 74–84. https://doi.org/10.1016/j.spc.2017.01.001

Rathnasiri, M. S. H., Dewasiri, N. J., & Kumar, A. A. (2024). Policy framework and implementation strategies for Sri Lanka's transition to a net-zero economy. In R. Singh, S. Khan, A. Kumar, S. Luthra, & H. Chokshi (Eds.), *Net zero economy, corporate social responsibility and sustainable value creation. CSR, sustainability, ethics & governance.* Springer. https://doi.org/10.1007/978-3-031-55779-8_3

Ray, P. (2019). Renewable energy and sustainability. *Clean Technologies and Environmental Policy, 21*, 1517–1533.

Rice, C. W., & Reed, D. (2019). *Soil carbon sequestration and greenhouse gas mitigation: A role for American agriculture.* Bipartisan Policy Centre. Retrieved from https://bipartisanpolicy.org/download/?file=/wp-content/uploads/2019/03/cwrice_report_30907_final.pdf

Sharp, J. D., Jaccard, M. K., & Keith, D. W. (2009). Anticipating public attitudes toward underground CO_2 storage. *International Journal of Greenhouse Gas Control, 3*(5), 641–651. https://doi.org/10.1016/j.ijggc.2009.04.001

Singh, R., Ravesangar, K., Narayanan, S., & Singh, K. K. H. (2024). HRM practices in creating net zero: Supporting organizational sustainability. In R. Singh, S. Khan, A. Kumar, S. Luthra, & H. Chokshi (Eds.), *Net zero economy, corporate social responsibility and sustainable value creation. CSR, sustainability, ethics & governance.* Springer. https://doi.org/10.1007/978-3-031-55779-8_1

Sovacool, B. K., & Griffiths, S. (2020a). The cultural barriers to a low-carbon future: A review of six mobility and energy transitions across 28 countries. *Renewable and Sustainable Energy Reviews, 119*. https://doi.org/10.1016/j.rser.2019.109569

Sovacool, B.K., & Griffiths, S. (2020). Culture and low-carbon energy transitions. *Nature Sustainability, 3*, 685– 693. https://doi.org/10.1038/s41893-020-0519-4

State of Green. (2024). *Physical event carbon capture, storage and utilisation energy storage.* The Danish CCUS Summit. Retrieved from https://stateofgreen.com/en/events/the-danish-ccus-summit-2024/

Stephenson, J. (2018). Sustainability cultures and energy research: An actor-centred interpretation of cultural theory. *Energy Research and Social Science, 44*, 242–249. https://doi.org/10.1016/j.erss.2018.05.034

Tianyu Li, X.-G. Y.-C. (2024). Towards Paris Climate Agreement goals. The essential role of green finance and green technology. *Energy Economics.* https://doi.org/10.1016/j.eneco.2023.107273

United Nations. (2024). *United Nations Climate Action, education is key to addressing climate change*. Retrieved from https://www.un.org/en/climatechange/climate-solutions/education-key-addressing-climate-change

Verfuerth, C., Demski, C., Capstick, S., Whitmarsh, L., & Poortinga, W. (2023). A people-cantered approach is needed to meet net zero goals. *Journal of the British Academy, 11*(S4), 97–124. https://doi.org/10.5871/jba/011s4.097

Wenguang Tang, L. M. (2023). Green innovation and resource efficiency to meet net-zero emission. *Resources Policy*. https://doi.org/10.1016/j.resourpol.2023.104231

Yuanxiong Lin, M. K.-P. (2023). Assessment of renewable energy, financial growth, and accomplishing targets of China's cities carbon neutrality. *Renewable Energy*. https://doi.org/10.1016/j.renene.2022.11.026

Index

R. Singh and D. Crowther (eds.), *Transition Towards a Sustainable Future*, Approaches to Global Sustainability, Markets, and Governance,
https://doi.org/10.1007/978-981-97-5756-5

The manufacturer's authorised representative in the EU is Springer
Nature Customer Service Centre GmbH, Europaplatz 3, 69115 Heidelberg,
Germany. If you have any concerns regarding our products, please
contact ProductSafety@springernature.com

Printed and bound by CPI Group (UK) Ltd, Croydon, CR0 4YY
01/10/2026
02225859-0011